Fundamentals of Optical Networks and Components

Fundamentals of Optical Networks and Components

Partha Pratim Sahu

CRC Press
Taylor & Francis Group
Boca Raton London New York

CRC Press is an imprint of the
Taylor & Francis Group, an **informa** business

First edition published 2020
by CRC Press
6000 Broken Sound Parkway NW, Suite 300, Boca Raton, FL 33487-2742

and by CRC Press
2 Park Square, Milton Park, Abingdon, Oxon, OX14 4RN

© 2021 Taylor & Francis Group, LLC

CRC Press is an imprint of Taylor & Francis Group, LLC

ISBN: 978-0-367-26545-8 (hbk)
ISBN: 978-0-429-29376-4 (ebk)

Typeset in Times
by codeMantra

To my family:

My grand mother Sushila Sahu and
my parents Harekrishna Sahu and Jyotsana Sahu,

My wife Arpita Sahu, and
my daughters Prakriti and Ritushree Sahu.

Contents

Preface...xvii
Acknowledgements xxi
Author xxiii

Chapter 1   Introductory Concept 1

    1.1   Basic Communication Model 1
        1.1.1   Local Area Network..2
            1.1.1.1   OSI Model..2
            1.1.1.2   TCP/IP Protocol...3
        1.1.2   Wide Area Network...5
            1.1.2.1   Circuit Switching 5
            1.1.2.2   Packet Switching...5
            1.1.2.3   Frame Relay 6
            1.1.2.4   Asynchronous Transfer Mode 6
        1.1.3   VSAT Network via Satellite 6
        1.1.4   Integrated Services Digital Network.......................... 10
            1.1.4.1   Narrowband ISDN 11
            1.1.4.2   Broadband ISDN... 11
    1.2   Optical Fiber Principle 11
        1.2.1   Optical Fiber.. 12
            1.2.1.1   Optical Transmission in Fiber 12
            1.2.1.2   Difference between Single- and
                           Multimode Fibers... 15
        1.2.2   Attenuation in Fiber 17
            1.2.2.1   Absorption 17
        1.2.3   Scattering Loss.. 18
        1.2.4   Dispersion in Fiber 19
        1.2.5   Nonlinearities.. 21
        1.2.6   Nonlinear Refraction... 21
        1.2.7   Stimulated Raman Scattering 22
        1.2.8   Stimulated Brillouin Scattering 22
        1.2.9   Four-Wave Mixing... 22
    1.3   Optical Transmitters... 22
        1.3.1   Laser Action 23
        1.3.2   Semiconductor Diode Laser.. 24
        1.3.3   Multiple Quantum Well Laser..................................... 25
        1.3.4   Tunable and Fixed Lasers.. 25
            1.3.4.1   Laser Characteristics 25
            1.3.4.2   Mechanically Tuned Lasers.......................... 26
            1.3.4.3   Acoustooptically and Electrooptically
                           Tuned Lasers...27

                1.3.4.4   Injection-Current-Tuned Lasers 27

        1.3.5   Laser Arrays 28

   1.4   Optical Receivers and Filters 29

        1.4.1   Photodetector 29

                1.4.1.1   PIN Photodiode 30

                1.4.1.2   Avalanche Photodiode 31

        1.4.2   Tunable Optical Filters 32

                1.4.2.1   Filter Characteristics 32

                1.4.2.2   Etalon 33

                1.4.2.3   Mach–Zehnder Chain 34

                1.4.2.4   Acousto-optic Filters 34

                1.4.2.5   Electrooptic Filters 35

                1.4.2.6   Liquid Crystal Fabry–Perot Filters 35

        1.4.3   Fixed Filters 35

                1.4.3.1   Grating Filters 35

                1.4.3.2   Fiber Bragg Gratings (FBG) 35

                1.4.3.3   Thin-Film Interference Filters 36

        1.4.4   Comparison between Different Filters 36

   1.5   Optical Modulation 36

        1.5.1   Digital-to-Digital Modulation 37

                1.5.1.1   NRZ 37

                1.5.1.2   Bipolar AMI 37

                1.5.1.3   Pseudo Ternary AMI 38

                1.5.1.4   Biphase Coding 38

                1.5.1.5   B8ZS Code 39

                1.5.1.6   HDB3 Code 39

        1.5.2   Digital-to-Analog Modulation 40

        1.5.3   Analog-to-Analog Modulation 41

                1.5.3.1   Amplitude Modulation 41

                1.5.3.2   Frequency Modulation 41

                1.5.3.3   Phase Modulation 42

   Summary 43

   Exercises 43

   References 45

Chapter 2   Different Optical Network Node 49

   2.1   Non-Reconfigurable Node 49

        2.1.1   Non-Reconfigurable Wavelength Router Node 49

        2.1.2   Arrayed Waveguide Grating-Based Node 50

        2.1.3   Node Architecture of a Passive-Star WDM

                Network 51

   2.2   Reconfigurable Wavelength-Routing Node 53

        2.2.1   Add/Drop Multiplexer-Based Reconfigurable

                Node in a Ring WDM Network 54

        2.2.2   Wavelength Convertible Node Architecture 57

2.2.3   Reconfigurable Node Architecture in
        WDM-Based Mesh Optical Network 58
        2.2.3.1   Wavelength-Router–Based
                  Reconfigurable Node 59
        2.2.3.2   Fully Wavelength Convertible Node
                  Architecture of a WDM Mesh Network 59
2.2.4   SONET over WDM Node Architecture for a
        Mesh Optical Network 59
2.2.5   Transport Node of a WDM Optical Network.............62
2.2.6   IP over WDM Network Node Architecture 62
2.2.7   Node Architecture for Multicasting Optical
        Network 62
2.2.8   Traffic Grooming Node Architecture for an
        Optical Mesh Network 64
2.2.9   Node Architecture of Optical Packet-Switched
        Network 66
2.3   Network Node Based on Delivery and Coupling Switch 68
2.4   Multihop Network Node Architecture 68
Summary 70
Exercises..70
References 71

Chapter 3   Devices in Optical Network Node 75

3.1   Basic Components of Integrated Waveguide Devices.............75
      3.1.1   Directional Coupler...76
              3.1.1.1   Coupled Mode Theory................................77
              3.1.1.2   Power Transferred between Two
                        Waveguides Due to Coupling 77
              3.1.1.3   Coupling Coefficient..................................79
      3.1.2   MMI Coupler..79
              3.1.2.1   Guided Mode Propagation Analysis...........81
              3.1.2.2   Power Transferred to the Output
                        Waveguides...82
      3.1.3   TMI Coupler..82
              3.1.3.1   Power Transferred to Output Waveguides 83
      3.1.4   Array Waveguide Grating 83
      3.1.5   MZ Active Device 85
              3.1.5.1   *TE* Polarization 85
3.2   Wavelength Division Multiplexer/Demultiplexer-Based
      Waveguide Coupler...88
      3.2.1   WDM-Based TMI Coupler 88
3.3   Optical Switching 90
      3.3.1   MZ Switch...91
              3.3.1.1   TOMZ Switch-Based DC 91
              3.3.1.2   *TE* Polarization 92

                          3.3.1.3   EOMZ-Based DC 93
                          3.3.1.4   MMI Coupler-Based MZ Switch 94
                          3.3.1.5   TMI Coupler-Based MZ Switch 94
              3.3.2   X-Junction Switch..95
              3.3.3   DC-Based Electrooptic Switch...................................96
              3.3.4   Gate Switches 97
      3.4   Optical Crossconnect (OXC) 98
              3.4.1   Architecture-Based Crossconnect............................99
              3.4.2   Micro Electro Mechanical Systems (MEMS)...........99
      3.5   Optical ADM (OADM) 100
              3.5.1   Thermooptic Delay Line Structure 103
      3.6   SONET/SDH 106
              3.6.1   Transmission Formats and Speeds of SONET......... 106
              3.6.2   SONET/SDH Rings.. 109
      3.7   Optical Regenerator... 109
              3.7.1   Optical Amplifiers.. 109
              3.7.2   Optical Amplifier Characteristics 110
              3.7.3   Semiconductor Laser Amplifier 111
              3.7.4   Doped Fiber Amplifier 112
              3.7.5   Raman Amplifier.. 116
      3.8   Channel Equalizers 117
      3.9   Wavelength Conversion 122
              3.9.1   Opto Electronic Wavelength Conversion 123
              3.9.2   Wavelength Conversion Using Coherent Effects...... 124
              3.9.3   Wavelength Conversion Using Cross Modulation.... 125
                          3.9.3.1   Semiconductor Laser Based
                                          Wavelength Conversion 126
                          3.9.3.2   All-Optical Wavelength Conversion
                                          Based on CPM in Optical Fiber................ 126
      3.10   High-Speed Silicon Photonics Transceiver 127
              3.10.1 Silicon Photonics Transceiver Architecture............. 127
              3.10.2 Performance 128
      Summary 129
      Exercises.. 129
      References 130

Chapter 4   Processing of Integrated Waveguide Devices for Optical
              Network Using Different Technologies... 135

      4.1   Fabrication and Characteristics of Silica (SiO₂)/Silicon
              Oxynitride (SiON)-Based Devices 135
              4.1.1   Deposition of Thin Film SiON Layer by Using
                          LPCVD.. 136
              4.1.2   Deposition of SiO₂/SiON Layer by Using PECVD... 137
                          4.1.2.1   Silicon Dioxide (SiO₂) 138
                          4.1.2.2   Silicon Nitride.. 138

        4.1.2.3   SiON Layer 139

    4.1.3   Tuning of Refractive Index Using Thermooptic
           Effect 144

    4.1.4   Devices Fabricated and Demonstrated by Using
           SiO_2/SiON Material.. 144

    4.1.5   Properties of SiO_2/SiON... 145

4.2   Fabrication and Characteristics of
    SiO_2/GeO_2-SiO_2 Waveguide Material................................... 145

    4.2.1   Deposition of SiO_2/GeO_2-SiO_2 Layer Using
           PECVD.. 146

    4.2.2   Deposition of SiO_2/GeO_2-SiO_2 Material Using
           Flame Hydrolysis.. 147

    4.2.3   Tuning of Refractive Index Using Thermooptic
           Effect 148

    4.2.4   Devices Fabricated and Demonstrated by
           Previous Authors Using SiO_2/GeO_2-SiO_2 Material ... 149

    4.2.5   Properties of SiO_2/GeO_2-SiO_2 149

4.3   Fabrication and Characteristics of SOI
    Waveguide Material 150

    4.3.1   Fabrication of SOI Wafer 150

           4.3.1.1   BESOI Processing 150

           4.3.1.2   SIMOX Method 150

    4.3.2   Device Fabricated and Demonstrated by
           Previous Authors Using SOI Material..................... 151

    4.3.3   Properties of SOI.. 152

4.4   Fabrication and Characteristics of Ti:$LiNbO_3$ Waveguide
    Material 153

    4.4.1   Processing of $LiNbO_3$-Based Waveguide................. 153

           4.4.1.1   Thermal in Ti-Diffusion Method............. 153

           4.4.1.2   Proton Exchange Method 157

    4.4.2   Tuning of Refractive Index Using Electrooptic
           Effect 158

    4.4.3   Devices Fabricated and Demonstrated by
           Previous Authors Using $LiNbO_3$ Material 158

    4.4.4   Properties of $LiNbO_3$ 158

4.5   Fabrication and Characteristics of InP/GaAsInP
    Waveguide Materials 159

    4.5.1   Processing of InP/InGaAsP Waveguide................... 159

           4.5.1.1   Deposition of GaAsInP and InP Layers
                    Using MBE Growth System 160

           4.5.1.2   InP/GaAsInP Waveguide Fabrication....... 163

    4.5.2   Tuning of Refractive Index of InP/GaAsInP
           Waveguide 163

    4.5.3   Devices Fabricated and Demonstrated by
           Previous Authors Using InP/GaAsInP Material 163

    4.5.4   Properties of InP/GaAsInP...................................... 164

4.6   Fabrication and Characteristics of Polymeric Waveguide
        Material 164
        4.6.1   Fabrication of Polymeric Waveguides...................... 165
        4.6.2   Tuning of Refractive Index Using Thermooptic
                 Effect 166
        4.6.3   Devices Fabricated and Demonstrated by
                 Previous Authors Using Polymer Technology.......... 166
        4.6.4   Properties of Polymeric Material 167
4.7   Comparative Study of Integrated Waveguide Materials 167
Summary 169
Exercises... 169
References 169

Chapter 5   Data Link Control for Optical Network 173

5.1   Frame Synchronization 173
        5.1.1   Asynchronous Transmission 173
        5.1.2   Synchronous Transmission...................................... 174
5.2   Flow Control.. 175
        5.2.1   Stop and Wait Flow Control 175
        5.2.2   Sliding Window Flow Control 176
5.3   Error Detection and Control.. 179
        5.3.1   Error Detection.. 179
                 5.3.1.1   Vertical and Horizontal Redundancy
                            Check 179
                 5.3.1.2   Cyclic Redundancy Check...................... 181
        5.3.2   Error Control 185
                 5.3.2.1   Stop and Wait ARQ 186
                 5.3.2.2   Go-Back-N ARQ..................................... 188
                 5.3.2.3   SREJ ARQ.. 189
5.4   High-Level Data Link Control (HDLC)................................. 191
        5.4.1   Types of Station... 191
        5.4.2   Types of Configurations 191
        5.4.3   Types of Data Transfer Modes 191
        5.4.4   HDLC Frame Format.. 192
        5.4.5   Operation of HDLC.. 194
                 5.4.5.1   Initialization... 194
                 5.4.5.2   Data Transfer 195
                 5.4.5.3   Disconnect 196
        5.4.6   Examples of HDLC Operations 196
5.5   Other Link Control Protocol 197
        5.5.1   LAPB... 197
        5.5.2   LAPD 198
        5.5.3   LLC/MAC 198
        5.5.4   LAPF... 198

          5.5.5   ATM 199
                  5.5.5.1   ATM Protocol 200
                  5.5.5.2   ATM Logical Connections 201
                  5.5.5.3   Transmission of ATM Cells.....................206
      Summary 208
      Exercises..208
      References 210

Chapter 6   Data Communication Networks Having No Optical Transmission..... 213

      6.1   History and Background of Networking-Different
            Generations...213
      6.2   First Generation of Network..214
            6.2.1   Protocol Architectures 214
            6.2.2   Topologies 216
                    6.2.2.1   Bus Topology 216
                    6.2.2.2   Tree Topology 218
                    6.2.2.3   Ring Topology 218
                    6.2.2.4   Star Topology................................220
                    6.2.2.5   Mesh Topology 221
            6.2.3   Medium Access Control..221
                    6.2.3.1   Round Robin 221
                    6.2.3.2   Reservation 224
                    6.2.3.3   Contention....................................225
            6.2.4   Logical Link Control...229
            6.2.5   Wireless LANs 230
                    6.2.5.1   Medium Access Control (MAC) 231
            6.2.6   Asynchronous Transfer Mode (ATM) LAN 232
      Summary 233
      Exercises..233
      References 235

Chapter 7   Fiber-Optic Network without WDM..237

      7.1   Bus Topology...237
            7.1.1   Fasnet..238
            7.1.2   Expressnet 239
            7.1.3   Distributed Queue Dual Bus (DQDB) 241
      7.2   Ring Topology: FDDI...242
            7.2.1   MAC Frame...243
            7.2.2   MAC Protocol of FDDI..244
      7.3   Star Topology 245
            7.3.1   Fibernet...246
            7.3.2   Fibernet-II..248
      7.4   Wavelength Routed Networks without WDM......................250

Summary 252
Exercises... 252
References 253

Chapter 8   Single-Hop and Multihop WDM Optical Networks 255

8.1   Single-Hop Networks 255
        8.1.1   Characteristics of a Basic Single-Hop WDM Star
                 Network 257
8.2   Different Single-Hop Optical Networks................. 260
        8.2.1   SONATA 260
        8.2.2   LAMBDANET....................................... 261
        8.2.3   Rainbow 261
                 8.2.3.1   Rainbow Protocol 262
                 8.2.3.2   Model of Rainbow 263
        8.2.4   Fiber-Optic Crossconnect (FOX)-Based
                 Single-Hop Network................................ 269
        8.2.5   STARNET.. 269
        8.2.6   Other Experimental Single-Hop Systems 269
8.3   Coordination Protocol for a Single-Hop System 270
        8.3.1   Non Pre-transmission Coordination............. 270
                 8.3.1.1   Fixed Assignment 270
                 8.3.1.2   Partial Fixed Assignment Protocols 271
                 8.3.1.3   Random Access Protocol I 272
                 8.3.1.4   Random Access Protocol II............ 272
                 8.3.1.5   The PAC Optical Network............. 272
        8.3.2   Pre-transmission Coordination Protocols 273
                 8.3.2.1   Partial Random Access Protocols 273
                 8.3.2.2   Improved Random Access Protocols 275
                 8.3.2.3   Receiver Collision Avoidance (RCA)
                           Protocol.................................... 275
                 8.3.2.4   Reservation Protocols 276
8.4   Multihop Optical Network 277
        8.4.1   Optimal Virtual Topologies Using Optimization..... 279
                 8.4.1.1   Link Flow 279
                 8.4.1.2   Delay-Based Optimization............. 280
        8.4.2   Regular Structures................................. 281
                 8.4.2.1   ShuffleNet 281
                 8.4.2.2   de Bruijn Graph 284
                 8.4.2.3   Torus (MSN)............................ 285
                 8.4.2.4   Hypercube.............................. 286
                 8.4.2.5   GEMNET 286
8.5   SC Multihop Systems 292
        8.5.1   Channel Sharing in Shuffle Net 292
        8.5.2   Channel Sharing in GEMNET.................... 293
Summary 295

Exercises...295
References 299

Chapter 9   Optical Access Architecture 303

    9.1   Performance Measures and Notation of Access
          Architecture...303
          9.1.1   Random-Access Methods...304
                9.1.1.1   ALOHA 305
                9.1.1.2   Slotted ALOHA..............................307
          9.1.2   Carrier Sense Multiple Access (CSMA) 308
                9.1.2.1   Non-Persistent CSMA 308
                9.1.2.2   Slotted Non-Persistent CSMA 311
                9.1.2.3   1-Persistent CSMA 313
                9.1.2.4   *p*-Persistent CSMA 317
          9.1.3   CSMA/CD: IEEE Standard 802.3 318
                9.1.3.1   Throughput Analysis for
                            Non-Persistent CSMA/CD................ 320
                9.1.3.2   Throughput Analysis for 1-Persistent
                              CSMA/CD 322
          9.1.4   Stability of CSMA and CSMA/CD.........................324
          9.1.5   Controlled-Access Schemes 325
                9.1.5.1   Token Ring: IEEE Standard 802.5 326
                  9.1.5.2   Token Bus: IEEE Standard 802.4 327
    9.2   Optical Access Network 330
          9.2.1   Issues in Optical Access Architecture 331
    9.3   Simple Fiber-Optic Access Network Architectures 331
    9.4   Components of PON Technologies...332
          9.4.1   Optical Splitters/Couplers 332
          9.4.2   PON Topologies 333
          9.4.3   Burst-Mode Transceivers 334
    9.5   EPON Access Architecture 334
          9.5.1   Operation of EPON 334
    9.6   Multi-Point Control Protocol (MPCP) 336
          9.6.1   Discovery Processing 336
          9.6.2   Report Handling 337
          9.6.3   Gate Handling 338
          9.6.4   Clock Synchronization 338
    9.7   Dynamic Bandwidth Allocation (DBA) Algorithms in
          EPON...339
          9.7.1   IPACT..340
          9.7.2   Services 341
    9.8   IP-Based Services over EPON 342
          9.8.1   Slot-Utilization Problem...342
          9.8.2   Circuit Emulation (TDM over IP)...........................343
          9.8.3   Real-Time Video and VoIP 344

        9.8.4   Performance of CoS-Aware EPON 345
        9.8.5   Light-Load Penalty 345
    9.9   Other Types of PONs 346
        9.9.1   APON 346
        9.9.2   GFP-PON 347
        9.9.3   WDM-PON 347
            9.9.3.1   Need for WDM in PONs 347
            9.9.3.2   Arrayed Waveguide Grating
                      (AWG)-Based WDM-PON 348
            9.9.3.3   WDM-PON Architectures 349
            9.9.3.4   Scalability of WDM-PON 351
        9.9.4   Deployment Model of WDM-PONS 352
            9.9.4.1   Open Access 352
    Summary 354
    Exercises 355
    References 358

Index 361

Preface

Before writing the book's topics for the intended audience, many books on optical networks have been consulted. In the fast-moving technology of optical networks, optical network security, reliability and survivability, optical fiber transmission media and its design codes, node components and their fabrication technology, apart from virtual topology design and wavelength routing in optical network have to be included. The basics of networks (specially data communication networks) for audiences who wants to start learning and initiate the research on optical networks need to be discussed.

In fact, the approach of this book is also to help the readers, including students, researchers, engineers, etc., to know and understand the basics of networking before going to advanced topics of optical network. Apart from the basics, the book discusses the basic theory of optical transmission media, integrated optics devices used as preferred hardware components in optical network for helping the design engineer for implementation of a nationwide high-speed optical network. The book is designed to be used at a number of levels, varying from a senior undergraduate elective to a more advanced graduate course, to a reference work for the designers and researchers in the field. The book has two volumes – *Optical Networks and Components: Fundamentals and Advances:* Volume 1 is entitled *Fundamentals of Optical Networks and Components* and Volume 2 is entitled *Advances in Optical Networks and Components*. Volume 1 consists of 9 chapters having mostly hardware components used in optical backbone and the basics of data transmission and optical access used in optical networks, whereas volume 2 comprises 11 chapters having prospectives and advances in optical networks. The chapters are as follows:

CHAPTERS IN VOLUME 1:

Chapter 1: Introductory Concept

This chapter describes the overview of different communication networks such as local area networks, VSAT networks, Integrated Services Digital Networks (ISDN), Broadband ISDN and Basic principles of optical transmission through optical fibers

Chapter 2: Different Optical Network Node

This chapter mentions WDM-based node architectures, Wavelength convertible node architectures, traffic groomed node architectures and multicast node architectures for flexible operation of optical networks

Chapter 3: Devices in Optical Network Nodes

To make it more compatible with modern technology, integrated optics concept emerged. Chapter 3 discusses integrated waveguide devices used in optical network nodes.

Chapter 4: Processing of Integrated Waveguide Devices for Optical Network Using Different Technologies

This chapter addresses processing technologies with $SiON/SiO_2$, SiO_2-GeO_2/SiO_2, SOI, $LiNbO_3$, GaAsInP/InP and polymeric materials used for fabrication of integrated waveguide devices.

Chapter 5: Data Link Control for Optical Network

Since Optical network deals with data transmission, this chapter discusses data link control schemes along with error detection in data communication for optical networks

Chapter 6: Data Communication Networks Having No Optical Transmission

This chapter addresses data transmission in networks having no optical transmission which is treated as the first generation of networks

Chapter 7: Fiber-Optic Network without WDM

This chapter deals with optical transmission network having transmission in optical domain and multiplexing and switching in electrical domain and network is treated as second generation of networks

Chapter 8: Single-Hop and Multihop WDM Optical Networks

This chapter describes single-hop and multihop transmission network using WDM treated as a third generation of networks which deals with both transmission and node device operation in optical domain.

Chapter 9: Optical Access Architecture

This chapter discusses optical access techniques having both optical access and electrical access technologies.

CHAPTERS IN VOLUME 2:

Chapter 1: Optical Ring Metropolitan Area Network

This chapter describes mainly optical ring topology used in metropolitan area networks

Chapter 2: Queuing System and Its Interconnection with Other Networks

This chapter discusses queuing theories used to analyze the performance of computer communication networks along with optical networks.

Chapter 3: Routing and Wavelength Assignment

This chapter discusses different static and dynamic routing and wavelength assignment approaches used in optical network.

Chapter 4: Virtual Topology

This chapter addresses formulation of virtual topology and its design for optical network.

Chapter 5: Wavelength Conversion in WDM Network

This chapter mentions various aspects and benefits of wavelength conversion in the networks and its incorporation in a wavelength routed network design for efficient routing and wavelength assignment.

Chapter 6: Traffic Grooming in Optical Network

This chapter addresses the different schemes of static and dynamic traffic grooming in optical networks.

Chapter 7: Survivability of Optical Networks

This chapter presents the fault management schemes such as protection deployed in a survivable network for SONET/SDH rings and mesh optical networks.

Chapter 8: Restoration Schemes in Survivability of Optical Networks

This chapter discusses the different restoration schemes in survivable optical networks.

Chapter 9: Network Reliability and Security

This chapter discusses the optical signal security schemes used in optical network having WDM and WDM apart from basic theory of network security.

Chapter 10: FTTH Standards, Deployments and Issues

This chapter presents different FTTH standards, deployments and research issues.

Chapter 11: Math Lab Codes of Optical Communication

This chapter discusses mathematical simulation codes of optical fiber link design and the codes help in designing the physical link in optical networks.

IMPORTANCE OF THE BOOK

Optical networks have become essential to fulfill the skyrocketed demands of bandwidth in present day's communication networks. In these networks, flexible operation such as routing, restoration and reconfiguration are provided by the nodes, where wavelength division multiplexing (WDM), optical matrix switches, add/drop multiplexers, EDFA, SONET wavelength router, etc. are the key devices. This book attempts to emphasize optical networks and these key devices are used in these networks. The main purpose of the book is to provide students, researchers and practicing engineers with an expert guide to the fundamental concepts, issues and state-of-the-art developments in optical networks. The features of the book are that it provides the concept of these devices along with its fabrication processes, optical encryption, etc. Since optical networks mostly handle data communication, it also provides data transmission control and protocols to make data communication interlinked with optical network. One of its special characteristics is that each optical network topology has a node architecture with its device operations. The book is organized into eighteen chapters, covering the basic principles and fundamental importance concerning the technology and latest development. Unlike other books in the area, this book covers a description of both hardware components and routing software.

1. We have mentioned 50 examples and 300 practices distributed uniformly throughout all the chapters of the book. These examples are very much helpful in understanding the basic problems and its solution, specially for undergraduate and postgraduate students.
2. These examples are helpful in solving problems in development of high-speed optical networks for network software development companies/vendors.
3. The practices mentioned in the books are useful in learning for formulation and modeling of conventional network topologies and nationwide mesh network topologies.
4. Both these examples and practices are also helpful for researchers to initiate research in this field and to develop research problem-solving capability in this field as all the chapters start from basics to latest developments.

The basic purpose of our book is how optical communication is used to provide high bandwidth demand due to a skyrocketed increase of number of users and services.

Actually more users start to use high-speed optical data networks, and their usage patterns evolve to include more bandwidth-intensive networking applications such as data browsing on the World Wide Web, Java applications and video conferencing. So, there is an acute need for very high-bandwidth transport network facilities, whose capabilities are much beyond those that current networks (e.g., today's Internet) can provide. Research and development on optical WDM networks have matured considerably over the past decade, with commercial deployment of WDM systems. In most cases WDM optical transmission systems have wavelength channel counts of **32** to 64. In future, channel counts may be increased to **320** per fiber strand. All working in the field of optical networking (researchers, government funding agencies, telecom network operators, equipment vendors, etc.) requires hybrid system in which optics, electronics and software play in building a successful optical network. So it needs the importance of cross-layer design issues involving the physical layer (optics and electronics).

The audience may need to understand the issues and challenges in designing such networks. It is anticipated that the next generation of the Internet will employ WDM-based optical backbones. In all chapters, we discuss these issues and challenges.

INTENDED AUDIENCE

The intended audience of this book are researchers, industry practitioners and graduate students (both as a graduate textbook and for doctoral students' research). Many electrical engineering, computer engineering and computer science programs around the world have been offering a graduate course on this topic. That is, research and development on optical communication networks have matured significantly to the extent that some of these principles have moved from the research laboratories to the formal (graduate) classroom setting.

Each chapter is typically organized in a stand-alone and modular fashion, so any of them can be skipped, if so desired.

I also hope that industry professionals will find this book useful as a well-rounded reference. Through my own industry relationships, I find that there exists a large group of people who are experts in physical-layer optics, and who wish to learn more about network architectures, protocols, and the corresponding engineering problems in order to design new state-of-the-art optical networking products.

Acknowledgements

Although my name is visible on the cover page, a large number of people are involved to produce a quality book.

First and foremost, I wish to thank my research and project students for the effortin getting the book to its current form. Much of the book's material are based on research that I have conducted over the years with my graduate students, research scientists and my research group member visiting my laboratory, and I would like to acknowledge them as follows: Dr Bidyut Deka for optical network hardware devices (Chapters 3 and 4), Dr Bijoy Chand Chatterjee for some portion of wavelength-routing material (Chapter 12), Dr Rabindra Pradhan for wavelength routing, traffic grooming and protection (Chapters 12–15); Dr Mahipal Singh Queuing system, network security (Chapters 11 and 18);

A number of additional individuals who guide and discuss with over the years and who I would like to acknowledge are the following: Prof. Alok Kumar Das, Prof. Mrinal Kanti Naskar, Prof. Debasish Datta, Prof. Utpal Biswas and Prof. Asis Kumar Mal and Prof. S Choudhury.

I like to acknowledge the people at CRC who work with me – Marc Gutierrez, Assunta Petrone and Nick Mould for their assistance during the book's production.

Finally, I like to thank my family members for their constant encouragement, specially from my father Harekrishna Sahu, my mother Jyotsana Sahu, my wife Arpita and my daughters Prakriti and Ritushree. Without their support, it was impossible to complete this project.

Author

P. P. Sahu received his M.Tech. degree from the Indian Institute of Technology Delhi and his Ph.D. degree in engineering from Jadavpur University, India. In 1991, he joined Haryana State Electronics Development Corporation Limited, where he has been engaged in R&D works related to optical fiber components and telecommunication instruments. In 1996, he joined Northeastern Regional Institute of Science and Technology as a faculty member. At present, he is working as a professor in the Department of Electronics and Communication Engineering, Tezpur Central University, India. His field of interest is integrated optic and electronic circuits, wireless and optical communication, clinical instrumentation, green energy, etc. He has received an INSA teacher award (instituted by the highest academic body Indian National Science Academy) for high level of teaching and research He has published more than 90 papers in peer-reviewed international journals, 60 papers in international conference and has written two books published by Springer-Nature and two books by McGraw-Hill, India. Dr Sahu is a Fellow of the Optical Society of India, Life Member of Indian Society for Technical Education and Senior Member of the IEEE.

1 Introductory Concept

Due to the skyrocketed increase of internet users and services, high-speed communication is required to to fulfill enormous demand of bandwidth. In this direction, all optical networks with wavelength divisional multiplexing (WDM) technology have become essential to develop such high-speed communication. This book deals with the principles and fabrication of optical network devices such as wavelength router [1–6], WDM [7–11], add/drop multiplexer (ADM) [12–17], photonic switch [18–23], Erbium-doped fiber amplifier (EDFA) [24–26], and EDFA gain equalizer [27–30].

1.1 BASIC COMMUNICATION MODEL

Before discussing optical networks, one should know the basic communication model. Figure 1.1 shows the general block diagram of a communication system having a source system, destination system and transmission media. The source system has a source device that generates raw signals such as data, voice/video and information, and a transmitter that transforms and encodes raw signals in such a way as to produce electromagnetic signals that can be transmitted through a transmission system, which is a complex network connecting source and destination. The destination system has

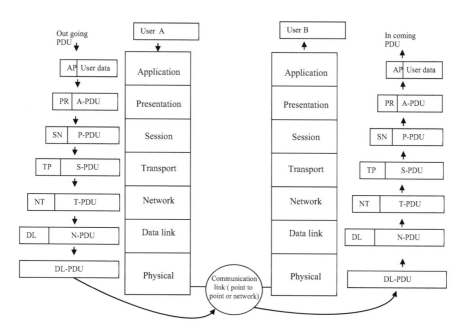

FIGURE 1.1   Transport of data under OSI environment.

1

a receiver that receives signal from a transmission system and converts it into a raw signal, and from the raw signal, the information/message is recovered. In this section, we provide an overview of different communication networks.

1.1.1   Local Area Network

Local area network is a communication network that covers a small geographical area (typically a building or a cluster of buildings) and provides a means for information exchange among the devices/nodes attached to it. The communication between different nodes of the network is mainly based on open system interconnection (OSI) model or transport control protocol/internet protocol (TCP/IP).

1.1.1.1   OSI Model

Since the origin of communication, its operation varies from vendor to vendor. So standards are needed to promote interoperability among vendor equipment and to encourage economics of scale. Because of the complexity of communication tasks, no single standard will be sufficient. It is better to form a framework for standardization rather than breaking the operation into manageable parts. In 1977, the International Standard Organization (ISO) had started to establish a subcommittee for developing the architecture of the framework. As a result, OSI has been developed [31,32]. The OSI model is a seven-layer architecture in which each layer performs definite functions, namely physical layer, data link layer, network layer, transport layer, session layer, presentation layer and application layer.

1. Physical layer: It permits interconnection with different control procedures such as V.24 and V.25 for various physical media.
2. Data link layer: It controls data transmission through the system having high error rate (i.e., an error rate not acceptable for a great majority of applications). It works in the framework of high-level data link control (HDLC). It is just above the physical layer.
3. Network layer: It selects a connection path or provides a rout (where the intermediate nodes may be present) for data transmission from one node to the other.
4. Transport layer: It controls successful transportation of data from the source to the destination node. It provides totality of transmission service and ensures that data are delivered error-free, in sequences with no loss and delicacy.
5. Session layer: It provides synchronization or organization dialog between the source and the destination before data transmission. It does function above the transport layer. It provides a mechanism for recovery and permits backup.
6. Presentation layer: It does general interest functions related to representation and manipulation of structured data just before the application layer. It defines the format of the data to be exchanged between different applications.
7. Application layer: It performs management functions and generally useful mechanisms that support distributed applications.

Figure 1.1 shows how data are transmitted in an OSI architecture with the use of a protocol data unit (PDU). When user A has a message to send to user B, it transfers these data to the application layer, where a header is added to the data making it A-PDU. Then, it is passed to the presentation layer. In the same way, these PDU goes through the layers as per the figure (by using HDLC format) to a data link layer. The data link layer unit, also called as a frame, is then passed to a communication path/link in the network by using a physical link. When the frame is received in the destination node/target node, a reverse process occurs. As the PDU ascends, each layer strips off the outermost header, acts on the protocol information contained therein, and passes the remainder up to the next layer.

1.1.1.2   TCP/IP Protocol

Since 1990, TCP/IP has become more popular than the OSI model because of its simplicity and interoperability over different networks, thus providing different services through its IP layer. In an OSI model, protocols at the same level of hierarchy have certain features in common. In this direction, TCP/IP architecture is better than that of the OSI model. The TCP/IP has five layers [33,34] – physical layer, network access layer, internet layer, transport layer and application layer.

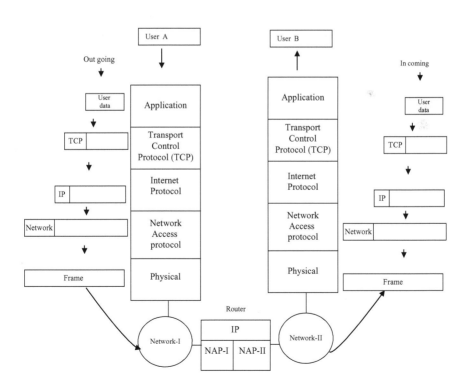

FIGURE 1.2   Transport of data under TCP/IP environment.

1. Physical layer: It defines the characteristics of transmission medium, signaling rate and encoding scheme.
2. Network access layer: It makes a logical interface between an end system and a subnetwork where a connection path is selected.
3. Internet layer: It does the function of routing data from the source node to the destination host through one or more networks connected by routers.
4. Host-to-host transport layer.
5. Application layer.

Figure 1.2 shows the transport layer through different layers in TCP/IP protocol [34]. When user A has a message to send to user B via different applications, as given in Figure 1.3a, and transfers these data to the application layer, a header of TCP is added to the data. In the same way, these user data go through the layers as per Figure 1.2, adding with different header files in these layers. In the IP layer, IP header files are added. Finally it is passed to the communication path/link in the network-I and via router having NAPI and II and IP interface, and the frame is received in the destination node/target node, where a reverse process occurs. The PDU ascends, and each layer strips off the outermost header, acts on the protocol information contained therein, and passes the remainder up to the next layer. Finally, it goes to user B. Figure 1.3 shows the different applications provided by the TCP/IP architecture, in which simple mail transfer protocol (SMTP), hypertext transfer protocol (HTTP), file transfer protocol (FTP), TELNET and broader gateway protocol (BGP) make use of TCP via IP layer to get connection to another host of corresponding module, whereas user datagram protocol (UDP), internet control message protocol (ICMP) and open shortest path first (OSPF) make use of IP directly for connection.

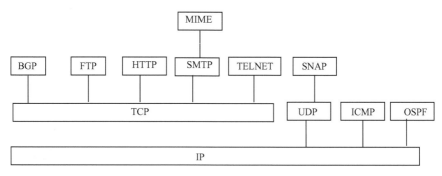

BGP – Border gate wayProtocol   MIME – Multi-purpose internet mail extension
FTP – File transfer protocol   SMTP – Simple mail transfer protcol
HTTP – Hyper text transfer protocol   SNMP – Simple network management protocol
ICMP – InternetControlMessageProtocol   TCP – Transmission control protocol
IP – Internet Protocol   UDP – User datagram protocol
OSPF – Open shortest path first

FIGURE 1.3   (a) Application protocol of TCP/IP.

(Continued)

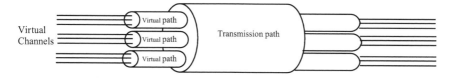

FIGURE 1.3 (CONTINUED)   (b) ATM connection relationship.

1.1.2 Wide Area Network

Wide area network (WAN) has been traditionally considered to be a network that covers large geographical area. It consists of a number of interconnected switched nodes. Here, the transmission from one device to the other is routed through these internal nodes to the specified destination device. For this purpose, a switching facility is used to move the data from one node to the other until it reaches the destination. The WANs are mainly implemented by using two switching technologies – circuit switching [35,36] and packet switching [35,36]. Apart from that, WANs may use asynchronous transfer mode (ATM) and frame relay architectures.

1.1.2.1 Circuit Switching

In a circuit-switched network [33], a dedicated communication path is established between two stations through the nodes of the network. There are three steps – circuit establishment, data transfer and circuit disconnect after data transfer is over.

1. Circuit establishment: Before signal for data is transmitted, an end-to-end (source-to-destination) circuit must be established. For this establishment, a signal path must be selected by routing. Routing in circuit switching is done in three ways: fixed routing, alternate path routing and adaptive routing. These routing approaches are discussed in the next chapter. After selection of path will be dedicated for transmission of data for this connection.
2. Data transfer: After establishment of connection, data are transmitted through a dedicated path selected for routing. This path cannot be shared by other stations or nodes till the data transfer is over.
3. Circuit disconnect: After data transfer is over, the circuit is disconnected and transferred to another user.

1.1.2.2 Packet Switching

A different approach is used in packet switching. Here, data are sent out in a sequence of small chunks with destination and route address, and these chunks are called as packets. Each packet is passed through the network from node to node along the path leading from source to destination. At each node, the entire packets is received store briefly and then transmitted to the next node. The links used in the path are shared by other users to send their packets. There are two types of packet switching – virtual circuit packet switching [36] and datagram packet switching [33].

Like circuit switching, in virtual circuit packet switching, there is a requirement of connection call setup, and after connection is set up for a path, all the packets should flow through the same path to the destination. But the links used in the path are shared by the packets of other users.

In datagram packet switching, it does not require to set up connection previously, and the individual datagram packets are routed independently to destination not by single path but maybe by a number of paths. The datagram packet switching is faster than virtual circuit packet switching because of less queue time delay.

1.1.2.3 Frame Relay

The packet switching relatively exhibits a high bit error rate while it is implemented for long-distance communication. To control these errors, it requires more overload; besides, extra time is required to process these overloads at each intermediate node. This overload is unnecessary and counterproductive. The frame relay networks are used for operating efficiently at user data rates of up to 2 Mbps. The purpose of achieving these high data rates is to strip out most of the overload involved with error control. In this direction, frame relay was developed by reducing overload, with consideration of a smaller number of layers [37].

1.1.2.4 Asynchronous Transfer Mode

ATM is another approach in which a fixed length packet called cell is transmitted for data transfer [38]. These ATM cells have little overload of error correction. Due to fixed length, the processing time is also reduced. In ATM, multi-virtual channels of fixed length are available for transmission of data. As per the demand for bandwidth, the numbers of virtual channels are set dynamically for maintaining the quality of service. After selection, the virtual channels are put into a virtual path. So, it is required to set virtual channel connections, and finally, a virtual path connection is set. The overload in ATM is less than that of frame relay. As a result, the data rate is more in ATM (10–100 Mbps) than that of frame relay (2 Mbps). Figure 1.3b shows an ATM connection concept, how the virtual channel connections construct virtual paths and finally how groups of virtual paths make a transmission path of ATM network.

1.1.3 VSAT Network via Satellite

Before discussing VSAT network, one should know about satellite microwave. A communication satellite is basically a microwave relay station in which satellite is used as a transponder to connect two or more ground-based microwave transmitter/receivers which are basically very small aperture terminals (VSAT) [33]. The satellite transmits one frequency band named as downlink frequency, whereas it receives transmission on one frequency band named as uplink frequency. For a satellite to get a communication effectively, it is needed to make it stationary relative to its position over earth because offline -of-sight communication with the users/ stations at all times. The coverage area for elevation angle I and altitude H is derived by considering the distance between two users stationed at two extreme points of coverage area, D_M written as [39] from Figure 1.4a

$$D_M = 2R_e \left[\pi/2 - I - \sin^{-1} \left\{ R_e \cos(I)/(R_e + H) \right\} \right]$$

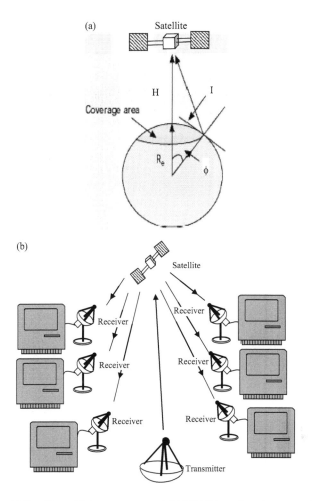

FIGURE 1.4   (a) Satellite with its coverage and (b) VSAT broadcasting network.

where R_e = radius of earth. Considering velocity of propagation of signal, C, the maximum propagation delay is written as

$$T_D = 2\sqrt{\left\{H + R_e - R_e \cos\left(D_M/2R_e\right)\right\}^2 + \left\{R_e \sin\left(D_M/2R_e\right)\right\}^2}\Big/C$$

The total number of satellites required for global coverage can be written as

$$N = 4\sqrt{3n\pi^2}\Big/\left[\pi/2 - I - \sin^{-1}\left\{R_e \cos(I)/\left(R_e + H\right)\right\}\right]^2 \qquad (1.1a)$$

where n = minimum number of satellites seen from any point at any time, where $n = 1$ for single-fold coverage and $n = 2$ for double-fold coverage.

There are three types of polar orbit satellites:

- Low earth orbit (LEO) – altitude 500–1500 km,
- Medium earth orbit (MEO) – altitude 5000–15,000 km,
- Geostationary orbit– altitude 35,784 km.

LEO and MEO satellite networks provide a wide service area where there is less non-telecommunication infrastructure, especially on rural and hilly regions of Asia, Africa, Eastern Europe, South America, and the polar areas [40]. These LEO and MEO satellite networks also cover global coverage to their users, which a typical GEO satellite system cannot accommodate. In this direction, LEO satellite system, which is Motorola's IRIDIUM system, was employed in May 1998 for global coverage [40].

The IRIDIUM system is the first initiative of global coverage of wireless communication system to provide voice, data, fax, and paging services to the world. At an altitude of 780 km above the earth, 66 satellites derived by using equation (1.1a) are required in six planes. Each plane has 11 satellites. Planes have a near-circular orbit, with co-rotating planes spaced 31.6° apart and counter-rotating planes (one and six) spaced 22° apart [40]. The minimum elevation angle normally for an earth station is considered to be 8.2°, which maximizes the coverage area of the satellite and improves the link quality compared with lower elevation angles. The average satellite in-view time is approximately 10 minutes.

Further, we can increase the coverage area with MEO satellite, but the propagation time delay increases due to higher altitude. Further, both LEO and MEO satellites have lower lifetime in comparison to GEO satellites

Out of three satellites, GEO satellite is mostly used for VSAT network, mainly because of higher coverage area than other two satellites and high lifetime, although it has a high cost of installation and maintenance. Two satellites using same frequency band will interfere with each other while they come closer. To stay away from the problems between two satellites, 4-degree spacing of one satellite is used for 4/6 GHz and a 3-degree spacing of other satellite for 12/14 GHz. There are two types of transmission of signal via satellite – broadcasting and point-to-point transmission.

In broadcasting transmission, data are transmitted by one user treated as transmitter, whereas other VSAT receives the data as a receiver as shown in Figure 1.4b. Among applications in this direction are television distribution, long-range radio broadcasting and private business broadcasting. In the case of public broadcasting service (PBS), the television programming is distributed by the use of satellite channels.

In point-to-point transmission, signal transmissions are in both directions via satellite. There are two types of VSAT network which uses point-to-point transmission – centralized VSAT network and distributed VSAT network. Two users in two different VSATs in centralized network transmit their data via a central HUB as shown in Figure 1.5a. In distributed network, there is no central HUB, and two users in two different VSATs transmit directly without going through via as shown in Figure 1.5b.

The satellite transmission uses a frequency range of 1–15 GHz. In fact, there is significant noise from natural sources including galactic, solar and atmospheric noise and human-made interference from various electronic devices below 1 GHz, whereas above 15 GHz, the signal power is heavily attenuated in atmosphere. There

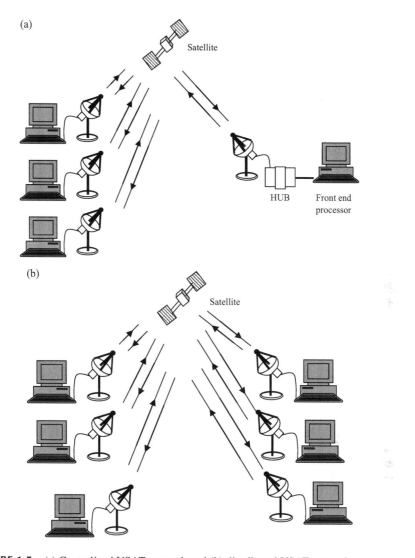

FIGURE 1.5 (a) Centralized VSAT network and (b) distributed VSAT network

(Continued)

are two frequency bands used for satellite communications in the range – C band and Ka band. For C band, the satellites provide point-to-point transmission with frequency ranges 5.925–6.425 GHz from earth to satellite (uplink) and 3.7–4.2 GHz from satellite to earth (downlink). Since this frequency ranges are saturated due to having so much traffic, there are two frequency ranges beyond 10 GHz used in Ka. In Ka band, the uplink and downlink frequency ranges are 14–14.5 and 11.7–12.2 GHz, respectively. Due to tremendous demand of bandwidth, there is another band Ku used in satellite communication even after 5 GHz. The ranges for Ku band are 27.5–31 and 17.7–21.2 GHz, respectively. But in the case of Ku band, the bandwidth used for both uplink and downlink is 3.5 GHz and those for other C and Ka band is 0.5 GHz. There are several characteristics of satellite communication to be considered:

- A propagation time of one quarter of second is taken for data transmission from one earth station to another station via satellite. Almost the same order of time delay is required for telephonic transmission.
- There are problems of error control and flow control that will be discussed later in this book.
- Satellite communication is mainly broadcasting in nature, but it can be used as point-to-point bidirectional communication.

1.1.4 Integrated Services Digital Network

Rapid development of communication technologies has resulted in an increasing demand of worldwide public telecommunication networks in which a variety of services such as voice and data (computer communication) are distributed. The Integrated Services Digital Network (ISDN) is a standard network having user interfaces and is also realized as digital switches and paths accommodating a broad range of traffic types and proving value-added processing services [41]. Standards of ISDN are made by ITU-T (formerly CCITT). The ISDN is a single worldwide uniformly accessible network having multiple networks connected within national boundaries. It has both circuit-switching and packet-switching connection at 64 kbps. There are mainly two types of services – voice communication and non-voice (data) communication. Figure 1.5c shows the ISDN architecture in which different services such as voice, PBX signal and data are connected to ISDN through its interface, and then all interfaces are connected to ISDN central office through digital transmitted media/pipe. The central ISDN office is connected to different networks such as circuit-switched network and packet-switched network with digital pipes at a certain bit rate. There are two generations of ISDN – narrowband ISDN (N-ISDN) [41] and broadband ISDN (B-ISDN) [41].

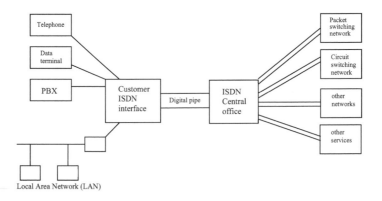

FIGURE 1.5 (CONTINUED)   (c) ISDN architecture.

1.1.4.1   Narrowband ISDN

The first generation of N-ISDN referred to as narrowband ISDN is based on 64 kbps channel as the basic unit of switching and has a circuit-switching orientation. It is developed by using a frame relay concept. All traffic in this system use D channel employing link access protocol-D (LAPD) which has two forms of service to LAPD users: the unacknowledged information-transfer service and acknowledged information-transfer service. The unacknowledged information-transfer service provides for the transfer of frames containing user data with no acknowledgment, whereas the acknowledged information-transfer service is a more common service similar to link access protocol-B (LAP-B) and HDLC [33]. Table 1.1a shows different narrowband channels included for transmission in N-ISDN with their application services.

1.1.4.2   Broadband ISDN

The second generation of ISDN referred to as B-ISDN provides services to both narrowband and broadband channels having a data speed of more than 100 Mbps. Initially, it uses ATM-based network, and later as the number of services increases, it uses optical network to provide services. We will discuss optical network thoroughly in this book.

Table 1.1b shows different broadband signals included for transmission in B-ISDN with their application services, apart from inclusion of narrowband signals.

1.2   OPTICAL FIBER PRINCIPLE

Optical network is mainly based on an optical fiber communication principle which requires optical fiber, optical transmitter and receiver. This section discusses about optical fiber, optical transmitter and receiver devices [25]. We attempt to know the physics behind the principles of optical transmission in fiber in order to provide some background.

TABLE 1.1A
Narrowband Signals

Channel Type	Data Rate	Application Services
D type	64 kbps	Control signal in network
B type	64 kbps	Voice, low-speed data
H_0 type	354 kbps	Compressed video, multiple voice, medium-speed data
H_{11} type	1.536 Mbps	Compressed video, multiple voice, EPBX signal medium-speed data
H_{12} type	1.920 Mbps	Compressed video, EPBX signal medium-speed data

TABLE 1.1B
Broadband Signals

Channel Type	Data Rate (Mbps)	Application Services
H_2 type	30–45	High-speed data, Full motion video, video telephony
H_3 type	60–70	High-speed data
H_4 type	120–140	Bulk text, Facsimile, enhanced video

1.2.1 Optical Fiber

Optical fiber is a circular waveguide consisting of a two-layered solid cylinder in which the inner layer is called as a core and the outer concentric layer is called as cladding [42]. It is protected by a thin plastic jacket as shown in Figure 1.6. Optical fiber having many properties is suited to an ideal transmission medium for high-speed networking. Figure 1.7 shows attenuation and dispersion characteristics of optical fiber. Normally, for traditional optical fiber made up of glass, there are two windows of wavelengths, mainly used for communication – one window is cantered at ~1310 nm with a bandwidth of 200 nm and an attenuation of less than 0.5 dB/km, and the total bandwidth in this region is about 25 THz [25], and the second window is centered at 1550 nm with a bandwidth of similar size, and an attenuation of ~0.2 dB/km having three bands, i.e., S band (1460–1530 nm), C band (1530–1560 nm) and L band (1560–1630 nm). The prominent loss is due to Rayleigh scattering, and the peak in loss in the 1400 nm neighborhood is mainly due to hydroxyl ion (OH–) impurities in the fiber [42]. Other losses are material absorption and radiation loss. The Er^{++}-doped optical amplifier is used for C band for long-haul (over 80 km) wide-area applications [25,42]. Apart from it enormous bandwidth and low attenuation, fiber also provides low error rates [25].

Apart from the high bandwidth of ~50 THz and low BER, it has the following advantages [25]:

1. small size and thickness,
2. flexible and light weight and less corrosive in different environments,
3. immunity to electromagnetic interference,
4. cheapest and most readily available substances available on earth, i.e., silica as fiber materials.

1.2.1.1 Optical Transmission in Fiber

The characteristics of the Optical fiber are based on the glass material, which is the main material of this circular waveguide. A waveguide is used as a path that allows the propagation of electromagnetic waves (light waves). As discussed earlier, it has an inner cylinder, core and outer layer, and a cladding in which the refractive index of the core is higher than that of the cladding. The ratio of the refractive indices of the cladding and core provides critical angle $\theta_c = \sin^{-1}(n_2/n_1)$. As shown in

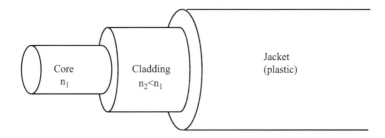

FIGURE 1.6 Optical fiber.

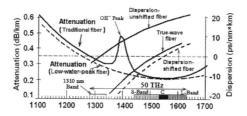

FIGURE 1.7 Attenuation characteristics of an Optical Fiber.

Figure 1.8, the light ray is launched to optical fiber at an angle θ_{in} which is more than $90°-\theta_c$. Once the light launches into the core, the light ray approaches towards the core-cladding interface at an angle more than critical angle $(\theta_i > \theta_c)$ inside the core and the light makes a total internal reflection. Due to laws of total internal reflection, light is completely reflected back into the interface, and similarly it will be reflected back in another same type of interface. Hence, it is transmitted along the length of a fiber without loss, as shown in Figure 1.9. Light travels through vacuum at a speed of $c = 3 \times 10^8$ m/s light can also travel through any transparent material, but the speed of light will be slower in the material than that in vacuum. The ratio of the speed of the signal inside the core is given by $C_{mat} = C/n_1 \sim 2 \times 10^8$ m/s, where refractive index of the core $n_1 = 1.5$ contributing propagation delay of 5ms/km.

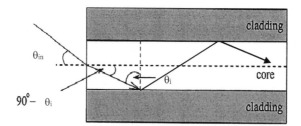

FIGURE 1.8 Launching of light ray into optical fiber.

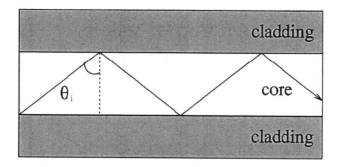

FIGURE 1.9 Light propagation in optical fiber with little loss.

The optical signal light is incident at an angle in which their fraction at the air-core boundary provides launching of the transmitted light. After launching the light signal, total internal reflection can take place at the core-cladding boundary. Figure 1.10 shows how a numerical aperture is found.

From Snell's law, it is written as

$$n_{air} \sin\theta_{air} = n_1 \sin\left(90° - \theta_c\right) = n_1 \sqrt{1 - \sin^2\theta_c} = n_1 \sqrt{1 - \frac{n_2^2}{n_1^2}}$$

$$= \sqrt{n_1^2 - n_2^2}$$

The quantity $n_{air} \sin\theta_{air}$ refers to the numerical aperture of optical fiber.

The two types of fiber are mainly step-index fiber and graded-index fiber. The step-index fiber having a constant refractive in the core and cladding is already discussed.

For light to launch to a fiber, the light must be incident on the core-cladding surface at an angle larger than critical angle, and for the step-index fiber, it is larger. The reduction of the critical angle is required to for more light to get total internal reflection and to reduce the same, the refractive index of the core has to be increased. It is difficult to increase the core index in glass fiber due to fabrication limitation. In this direction, a *graded-index* fiber [43] reduces the critical angle required for more light to get total internal reflection, and at the same time, it also reduces the inter modal is person in the fiber. In a *graded index*, the interface between the core and the cladding undergoes a gradual change in refractive index with $n_1(r_i) > n_1(r_i + 1)$ (Figure 1.11). The refractive index of the core at a position r from the center of the core is written as [25].

$$n_1(r) = \sqrt{n_1^2(0) - \left(\frac{r}{a}\right)^\alpha \left\{n_1^2(0) - n_2^2\right\}} \quad \text{for } r \leq a$$

$$= 0, \text{ for } r > a$$

where a = radius of the core, $n_1(0)$ = refractive index at the center of the core, n_2 = refractive index of the cladding, and α = profile parameter. The numerical aperture of the core at a position r from the center of the core is written as

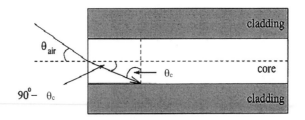

FIGURE 1.10   Numerical aperture of optical fiber.

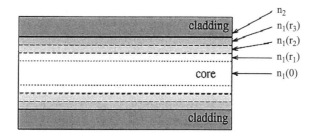

FIGURE 1.11   Graded-index fiber.

$$NA(r) = \sqrt{n_1^2(r) - n_2^2} = \sqrt{n_1^2(0) - \left(\frac{r}{a}\right)^\alpha \left\{n_1^2(0) - n_2^2\right\} - n_2^2}, \quad \text{for } r \le a$$

$$= 0, \text{ for } r > a$$

Example 1.1

A step-index fiber has a normalized frequency $V = 26.6$ at 1300 nm wavelength. If the core radius is 25 μm. What will be the value of numerical aperture?

$$V = \frac{2\pi a}{\lambda} NA$$

$$= NA = 26.6 \times 1.3 / (2 \times 3.14 \times 25) = 0.22$$

1.2.1.2   Difference between Single- and Multimode Fibers

A mode in an Optical Fiber representing a wave with a fixed propagation constant transmits through the fiber. It can also be considered as a standing wave in the transverse plane of the fiber. More formally, a mode corresponds to a solution of the wave equation which is derived from Maxwell's equations [42].

An electromagnetic wave is constructed by an electric field vector, E, and a magnetic field vector, H, each of which has three components. Since the optical fiber is represented well by cylindrical component, these components in the cylindrical co-ordinate system are E_r, E_ϕ, E_z and H_r, H_ϕ, H_z, where component of field is normal to core-cladding boundary of the fiber, ϕ component is the tangential to the boundary and z component is along the direction of propagation. There are two types of modes – transverse electric (TE) mode and transverse magnetic (TM) mode. For TE mode, $E_z = 0$, whereas for TM mode, $H_z = 0$.

Although total internal reflection in the core-cladding boundary can take place at an angle more than the critical angle, the light incident at an angle less than the critical angle will not propagate due to destructive interference between the incident and the reflected light at the core-cladding interface with in the fiber. For other angles of incidence, the incident wave and the reflected wave at the core-cladding interface constructively interfere in order to maintain the propagation of the wave. Fiber having propagation of more than one mode is called multimode fiber, whereas fiber having propagation of one mode is known as single-mode fiber (SMF). The number of modes supported by an Optical Fiber depending on normalized frequency V is written as

$$V = k_0 a \sqrt{\left(n_1^2 - n_2^2\right)}$$

where $k_0 = 2\pi/\lambda$, a is the radius of the core and λ is the wavelength of the propagating light in vacuum. For SMF $V \leq 2.404$. For multimode step-index fiber, the number of modes, m, is approximately written as [25]

$$m \approx \frac{1}{2} V^2$$

In multimode graded-index fiber of profile parameter α, the number of modes is given approximately by [25]

$$m \approx \frac{\alpha}{\alpha + 2} a^2 k^2 n_1^2 \Delta$$

where $\Delta = (n_1 - n_2)/n_1$. The advantage of multimode fiber is that its core diameter is relatively large as a result, injection of light in to the fiber with low coupling loss can be accomplished by using in expensive, large-are a light source, such as light-emitting diodes (LEDs).

The disadvantage of multimode fiber is the occurrence of intermodal dispersion [5]. In multimode fiber, each mode propagates with different velocity due to having different angles of incidence at the core-cladding boundary. This effect results in different rays of light from the same source to reach at the other end of the fiber at different times, contributing to spreading of pulse in the time domain. The intermodal dispersion resulting in spreading of pulse enhances with increase of propagation distance. To reduce this intermodal dispersion, graded-index fiber is used because having gradual changes in refractive index tries to make the same angle of incidence at the core-cladding boundary. For graded-index multimode fiber, intermodal dispersion constraints the bit rate of the transmitted signal and the distance of transmission of the signal.

For restriction of intermodal dispersion, the number of propagated modes in fiber is reduced. The reduction in the number of modes can be obtained by reducing the core diameter to make $V \leq 2.404$ by reducing the numerical aperture or by increasing the wavelength of light.

By reducing the fiber core diameter V value is made lower than 2.404 and a single mode is propagated in the fiber. This single mode is HE_{11} mode, also known as the fundamental mode. The SMF usually has a core size of about or less than $10\,\mu m$, while multimode fiber typically has a core size of more than 10 pm (Figure 1.12). Since SMF transmits light power with only one mode, it eliminates intermodal dispersion, supporting transmission over much longer distances. Due to small size

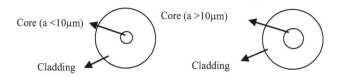

FIGURE 1.12   Single-mode and multimode Optical Fiber.

of SMF, LEDs having comparatively wide beam cannot couple enough light in to an SMF to facilitate long-distance communications. In order to get more coupling in SMF, semiconductor laser can be used, as it provides a high concentration of light energy with a narrow beam of light.

1.2.2   ATTENUATION IN FIBER

Attenuation of optical power In Optical Fiber reduces the signal. Before designing an optical fiber link, the propagation constant and attenuation of the optical fiber should be known. $P(L)$ is the power of the signal transmitted through fiber at a distance L km from the transmitter and α be the attenuation constant of the fiber. Attenuation is characterized by [25]

$$P(L) = P(0)e^{-\alpha L}$$

where $P(0)$ = optical power at the transmitter. To receive the signal by receiver, $P(L)$ should be greater than the receiver sensitivity P_r which is the minimum power required for detection at the receiver. So, the maximum distance L_{max} traversed by the optical signal without amplification in optical fiber is written as [25]

$$L_{max} = \frac{10}{2.343\alpha} \log_{10} \frac{P(0)}{P_r}$$

So L_{max} depends more heavily on the constant α than on the optical power launched by the transmitter. Figure 1.7 shows that the lowest attenuation is 0.2 dB/km at approximately wavelength region 1.5–1.6 µm. In optical communication system, the normal distance of optical propagation is 80 km without amplification. Using low loss fibers, amplifier spacing is enhanced from 80 to 160 km. The basic attenuations in a fiber are material absorption, scattering, bending losses and radiative losses of optical energy. The scattering loss is associated with structural imperfection and non-uniformity of fiber material composition. Radiative effects in fiber are due to fiber geometry which is bending (microscopic and macroscopic).

1.2.2.1   Absorption

Absorption has three mechanisms which are given below:

- Absorption due atomic defects in glass composition,
- Extrinsic absorption by impurity atoms in glass material,
- Intrinsic absorption by basic constituent atoms in fiber materials.

Atomic defects are imperfections of atoms arranged in fiber materials such as missing molecules, higher density cluster of atoms or oxygen defects. These defects are less and can be removed by proper treatment of materials.

The dominant factor of absorption is due to the presence of impurity in fiber materials. These impurities are transition metals such as iron, cobalt, chromium and copper which are present in glass and OH ions which may be introduced during fabrication. These impurities create the absorption energy levels due to which light wavelength can be absorbed.

Intrinsic absorption is related to the basic fiber materials and is due to electronic absorption bands in the ultraviolet (UV) region and form atomic vibration bands in the near-infrared region. The absorption coefficient in UV region can be expressed using the Orbach rule as

$$\alpha_{uv} = Ce^{E/E_0}$$

where C and E_0 are empirical constants and E is the photon energy.

1.2.3 Scattering Loss

Scattering loss is caused due to microscopic variations in the material density, composition fluctuations and structural in homogeneities/defects occurred during fabrication of fiber. Glass is made up of several oxides such as SiO_2, GeO_2 and P_2O_5, and there may be compositional fluctuation occurred during mixing/doping of these material in preparation of fiber materials. This gives rise to refractive index fluctuation. There is microscopic variation of concentration of materials also providing refractive index fluctuation. Finally, inhomogeneity in structure such as bubbles inside optical fiber also causes refractive index fluctuation. This index fluctuation causes a Rayleigh scattering in glass. The scattering loss can be formulated approximately as

$$\alpha_{scatt} = \frac{8\pi^3}{3\lambda^4}\left(\delta n^2\right)^2 \delta V \cdot K_B T \cdot \beta_T$$

where λ = wavelength, K_B = Boltzmann constant, T = temperature, β = isothermal compressibility, δn^2 = mean square refractive index fluctuation over volume δV and written as

$$\left(\delta n^2\right)^2 = \left(\frac{\delta n^2}{\delta\rho}\right)^2 (\delta\rho)^2 + \sum_{i=0}^{m}\left(\frac{\delta n^2}{\delta C_i}\right)^2 (\delta C_i)^2$$

where $\delta\rho$ = density fluctuation and δC_i = concentration fluctuation of ith glass composition of fiber. The structural in homogeneities are due to defects created in the form of bubbles of unreacted materials and crystalline regions during fabrication.

1.2.4 DISPERSION IN FIBER

Dispersion causes widening of a pulse duration when it transmits through a fiber. Due to widening of pulse, it interferes with neighboring pulses in the fiber, contributing inter symbol interference (ISI), as shown in Figure 1.13. Dispersion restricts the bit rate, i.e., the maximum transmission rate through a fiber-optical channel. There are three dispersions occurring in the optical fiber – intermodal dispersion [25], chromatic dispersion [25,44] and polarization mode dispersion (PMD) [25,44].

The intermodal dispersion arises when multiple modes of the same signal are transmitted with different velocities along the fiber. Intermodal dispersion does not appear in an SMF.

Chromatic dispersion arises due to the fact that different colors or wavelengths of signal transmit with different speeds, with in the same mode of the signal. This is due to the fact that in an optical fiber material, the refractive index is different for different wavelength material dispersion arises due to having different velocities of each wavelength in a material. Since any information-carrying signal will have an on-zero spectral width (range of wavelengths/frequencies in the signal), material dispersion will occur in most of the systems. Waveguide dispersion is due to the fact that the propagation of different wavelengths is a function of waveguide characteristics – the indices and shape of the fiber core and cladding. Profile dispersion is due to the variation of refractive index with the wavelength. Although SMF removes intermodal dispersion, chromatic dispersion and PMD cannot be eliminated in SMF.

PMD is due to the effect that SMFs produce two perpendicular polarizations of original transmitted signal, and these two polarizations of mode transmits with different

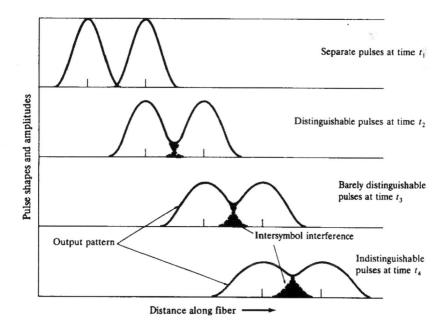

FIGURE 1.13   ISI between two pulses.

velocities in optical fiber [45]. In a perfectly circular fiber free from all stresses both polarization modes transmit at exactly the same speed, providing zero PMD. The practical fibers are not perfectly circular. The two polarizations transmit at different speeds and hence arrives at the end of the fiber at different times, where one perpendicular polarization direction is the fasted is and the other one is the slowed is The difference in arrival times between axes is known as PMD. The PMD results digitally transmitted pulses to spread out in polarization r no researched at their destination at different times. This effect of PMD contributes to bit errors at the receiver. The limit of acceptable dispersion penalty is usually 2 dB, which can be tolerate optical attenuation is low.

Currently, there are a number of special designs of optical fibers available, which offer lower dispersion than the dispersion-unshifted SMF. In a dispersion-shifted fiber (e.g., non-zero dispersion-shifted fiber (NZDSF) [5,45]), the index profile of core and cladding is designed as shown in Figure 1.14.

To reduce the dispersion, the dispersion flattening approach is used, where dispersion is spread minimum out over a wide range of optical fiber and fibers are called as dispersion-flattened fiber [5,44]. The refractive index profiles of dispersion-flattened fiber are designed using double cladding as shown in Figure 1.15.

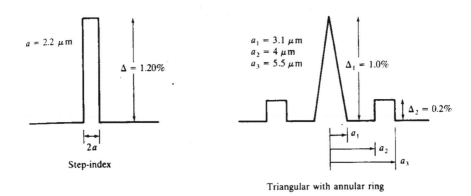

FIGURE 1.14 Design of refractive index profile for a dispersion-shifted fiber.

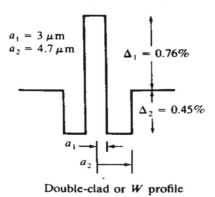

FIGURE 1.15 Design of refractive index profile for a dispersion-flattened fiber.

1.2.5 NONLINEARITIES

Nonlinearities effect of optical fiber provides attenuation, distortion, and cross-channel interference in WDM-based optical network [46]. In a WDM system, these effects show constraints of spacing between adjacent wavelength channels, the maximum power per channel and the maximum bitrate are limited. Nonlinear effects degrade the performance of WDM optical networks. Such nonlinearities restrict the optical power on each channel, and the maximum number of channels and the maximum transmission rate is constrained by the spacing between different channels.

In a WDM system using channels spaced 10 GHz a part and a transmitter power of 0.1 m W per channel, a maximum of about 100 channels are obtained in the 1550-nm low-attenuation region. The nonlinear effects are used for Raman amplification of the signal and also for wavelength conversion by using four-wave mixing (FWM). Nonlinearities also contribute to optical soliton propagation having a pulse with specific shape, power, and duration.

An optical soliton system is also used for compensation of dispersion. The main limitation of soliton systems is fiber loss. For low-power signal, compensation between dispersion and nonlinearities is not obtained.

The details of optical nonlinearities are very complex. The existence of these nonlinearities limits the number of nodes and the number of channels. There are different nonlinear effects providing signal transmission in optical fiber.

1.2.6 NONLINEAR REFRACTION

In an Optical fiber, the refractive index controls the optical intensity of signals propagating through the fiber [47]. Thus, the phase of the light depends on the length of the fiber and optical intensity. There are two types of nonlinear effects for phase change in refraction – self-phase modulation (SPM) and cross-phase modulation (XPM).

The SPM arises due to the variations in the power of an optical signal and results in variations in the phase of the signal. The phase shift contributed by SPM is given in the equation:

$$\varphi_L = k_0 n_L L |E|^2$$

where n_L is the nonlinear coefficient for the index of refraction, $k_0 = 2/\lambda$, L is the length of the fiber, optical intensity $= |E|^2$.

In phase-shift keying (PSK), SPM degrade system performance. Instantaneous variations in as signal's phase arisen due to the changes in the signal's intensity cause instantaneous variations of frequency around the signal's central frequency. For very short pulses, the additional frequency components generated by SPM with the effects of material dispersion cause spreading or compression of the pulse in the time domain. XPM is a shift in the phase of a signal arisen due to the change in intensity of a signal transmitted at a different wavelength. Phase modulation (PM) provides asymmetric spectral broadening, and combined with dispersion, also distorts the pulse shape in the time domain. The XPM modulates a pump signal at one wavelength from a modulated signal at different wavelengths. So it is used for wavelength conversion.

1.2.7 Stimulated Raman Scattering

Stimulated Raman Scattering (SRS) [48] is caused by the interaction of light with molecular vibrations. Light incident on the molecules creates scattered light at a longer wavelength than that of the incident light. A portion of the light traveling at each frequency in Raman-active fiber is down shifted across are going of lower frequencies. The light wave generated at lower frequencies is called the Stokes wave. The range of frequencies occupied by the Stokes wave is determined by the *Raman gain* spectrum, which covers a range of around 40 THz below the frequency of the input light. In silica fiber, the Stokes wave has a maximum gain at a frequency of around 13.2 THz less than the input signal.

The fraction of power transferred to the Stokes wave grows rapidly as the power of the input signal is increased. Under very high input power, SRS causes almost all of the power in the input signal to be transferred to the Stokes wave. In multi-wavelength systems, the shorter-wavelength channels cause loss of some power to each of the higher-wavelength channels within the Raman gain spectrum. To reduce the amount of loss, the power on each channel needs to be below a certain level.

1.2.8 Stimulated Brillouin Scattering

Stimulated Brillouin scattering (SBS) provides shift frequency from the actual frequency of the input signal [48]. In the SBS, the frequency shift is arisen due to having sound waves rather than molecular vibrations. The Stokes wave propagates in the opposite direction to the input light, and SBS occurs at relatively low input powers for wide pulses (greater than 1 ps). So, it has a negligible effect for short pulses (less than 10 ns). The intensity of the scattered light is more in SBS than that in SRS, but the frequency range of SBS, on the order of 10 GHz, is much lower than that of SRS. Also, the gain bandwidth of SBS is only on the order of 100 MHz. The input power is below a certain threshold. The SBS may induce crosstalk between channels. Two counter-propagating channels differ in frequency by the Brillouin shift, which is around 11 GHz for wavelengths at 1550 nm. However, the narrow gain bandwidth of SBS makes SBS crosstalk which is avoided.

1.2.9 Four-Wave Mixing

As the bit rate of optical data streams in fibers increase, FWM is another nonlinear effect impulse propagation [47]. FWM provides worst-case interchannel crosstalk and for equally spaced WDM channels. FWM penalty is less ended by using fiber switch high local dispersion (SMF) or unequally spaced channels.

However, there are some significant motivations that are taking advantage of FWM in WDM networks. Also, FWM provides wavelength conversion.

1.3 OPTICAL TRANSMITTERS

Before discussing about tunable optical transmitters, the fundamental principles of lasers should be discussed. The principle of lasers and the various implementations of tunable lasers and their properties are mentioned here.

1.3.1 LASER ACTION

Laser is an amplified stimulated emission [49], which is an intense high-powered highly directional beams of constant phase. Before discussion of stimulated emission, we should know about the energy levels of atoms. Atoms are stable in the ground state with electrons in lower energy levels. In order for the electron to move to a higher energy level, the atom in the ground state must absorb energy. At higher energy levels, the electron is unstable and, usually, moves quickly back to the ground state by releasing a photon, i.e., spontaneous light. There are certain materials with quasi-stable state known as metastable state, where the substances are likely to stay in the excited state for longer periods of time. By applying enough energy (either in the form of an optical pump or in the form of an electrical current) to as substance with quasi-stable state for a long enough period of time, population inversion occurs, which means that there are more electrons in the excited state than in the ground state. Under population inversion, light energy is emitted with highly directional and intense beam of constant phase. This emission is called as stimulated emission. For getting amplification of stimulated emission to obtain LASER, it requires a special arrangement known as resonance cavity consisting of two mirrors which form the space between the mirrors. The cavity has aliasing medium as an excitation medium. The excitation medium applied current to the lasing medium, which is made of a quasi-stable substance. The applied current excites electrons in the lasing medium, and when an electron in the lasing medium drops back to the ground state, item it's a photon of light. The photon will reflect by means of the mirrors at each end of the cavity and will pass through the medium a gain. The cavity with mirror is called resonator. Stimulated emission occurs when a photon passes closely to an excited electron. Photons for which the frequency is an integral fraction of the cavity length will coherently combine to build up light at the given frequency within the cavity. The mirrors feed the photons back and forth, and stimulated emission can occur, and higher intensities of light can be produced. One of the mirrors is partially transmitting, so that some stimulated photons emerge from the cavity in the form of a narrowly focused beam of light as shown in Figure 1.16.

The wavelength of the emitted photon depends on its change in energy levels and is given by

$$\lambda = \frac{\hbar c}{E_q - E_g}$$

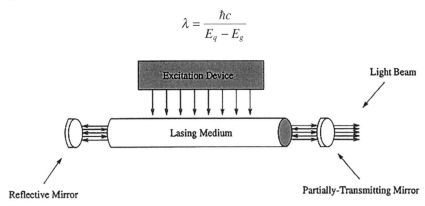

FIGURE 1.16 Basic structure of LASER having pumping and resonator mirror.

where $\hbar$ = Planck's constant, E_q = energy level of quasi-stable state of electron and E_g = energy level of the ground state. In a gas laser, the distribution for $E_q - E_g$ is given by an exponential probability distribution, known as the Boltzmann distribution, which changes depending on the temperature of the gas. Although many wavelength emissions are possible, only a single frequency determined by the cavity length is emitted from the laser. When the lasing medium has gases as excitation medium, it is called as gas laser. The examples are He Ne laser, Argon laser, CO_2, etc. If the lasing medium is a solid material, then it is called as solid-state laser. If the lasing medium is liquid, then it is called as liquid laser (e.g., dye laser).

1.3.2 SEMICONDUCTOR DIODE LASER

The most useful laser is a semiconductor diode laser for optical communication application [49,50]. The semiconductor laser is based on a double heterostructure semiconductor junction (shown in Figure 1.17) having an active region of direct bandgap semiconductor [25] in which stimulated emission is radiated due to conservation of momentum, and both sides of active region having indirect bandgap material of gap between valence band and conduction band higher than that of active region makes potential well in active region. To get laser action, two phenomena had occurred – charge carrier confinement and optical confinement. Due to the potential well in the active region, electron share confined in the conduction band for a while once itis exited from the valence band to the conduction band. When an electron jumps from the valence band to the conduction band, a vacancy or hole is generated

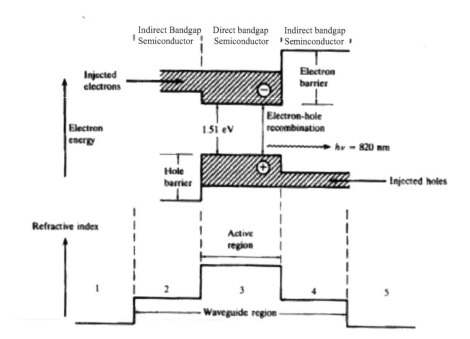

FIGURE 1.17 Double heterostructure laser diode band structure and index profile.

in the valence band. When the electron moves back from the conduction band to the valence band, it recombines with the hole, and as a result, it generates stimulated emission of a photon due to direct bandgap principle, maintaining momentum conservation. The refractive index of an active region is larger than that of both sides. So, the emitted radiation in the active region is confined. This confinement is called as an optical confinement. The amplification of stimulated radiation is made by mirror edges of the laser diode. The wavelength of the emitted photon depends on its change in energy levels and is given by

$$\lambda = \frac{\hbar c}{E_c - E_v}$$

where $\hbar$ = Planck's constant, E_c = energy level of the conduction band and E_v = energy level of the valence band. The distribution of energy levels occupied by electrons is given by the Fermi–Dirac distribution.

1.3.3 MULTIPLE QUANTUM WELL LASER

The bulk laser diode is a multiple quantum well (MQW) laser [25], where quantum well share made of thin alternating layers of semiconductor materials. The alternating layers generate potential barriers in the semiconductors confining the position of electrons and holes to a smaller number of energy states. The quantum wells are placed in the region of the *p-n* junction. By confining the possible states of the electrons and holes, higher-resolution, low-line width lasers are made.

1.3.4 TUNABLE AND FIXED LASERS

The transmitter lasers used in WDM networks should have the capability to tune to different wavelengths. This section discusses tunable and fixed, single-frequency laser designs [25].

1.3.4.1 Laser Characteristics

Physical characteristics of lasers affecting the performance of a laser are laser line width, frequency stability and a number of longitudinal modes [50]. The laser line width is the spectral width of the light puke made by the laser. The spacing of channels is related with the line width, which is affected by the amount of dispersion that occurs while light propagates along a fiber. The spreading of a pulse due to dispersion reduces the bitrate. Frequency in stabilities in lasers are due to the variations of the laser frequency. There are three causes of frequency variations – mode hopping, mode shifts and wavelength chirp.

Mode hopping is obtained in injection-current lasers. It provides a sudden jump in the laser frequency due to a change in the injection current above a given threshold. Mode shifts are changes in frequency due to temperature changes. Wavelength chirp is a change in the frequency obtained due to variations in injection current. In WDM systems, frequency instabilities restrict the assignment and spacing of channels. For removal of large shifts in frequency, variations in temperature or injection current must be controlled. One method for temperature compensation is laser packaged with a thermo electric cooling element providing cooling as a function of applied current which is obtained through thermistors.

The number of longitudinal modes in a laser is amplified in the cavity of lasers. In a simple cavity, wavelengths for which an integer multiple of the wavelength is equal to twice the cavity length will be amplified (i.e., wavelength λ for which $n\lambda = 2L$, where L is the length of the cavity and n is an integer). The unwanted longitudinal modes produced by a laser provide significant dispersion. So, it is required to produce only a single longitudinal mode.

The main characteristics for tunable lasers are tuning range and tuning time. The laser may be continuously tunable (over its tuning range) or discretely tunable (only to selected wavelengths). The tuning range is specified by a wavelength range which the laser emits, whereas the tuning time is the time required for the laser to tune from one wavelength to another.

1.3.4.2   Mechanically Tuned Lasers

Figure 1.18 shows mechanically-tuned lasers having a Fabry–Perot cavity [51], which is an optical resonator in which amplification/gain is obtained by optical feedback through two parallel mirrors with very high reflectivity and one mirror having slightly less reflectivity to emit the laser out. The external cavity is adjacent to the lasing medium to filter out unwanted wavelengths. Tuning occurs by physically adjusting the distance between two mirrors one it here and of the cavity where only a desired wavelength constructively interferes with its multiple reflections in the cavity. Here, at using range is covered by the entire useful gain spectrum of the semiconductor laser, but the tuning time is the order of milliseconds due to them mechanical tuning and the length of the cavity. The external cavity is used for tuning and has very good frequency stability. It has higher slope efficiency, lower noise, more linearity in tuning and much greater stability over temperature.

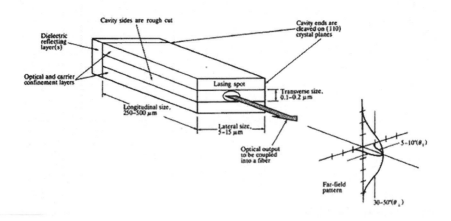

FIGURE 1.18   Mechanically tuned Laser diode with Fabry–Perot elation [25].

The Fabry–Perot tuning approach is of two types – Buried Hetero (BH) and MQW types. In this direction, MQW lasers have advantages over all former types of Fabry–Pero lasers. MQW lasers provide good performance margins and yields and reduce laser cost. One disadvantage of MQW lasers, however, is their tendency to be more susceptible to back reflections. Tuning range and tuning time of mechanically tuned laser diode are ~550 nm and ~1–10 ms, respectively.

1.3.4.3 Acoustooptically and Electrooptically Tuned Lasers

The tunable lasers use acoustic optical tuner and electro optical tuner [51]. In an acousto-optic tuner index of refraction in the external cavity is varied by using sound waves where as it is varied with electrical current in an electrooptical tuner. The change in the index provides a transmission of light at different frequencies. In the set unable lasers, the tuning time is restricted by the time required to obtain the cavity at the new frequency.

An acoustic optic laser has less tuning time. The packet switching with multi-gigabit-per-second channels is controlled by the tuning time which should be reduce din comparison with that of mechanically tuned lasers having millisecond tuning times. Electrooptically tuned lasers tune with few nanoseconds. Tuning range and tuning time of Acoustooptically tuned laser diode [52] are ~730 nm and ~9 μs, respectively, whereas the tuning range and tuning time of electrooptically tuned laser diode are ~ 9 nm and ~1–10 ns, respectively. Neither of these approaches shows a continuous tuning over a range of wavelengths.

1.3.4.4 Injection-Current-Tuned Lasers

Injection-current-tuned laser acts as a transmitter that permits wavelength selection via a diffraction grating. Figure 1.19 shows a Distributed Feedback (DFB) laser with a diffraction grating placed in the lasing medium [25]. The grating has a waveguide in which the index of refraction is changed alternatively and periodically between two values. Only wavelengths matching the period and indices of the grating are constructively reinforced. All other wavelengths are destructively interfered and are not transmitted through the waveguide. The condition for propagation of wavelengths is given by

$$D = \frac{\lambda}{2n}$$

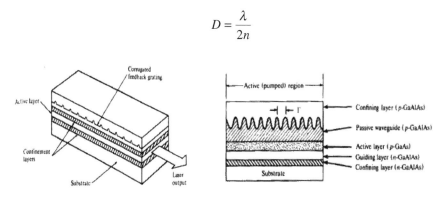

FIGURE 1.19 DFB laser with a cross-sectional structure made up of GaAlAs [25].

where D = the period of the grating and n = effective refractive index of grating region. The laser is tuned by injecting a current that changes the index of the grating region.

If the grating is placed outside of the lasing medium, the laser is called a Distributed Bragg Reflector (DBR) laser, which is shown in Figure 1.20. The power coupling becomes maximum at wavelengths close to Bragg's wavelength λ_B, which is written as [25]

$$\lambda_B = \frac{2Dn}{k}$$

Where k = order of the grating, D = grating period and n = effective refractive index.

The DBR-based laser operates over abroad range of biasing conditions and a large tuning range of emitted wavelength. There are three DBR lasers [52] – sampled grating DBR (SG-DBR), digital supermodel (DS-DBR) and super structure grating DBR. An SG-DBR laser has a tuning range more than 40 nm, whereas a tunable DS-DBR laser has a continuous tuning range of 0–45 nm. However, DBR devices have a major drawback of mode hopping that occurs due to the changes in the lasing-region index of refraction with increasing driven current.

Tuning range and tuning time of both DFB- and DBR-tuned laser diodes are ~48 nm and ~1–10 ns, respectively.

1.3.5 LASER ARRAYS

Another method to tune lasers is a laser array that has a set of fixed-tuned lasers integrated into a single component. The advantage of using a laser array is that if each of the wavelengths in the array is modulated independently, then multiple transmissions may take place simultaneously.

MEMS-based vertical-cavity surface emitting laser (VCSEL) [51] is a laser array approach, which is a new laser structure emitting laser light vertically from its surface. The VCSEL's principles of operation closely resemble those of the conventional edge-emitting semiconductor lasers. The core of the VCSEL has an electrically pumped

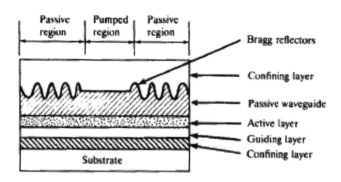

FIGURE 1.20   DBR laser [25].

gain region known as an active region emitting light. Layers of varying semiconductor materials above and below the gain region make mirrors. Each mirror reflects a wavelength back into the cavity causing light emission at a single wavelength. VCSELs are typically MQW devices with lasing occurring in layers only 20–30 atoms thick. In VCSELs, the Bragg reflectors with as many as 120 mirror layers form the laser reflectors. There are many advantages of VCSELs. The small-size and high-efficiency mirrors produce allow threshold current, which is less than 1 mA. The MEMS-based VCSEL has a faster tuning time of 1–10 ps in comparison to others.

1.4   OPTICAL RECEIVERS AND FILTERS

Tunable optical receiver/filter is one of the key devices in WDM optical networks. This section discusses about a tunable photodetector [53]. There are several types of photodetectors, photomultipliers, pyroelectric detectors, phototransistors and semiconductor-based photodetectors [25]. Photomultipliers having photocathode and electron multiplier packaged in a vacuum tube provide high gain and low noise, but it is not used in optical network because of large size and high voltage requirement. Pyroelectric photodetectors provide conversion of photon streams into heat energy, which results in a temperature change of dielectric material, i.e., variation of dielectric constant measured as a change of capacitance. Due to its lower speed, it is not suitable for high-speed optical communication. Semiconductor-based photodetectors work in principle of conversion of photon streams into electrical energy and are very much suitable for optical communication because of fast response time, high sensitivity and small size. Here we discuss only about semiconductor-based photodetector.

1.4.1   PHOTODETECTOR

The semiconductor-based photodetector is a *p-n* junction acting as a photodiode that converts the incoming optical signal (representing stream of photons) into electrical energy representing flow of electron streams. The electron stream generating electrical current is then amplified and passed through a threshold device. The two types of photodiode are PIN photodiode and avalanche photodiode (APD) [25].

The photodiode is basically are verse-biased *p-n* junction. Through the photoelectric effect, light incident on the junction generates electron–hole pairs in both the "*n*" and the "*p*" regions of the photo diode. The electrons released in the "*p*" region pass to the "*n*" region, and the holes created in the "*n*" region pass over to the "*p*" region providing a current flow.

In the fiber optic network, the photodiode is generally required to detect very weak signals. The detection of the weakest signal requires the photodetector, and an amplification circuit should be selected so that minimum signal-to-noise ratio (S/N) should be maintained. The S/N is written as

$$\frac{S}{N} = \frac{\text{signal power from photocurrent}}{\text{photo det ector noise} + \text{amplifier noise power}}$$

To achieve high S/N, the following conditions are obtained:

    i. The photodetector should have a high quantum efficiency to generate large signal current.
   ii. The photodetector and amplifier noises should be kept as low as possible.

The noise power determines the minimum optical power level that detected.

1.4.1.1 PIN Photodiode

Figure 1.21 shows the PIN photodiode having p and n regions separated by a very lightly n-doped intrinsic (i) region. Usually in its operation, a reverse-biased voltage is applied across the device so that the intrinsic region is fully depleted of carriers. The intrinsic n and p carrier concentrations are neglected in comparison to other impurity region concentrations. When a photon of energy greater than the bandgap energy of the semiconductor is incident, the photon can supply energy to excite the electron from valence band to conduction band as shown in Figure 1.22. This process generates free electron–hole pairs known as photo-carriers generated in the depletion region where most incident light is absorbed. The high reverse electric fields present in the depletion region causes the carrier to be separated and collected across the junction. This gives current flow by every pair generated. This current is called as photocurrent [25].

If the width of the depletion region is w, the total power absorbed within this region is written as

$$P(w) = P_{in}\left(1 - e^{-\alpha_s(\lambda)w}\right)$$

where P_{in} = incident power in the depletion region and $\alpha_s(\lambda)$ = absorption coefficient depending on wavelength. The absorption of a range of wavelengths depends on the bandgap energy (E_g) of semiconductor used in PIN photodiode. The upper cut wavelength absorbed is written as

$$\lambda_{cutoff} = \frac{hc}{E_g}$$

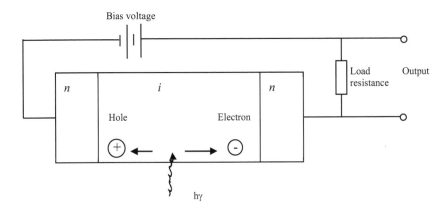

FIGURE 1.21 PIN photodiode and its circuit.

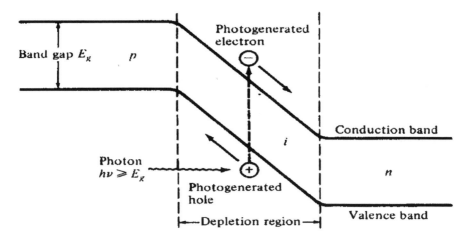

FIGURE 1.22   Energy band diagram of a PIN photodiode.

The cut off wavelength λ_{cutoff} for Si and Ge are 1.06 and 1.6 μm, respectively.

We consider the reflectivity R_f at the entrance surface of the photodiode, and the primary photocurrent I_p is written as

$$I_p = \frac{q}{h\nu} P_0 \left(1 - e^{-\alpha_s w}\right)\left(1 - R_f\right)$$

where P_0 = incident optical power on photodiode, q = electron charge and $h\nu$ = each photon energy. The important parameters are quantum efficiency, response time and responsivity and detectivity. The quantum efficiency is written as

$$\eta = \frac{\text{number of electron} - \text{hole pair generated}}{\text{number of incident photons}} = \frac{I_p/q}{P_0/h\nu}$$

The performance of a photodetector is measured by responsivity, which is written as $R = I_p/P_0$.

1.4.1.2   Avalanche Photodiode

APD is another photodetector [25] in which the photo generated carriers get multiplied before it enters the input circuitry of the amplifier. For carrier multiplication, these carriers must travel through a region where a high electric field is present. In this high-field region, photo generated carriers can gain enough energy so that it ionizes the bound electrons in valence band upon colliding with them. This ionization is called impact ionization. The newly generated carriers are also accelerated by high electric field and get enough energy to cause further ionization. This is called avalanche impact ionization. While the photocurrent is enhanced by multiplication, thermal noise signal current is also multiplied. It can be avoided by reach through avalanche construction as shown in Figure 1.23. It consists of a

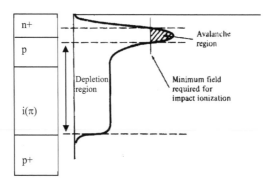

FIGURE 1.23 Operation of APD [25].

high resistivity intrinsic material deposited as an epitaxial layer on a *p*+ (heavily doped) substrate. The *p*-type diffusion is then made on high-resistive followed by an *n*+ material layer. This makes *p*+*pn*+ reach through construction. When a low reverse bias voltage is applied, the potential drop is obtained at a *p-n* junction. The depletion layer widens with increasing bias until a certain voltage is reached at which the peak electric field at *pn*+ junction is about 5%–10% below that needed to cause an avalanche breakdown. This is called just reach through the nearly intrinsic region. When it is fully depletion mode, the light enters the device through *p*+ region and is finally absorbed in the region, and then these photo generated carriers are accelerated by a high electric field. While accelerated, these carriers are multiplied through avalanche ionization [25]. The multiplication factor of APD is written as $M = I_M/I_p$, where I_M = average value of total multiplied output current and I_p is the primary current. The responsivity of APD is written as $R_{APD} = \dfrac{\eta q}{h\nu} M$.

1.4.2 Tunable Optical Filters

There are different types of tunable optical filters and fixed-tuned optical filters used in optical network in tunable transceiver. The viability of many local WDM networks depends on the speed and range of tunable filters [54].

1.4.2.1 Filter Characteristics

Tunable optical filters [55] are analyzed by their tuning range and tuning time. The tuning range is a range of wavelengths that are accessed by a filter. A wide tuning range permits the systems to utilize a greater number of channels. The tuning time of a filter is the time required to tune from one wavelength to another. Fast tunable filters are needed for WDM network architectures.

Some filters such as etalon are further analyzed by two parameters – free spectral range (FSR) and finesse. The transfer function or shape of the filter passband is repeated periodically. The period of such devices is referred to as the FSR. The filter permits every frequency separated from $n \times$ FSR from the selected frequency, where n is a positive integer. For example, in Figure 1.24, the finesse of a filter is a measure

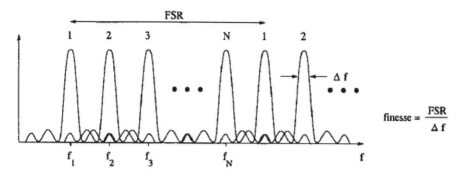

FIGURE 1.24 Filter characteristics with FSR and finesse [58].

of the width of the transfer function and determined by the ratio of FSR to channel bandwidth, where the channel bandwidth is considered to be 3-dBbandwidth of a channel.

The number of channels in an optical filter is restricted by FSR and finesse. All of the channels are obtained within one FSR. For high finesse, the transfer functions (pass band peaks) are thin, providing more channels within one FSR. With a low finesse, the channels are spaced more to avoid cross talk, giving less channels. Increasing the number of channels is made by cascading filters to enhance FSRs.

1.4.2.2 Etalon

Two types of etalon are used – planar etalon consisting of a single cavity made between two parallel flat mirrors and spherical etalon consisting of a cavity between two identical spherical mirrors with their concave sides facing each other and with the distance between the mirrors equal to each mirror's radius of curvature (confocal or spherical etalon) [56].

The planarization filter is based on a Fabry–Perot interferometer [56] and is also known as the Fabry–Perot etalon. Here, a monochromatic light ray travels back and forth between two mirrors, and the distance between mirrors equals an integral number of wavelengths. Finally, the light passes through the etalon. There are many modifications such as multicavity and multipass to the etalon made to improve the number of resolvable channels. In a multipass filter, the light travels back and forth through the cavity (made between two mirrors) multiple times, whereas a multi-cavity filter consists of multiple etalons of different FSRs cascaded to effectively increase the finesse.

The Fabry–Peroneal on virtually accesses the entire low-attenuation region of the fiber and provides narrow pass bands. The tuning time is the order of tens of milli-seconds due to its mechanical tuning. It is not useful for packet-switched applications where the packet duration is smaller than the tuning time.

The FSR of the Fabry–Perot et alon corresponds to the typical channel spacing of 100 or 50 GHz which are obtained with cavity of lengths 1–2 nm. All-fiber Fabry–Perot devices with finesse 240 covering a spectral band of 26 nm are presented.

1.4.2.3 Mach–Zehnder Chain

In Figure 1.25, a Mach–Zehnder (MZ) interferometer consists of a splitter that splits the incoming wave into two waveguides, and a combiner that recombines the signals at the outputs of the waveguide. An adjustable delay element controls the optical path length in one waveguide arm of an MZ structure, providing a phase difference between the two signals before these are combined. Wavelengths having a phase difference of 180 are filtered out.

MZ structures cascaded in a chain is a low-cost device, but its tuning time is still on the order of milliseconds, and its tuning control is complex. The high tuning time is due to thermal elements used to implement the delay elements. Recent advances have produced a fast tuning MZ filter, which exhibits a total FSR of 16 nm, an extinction ratio of 20 dB, and a 3-dB transmission bandwidth of 32 GHz.

1.4.2.4 Acousto-optic Filters

The acousto-optic filters have a fast tuning time [56]. Here radio frequency (RF) waves are passed through a transducer which is a piezoelectric crystal that converts sound waves to mechanical moves. The sound waves change crystal's index of refraction, making the crystal to act as grating. When light is incident upon the transducer, it is diffracted at an angle depending on the angle of incidence and the wavelength of light. By changing the RF wave frequency, a single optical wavelength is selected to pass through the material, whereas the rest of the wavelengths are destructively interfered.

The tuning time of the acousto-optic filter is mainly the flight time of the surface acoustic wave (SAW) which is 10 ps. However, the tuning range for acousto-optic filters is the entire 1300–1560 nm spectrum that includes almost low loss window of almost full frequency range of ~1520–1560 nm. This tuning range potentially allows more than 100 channels. If more than one RF wave is passed through the grating simultaneously, more than one wavelength is filtered out. This allows the filter to be effectively tuned to several channels at the same time.

An acousto-optic tunable filter (AOTF) has the following advantages for wavelength tunable lasers:

1. It has a tuning range of 0–100 nm and fast switching speed of several microseconds.
2. It has a stable operation against shock and vibration owing to its non-mechanical structure.

The disadvantage of acousto-optic filters is of their wide transfer function contributing crosstalk from adjacent channels if the channels are closely spaced.

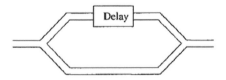

FIGURE 1.25 MZ interferometer (MZI) filter.

1.4.2.5 Electrooptic Filters

Since the tuning time of the acousto-optic filter is restricted by the speed of sound, crystals having indices of refraction changed by electrical currents are used. Electrodes are taken in the crystal to provide tuning electric current to change the crystal's index of refraction, which selects the wavelengths to pass through and others are destructively interfered. The tuning time is the order of several nanoseconds.

1.4.2.6 Liquid Crystal Fabry–Perot Filters

The liquid crystal (LC) Fabry–Perot filters consist of a cavity with an LC [55,56]. For tuning, the refractive index of LC is changed by an electrical current to filter out a desired wavelength, as in an electrooptic filter. These filters have low power requirements and lower fabrication cost. The filter speed of LC filter technology is enough to handle high-speed packet switching in WDM networks. The tunable FSR of the filters is 10 nm, loss is 2.2 dB, finesse is 150 and tuning speed is ~1 ps.

1.4.3 Fixed Filters

The fixed filters are important devices used in optical networks. Grating devices are typically fixed filters that filter out one or more different wavelength signals from a single fiber. Such devices are used to implement optical multiplexers and demultiplexers or receiver arrays.

1.4.3.1 Grating Filters

Diffraction grating is an example of fixed filters. The diffraction grating is made of a flat layer of transparent material (e.g., glass or plastic) with a row of parallel grooves cutting to it. The grating separates light into its component wavelengths by reflecting light incident with the grooves at all angles. At certain angles, only one wavelength is constructively interfered, whereas all others are destructively interfered. This allows us to select the wavelength(s) required by placing a filter tuned to the proper wavelength at a proper angle. Another grating is transmissive grating in which one wavelength is constructively interfered at a certain angle and other wavelengths are destructively interfered. These types of grating are normally used in tunable lasers.

1.4.3.2 Fiber Bragg Gratings (FBG)

The grating made with fiber is called as a fiber Bragg rating (FBG) [57] in which a periodical variation of the index of refraction is directly made by using photo induction in the core of an optical fiber. A given wavelength of light is reflected back to the source while passing the other wavelengths. Two primary characteristics of a Bragg rating are the reflectivity and the spectral bandwidth. Typical spectral bandwidths are on the order of 0.1 nm, while a reflectivity in excess of 99% is achievable. While inducing grating directly into the core of a fiber leads to low insertion loss, a drawback of Bragg rating is that the refractive index in the grating varies with temperature, with an increase in temperature resulting in longer wavelengths being reflected. An FBG is also used in the implementation of multiplexers, demultiplexers and tunable filters.

1.4.3.3   Thin-Film Interference Filters

Thin-film interference filters filter out one or more wavelengths from a number of wavelengths. The functions of these filters are similar to that of FBG devices. These filters are fabricated by depositing alternative layers of low-and high-index materials on to a substrate layer. There are many drawbacks of thin-film filter technology such as poor thermal stability, high insertion loss and poor spectral profile.

1.4.4   COMPARISON BETWEEN DIFFERENT FILTERS

Table 1.2a compares the tuning parameters of different tunable filters. It is seen that acousto-optic filter has a higher tuning range of 250 nm and a moderate tuning time of 10 µs. Table 1.2b compares the filter characteristics of different tunable filters. FBG is reasonably less channel spacing with less crosstalk.

1.5   OPTICAL MODULATION

In order to transmit data across an optical fiber, the information must be first encoded, or modulated, on to the laser signal. Figure 1.26 shows the block diagram of optical modulation in which message is converted into an analog optical signal. The message/data are converted into data bits/digital signal by using ASCII encoder. The digital signals maybe coded with a digital-to-digital encoder that includes not return to zero level (NRZ-L), Bipolar AMI, bipolar 8 zero substitution (B8ZS), high-density bipolar 3 zeros (HDB3), Manchester and differential Manchester encoder. The coded digital

TABLE 1.2A
Tunable Filters with Tuning Parameters

Tunable Receiver	Tuning Range (nm)	Tuning Time
Fabry–Perot	500	1–10 ms
Acousto-optic	250	~10 µs
Electrooptic	16	1–10 ns
LC Fabry–Perot	50	0.5–10 µs

TABLE 1.2B
Fixed Filters with Filter Characteristics

Technology	Loss(dB)	Channel Spacing (GH*)	Crosstalk (dB)	Tunability	Maturity
FBG	0.5	50	30	Yes	High
TFF	1	100	12	Yes	High
AWG	6	50	30	No	High
MZI	1	50	30	No	High
AOF	3	100	15–20	Yes	Low
DCE	5	50	30	Yes	High

FIGURE 1.26 Optical modulation for data communication.

signals are modulated with digital-to-analog modulation which include amplitude-shift keying (ASK), frequency-shift keying (FSK) and PSK. Analog-to-analog modulation technique is required to generate signals that can be transmitted through optical fiber. Analog-to-analog modulation includes amplitude modulation (AM), frequency modulation (FM) and PM.

1.5.1 DIGITAL-TO-DIGITAL MODULATION

Binary data bits after obtaining from ASCII coder are required to encode with the digital signal, which is a sequence of discrete, discontinuous voltage pulses. Binary data are transmitted by encoding each data bit into signal elements so that there should be one-to-one correspondence between bits and signal elements. Different kinds of digital to digital encoder [33] are – NRZ-L, not return to zero inverted (NRZI), bipolar AMI, Pseudo ternary AMI, Manchester, differential Manchester, B8ZS and HDB3 [33].

1.5.1.1 NRZ

The most common and easiest way to transmit digital signals is to use two different voltage levels for two binary digits. There are two types of NRZ – NRZ-L and NRZ-I. In NRZ-L code, the binary zero and one are represented by high and low voltage levels, respectively. Figure 1.27 shows an example of NRZ-L code in which there are no transition for both long strings of zero and one. This no transition problem makes loss of synchronization of bit because starting and end of bits are not detected.

For removal of no transition problem at long strings of one, the NRZ-I code is used for digital-to-digital modulation. The rules are

1. For binary zero, there is no transition at the beginning of the interval (one bit at time).
2. For binary one, there is a transition at the beginning of the interval.

1.5.1.2 Bipolar AMI

To remove the problems of no transition at long strings of zero and one the multilevel signals coding is used. Bipolar AMI is one of the multilevel binary coding where apart from zero level (no voltage) both positive and negative voltage pulses are used [33]. The rules are

1. The binary zero is represented by no line signal.
2. The binary one is represented by positive or negative voltage levels alternatively.

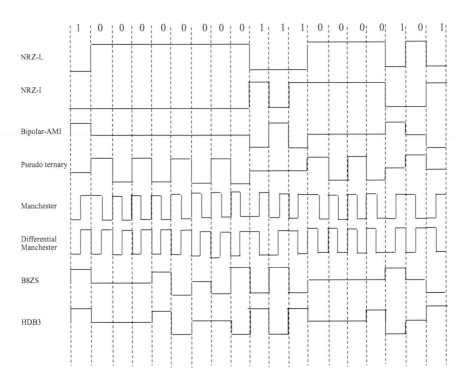

FIGURE 1.27   Different digital encoding schemes.

Figure 1.27 shows an example for bipolar AMI coding. It is seen that no transition problem in a long string of one is removed but that of zero still exists.

1.5.1.3   Pseudo Ternary AMI

To remove the problems of no transition at long string of zero, pseudo ternary AMI is one of the multilevel binary coding where apart from zero level (no voltage) both positive and negative voltage pulses are used [33]. The rules are

1. The binary one is represented by no line signal.
2. The binary zero is represented by positive or negative voltage levels alternatively.

Figure 1.27 shows an example for bipolar AMI coding. It is seen that no transition problem in a long string of zero is removed but that of one still exists.

1.5.1.4   Biphase Coding

Biphase coding is another coding technique that overcomes the limitations of no transition for long string of zeros and ones in the case of NRZ codes. In biphase coding, each bit has two phases. There are two types – Manchester and differential Manchester coding [25,33].

In Manchester code, there is a transition at the middle of each bit period. The mid bit transition serves as a checking mechanism and apart from those the following rules are followed:

1. In the case of binary zero, there is a transition from high to low in the middle of the interval.
2. In the case of binary one, there is a transition from low to high in the middle of the interval.

Figure 1.27 shows an example for Manchester coding. It is seen that there is a loss of synchronization from zero to one or from one to zero.

In differential Manchester code, there is always a transition at the middle of each bit period. Apart from those, the following rules are followed:

1. In the case of binary zero, there is a transit at the beginning of the interval.
2. In the case of binary one, there is no transition at the beginning of the interval.

Figure 1.27 shows an example for differential Manchester coding. It is seen that there is loss of synchronization from zero to one or from one to one.

1.5.1.5 B8ZS Code

Although biphase coding provides reduction of no transition problems at long string of binary one or zeros, cent percent removal of transition is not possible. In bipolar AMI, it is seen that there is a loss of synchronization for a long string of zeros. For reduction of no transition problem in bipolar AMI, a coding scheme used in North America is known as B8ZS [33]. This technique forces two code violations. To overcome this problem, the following rules are adapted apart from the rules of bipolar AMI.

1. If an octet of all zeros occurs and the last voltage pulse proceeding octet is positive, then the 8 zeros of the octet are encoded as 000+−0−+.
2. If an octet of all zeros occurs and the last voltage pulse proceeding octet is negative, then the 8 zeros of the octet are encoded as000−+0+−.

Figure 1.27 shows an example for B8ZS coding. It is seen that there is a loss of synchronization from zero to zero or for less than 8 zeros.

1.5.1.6 HDB3 Code

Till there is a loss of synchronization for a string of zeros less than eight. To reduce such loss of synchronization in B8ZS code, another coding scheme used in Europe and Japan is known as B8ZS. To overcome this problem, the following rules are adapted apart from the rules of bipolar AMI [33] as shown in Table 1.3.

Figure 1.27 shows an example for HDB3 coding. It is seen that there is a loss of synchronization from zero to zero or for less than 4 zeros.

TABLE 1.3
Rules of HDB3

	Number of Bipolar Pulses (Ones) Since Last Substitution Application Services	
Polarity of Proceeding Pulse	Odd	Even
−	000−	+00+
+	000+	−00−

1.5.2 DIGITAL-TO-ANALOG MODULATION

For transmission of digital signal in optical fiber, it is required to convert the same into analog signals. This modulation is called as digital-to-analog modulation. Another use of it is transmission of digital data/signal through public telephone network and wireless communication network/mobile communication network. This modulation involves operation of one or more of the three characteristics of a carrier signal – amplitude, frequency and phase. So there are three modulation techniques – ASK, FSK and PSK. In ASK, also known as on–off keying (OOK), the signal is switched between two power levels. The lower power level represents a "0" bit, while the higher power level represents a "1" bit. The binary 0 and 1 bits are encoded as follows [33]:

$$S(t) = \begin{array}{ll} A\cos(2\pi f_c t) & \text{binary } 1 \\ 0 & \text{binary } 0 \end{array}$$

where f_c = carrier frequency. Figure 1.28 shows an example of ASK waveform of a digital signal. In systems employing OOK, modulation of the signal can be achieved by simply turning the laser on and off (direct modulation). This OOK provides variations in the laser's amplitude and frequency, when the laser is turned on. A preferred approach for high bit rates (>2 Gbps) is to have an external modulator that modulates the light coming out of the laser. The modulator allows to pass light or not, depending on the current applied to it.

In FSK, the two binary values are represented by two different frequencies near the carrier frequencies:

$$S(t) = \begin{array}{ll} A\cos(2\pi f_1 t) & \text{binary } 1 \\ A\cos(2\pi f_2 t) & \text{binary } 0 \end{array}$$

where f_1 and f_2 are typically offset frequencies from central carrier frequency f_c and can be written as $f_1 = f_c + f$ and $f_2 = f_c - f$. Figure 1.28 shows an example of FSK waveform of a digital signal. The FSK is used for the modulation of a voice grade line. High frequencies of 3–30 MHz are used. For wireless local area network, even higher frequency can be used for FSK modulation.

In PSK, the phase of the carrier signal is shifted to represent data. There are different types of PSK – binary PSK (BPSK) and quadratic PSK (QPSK). In BPSK, the two binary values 0 and 1 are represented by the following analog signals with phase difference π.

$$S(t) = \begin{cases} A\cos(2\pi f_c t + \pi) & \text{binary 1} \\ A\cos(2\pi f_c t) & \text{binary 0} \end{cases}$$

Figure 1.28 shows an example for BPSK waveform of a digital signal. In QPSK, the two binary bit signal elements are represented with the following analog signals with phase difference π.

$$S(t) = \begin{cases} A\cos(2\pi f_c t + 45°) & \text{binary 11} \\ A\cos(2\pi f_c t + 135°) & \text{binary 10} \\ A\cos(2\pi f_c t + 225°) & \text{binary 00} \\ A\cos(2\pi f_c t + 315°) & \text{binary 01} \end{cases}$$

1.5.3 ANALOG-TO-ANALOG MODULATION

Sometimes it is difficult to have effective transmission with baseband signals, as obtained from digital-to-analog modulation such as ASK, FSK and PSK modulation. In order to get effective transmission with higher frequencies specially for wireless media and optical fiber, we can use analog-to-analog modulation in which an input signal $m(t)$ and a carrier of frequency f_c are combined to produce signal $s(t)$ whose bandwidth is centered on f_c. This modulation permits frequency division multiplexing to allocate a particular transmission frequency for an input signal $m(t)$. There are three types of analog-to-analog modulation – AM, FM and PM.

1.5.3.1 Amplitude Modulation

AM is the simplest form of modulation which mathematically can be expressed as [33]

$$s(t) = [1 + n_a m(t)]\cos(2\pi f_c t)$$

where $\cos(2\pi f_c t)$ = carrier signal with frequency f_c, n_a = modulation index and $m(t) = \cos(2\pi f_m t)$ = input signal of frequency f_m containing data. The above expression can be written as

$$s(t) = \cos(2\pi f_c t) + \frac{n_a}{2}\cos 2\pi(f_c + f_m)t + \frac{n_a}{2}\cos 2\pi(f_c - f_m)t$$

1.5.3.2 Frequency Modulation

FM is one of the angles modulations techniques in which the frequency is written as

$$s(t) = \cos\left[2\pi(f_c t + n_f \cos 2\pi f_m t) + \varphi\right]$$

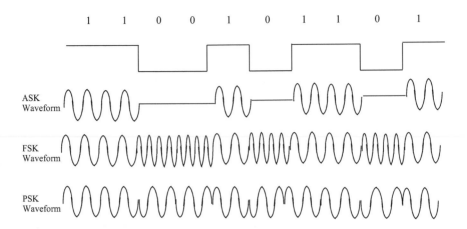

FIGURE 1.28   Digital-to-analog modulation waveform.

The frequency-modulated signal mathematically can be expressed as

$$s(t) = \cos\left[2\pi\left(f_c t + n_f \cos 2\pi f_m t\right) + \varphi\right]$$

where $\cos\left(2\pi f_c t\right)$ = carrier signal with frequency f_c and n_f = modulation index of FM and $m(t) = \cos\left(2\pi f_m t\right)$ = input signal of frequency f_m containing data. The above expression can be written as

$$s(t) = \sum_{n=-\infty}^{n=\alpha} J_n\left(n_f\right)\cos\left[2\pi\left(f_c + n f_m\right)t + n\pi/2 + \varphi\right]$$

where $J_n\left(n_f\right)$ = nth-order Bessel function of the first kind, and using the following property of the Bessel function $J_{-n}\left(n_f\right) = \left(-1\right)^n J_n\left(n_f\right)$, we can write $s(t)$ as

$$s(t) = J_0\left(n_f\right)\cos 2\pi f_c t + \sum_{\substack{n=-\infty \\ n\neq 0}}^{n=\alpha} J_n\left(n_f\right)\left[\cos 2\pi\left(f_c + n f_m\right)t + n\pi/2 + \varphi\right]$$

1.5.3.3   Phase Modulation

PM is one of the angles modulations techniques in which the frequency is written as

$$\varphi(t) = 2\pi n_p \cos 2\pi f_m t$$

The phase-modulated signal mathematically can be expressed as

$$s(t) = \cos\left[2\pi\left(f_c t\right) + 2\pi n_p \cos 2\pi f_m t\right]$$

where $\cos(2\pi f_c t) =$ carrier signal with frequency f_c, and $n_p =$ modulation index of PM and $\cos(2\pi f_m t) =$ input signal of frequency f_m containing data. The above expression can be written as

$$s(t) = \sum_{n=-\infty}^{n=\alpha} J_n(n_f)\cos\left[2\pi(f_c + nf_m)t + n\pi/2\right]$$

where $J_n(n_f) = n$th order Bessel function of the first kind, and using the following property of the Bessel function $J_{-n}(n_f) = (-1)^n J_n(n_f)$, we can write $s(t)$ as

$$s(t) = J_0(n_f)\cos 2\pi f_c t + \sum_{\substack{n=-\infty \\ n\neq 0}}^{n=\alpha} J_n(n_f)\left[\cos 2\pi(f_c + nf_m)t + n\pi/2\right]$$

In long-haul, high-speed WDM transmission links, a narrow spectral width, low susceptibility to fiber nonlinear effects, large dispersion tolerance and a simple and cost-effective configuration are needed for signal generation [33].

SUMMARY

This chapter is started with a description of a basic communication model. Since optical network is mainly used to transmit data, we have discussed local area network operated with an OSI model and TCP/IP protocol. Optical backbone is based on an optical network called WAN. So, this chapter provides basic concepts such as circuit switching, packet switching, frame relay and ATM switching. We have discussed about N-ISDN and B-ISDN which provide other services such as voice transmission and video transmission apart from data transmission. We have also mentioned about digital-to-digital, digital-to-analog and analog-to-analog modulation/demodulation used for data transmission in brief.

Since optical network uses basic optical fiber transmission, we have discussed basic devices such as optical transmitter, optical fiber and optical receiver. For comparison, we have also discussed other data transmission medium.

EXERCISES

1.1. Consider a step-index fiber which has a core refractive index of 1.495. What is the maximum refractive index of the cladding in order for light entering the fiber at an angle of 60° to propagate through the fiber? Air has a refractive index of 1.0.

1.2. Find the formula of the numerical aperture in a graded-index fiber with two layers. Compare the answer with the numerical aperture of the step-index fiber. Can we use geometric optics to deal with situations where the wavelength and core diameter are of the same order of magnitude (e.g., SMF)?

1.3. Consider a step-index multimode fiber in which the refractive indices of the cladding and core are 1.35 and 1.4, respectively. The diameter of the core is 50 μm approximately how many modes are supported by the fiber for a signal at a wavelength of 1550 nm?

1.4. Find the approximate number of modes in a 100 μm core step-index multimode fiber with a wavelength of 850 nm. Assume the refractive index of the core to be 1.5 and that of the cladding to be 1.47.

1.5. Consider an optical link in which power at the transmitter is 0.1 mW, and the minimum power required at the receiver is 0.08 mW. The attenuation constant for the fiber material is 0.033 dB/km. What is the maximum length of the optical link, assuming that there are no amplifiers?

1.6. Describe the various types of dispersion and explain how the effects of each type of dispersion can be reduced.

1.7. Consider a 1-mW 1550 nm signal that is transmitted across a 5 km fiber, through an 8 × 8 passives tar coupler, and through another 15 km of fiber before reaching its destination. No amplifiers are used. What is the power of the signal at the destination?

1.8. Draw the waveform of Manchester coding, B8ZS, HDB3 and differential coding of the following signals:
   i.   10100000000110100000101
   ii. 11010000000001000001101

1.9. Draw the waveform of NRZ-L, Bipolar AMI and pseudo ternary coding of the following signals:
   i.   10100000001110100110101
   ii. 11010000011010011001101

1.10. Draw the waveform of ASK, FSK and BPSK modulation signal of the following digital signals
   i.   10100000001110100110101
   ii. 11010000011010011001101

1.11. A sine wave is to be used for two different signalling schemes: (a) BPSK and (b) QPSK. The duration of signal the element is 10^{-5} s. If the received signal is of the following form

$$s(t) = 0.005\sin\left[2\pi10^{6}t + \varphi\right]V$$ and if the measured noise power at the

receiver is 2.5×10^{-8} W/Hz, determine the E_b/N_0 in dB for each case.

1.12. What SNR is required to achieve the bandwidth efficiency of 5 for ASK, FSK, BPSK and QPSK having a bit error rate of 10^{-5}.

1.13. How can step size and sampling time relate to reduce slope overload noise and quantization noise in case of delta modulation.

1.14. An NRZ-L signal is passed through a filter with $r = 0.5$ and then modulated onto a carrier. The data rate is 2400 bps. Evaluate the bandwidth for ASK and FSK. For FSK, the frequencies are 50 and 55 kHz.

1.15. Consider an audio signal with spectral components in the range of 300–3000 Hz. Assume that a sampling rate of 7 kHz will be used to generate a PCM signal. What is the data rate and the number of uniform quantization levels needed if SNR is 30 dB?

1.16. The density of fused silica is 2.6 g/cm^3. How many grams are needed to make 1 km 50 μm long fiber? If the core material is to be deposited inside the glass tube at 0.5 g/min, how much time is required to make the perform of the fiber.

1.17. A manufacturer wishes to make step-index fiber with normalized frequency $V = 75$ and numerical aperture NA = 0.3 and core index = 1.458. What should be the core diameter and cladding index.

1.18. Calculate the number of modes at 820 nm wavelength in a graded-index fiber with a parabolic index profile $\alpha = 2$ and 25 μm core radius and $n_1 = 1.48$ and $n_2 = 1.46$.

1.19. An optical signal has lost 60% of the power after traversing 5 km of the fiber. Find the propagation loss in dB of the fiber.

1.20 A continuous 12 km long optical fiber has 1.5 dB/km. What is the input power that should be launched to the fiber of 8 km length to maintain a power level of 0.3 μW at the receiving end?

1.21. The active region of LASER diode has a bandgap of $E_g = 1.1$ eV. Find the emitted wavelength of LASER diode.

1.22. The empirical formula of a direct bandgap semiconductor $In_x Ga_{1-x} As_y P_{1-y}$ is given by

$E_g(x, y) = 1.38 + 0.66x - 0.0021y + 0.22xy$. Find the bandgap and emission wavelength of LASER made of the same semiconductor with $x = 0.74$ and $y = 0.56$.

1.23. Find the maximum coupling wavelength of DBR-based LASER having grating of order 2, a grating period of 460 nm and a refractive index = 1.452.

1.24. Find the cut off wavelength of Ge and Si PIN diode detector.

1.25. Find finesse of a filter having 3 dB channel bandwidth of 1 MHz with FSR of 100 MHz Find the number of channels made with the same FSR if 3 dB channel spacing of 3 MHz is used.

1.26. Consider **1.5** mW, 1550 nm signal is transmitted across a 5 km fiber, through an 8×8 passive star coupler, and through another 15 km of fiber before reaching its destination. No amplifiers are used. What is the power of the signal at the destination?

1.27. Consider an optical link in which power at the transmitter is 0.1 mW and the minimum power required at the receiver is 0.08 mW. The attenuation constant for the fiber material is 0.033 dB/km. What is the maximum length of the optical link, assuming that there are no amplifiers?

REFERENCES

1. R. Ramamurthy and B. Mukherjee, "Fixed-alternate routing and wavelength conversion in wavelength-routed optical networks," *IEEE/A CM Transactions on Networking*, vol. 10, no. 3, pp. 351–367, June 2002.

2. D. Banerjee and B. Mukherjee, "Practical approaches for routing and wavelength assignment in large all-optical wavelength routed networks," *IEEE Journal on Selected Areas in Communications*, vol. 14, pp. 903–908, June 1996.

3. P. P. Sahu, "A new shared protection scheme for optical networks," *Current Science Journal*, vol. 91, no. 9, pp. 1176–1184, 2006.

4. P. P. Sahu, "New traffic grooming approaches in optical networks under restricted shared Protection," *Photonics Communication Networks*, vol. 16, pp. 233–238, 2008.

5. B. C. Chatterjee, N. Sarma, and P. P. Sahu, "Priority based dispersion-reduced wavelength assignment for optical networks," *IEEE/OSA Journal of Lightwave Technology*, vol. 31, no. 2, pp. 257–263, 2013.

6. B. C. Chatterjee, N. Sarma, and P. P. Sahu, "Priority based routing and wavelength assignment with traffic grooming for optical networks," *IEEE/OSA Journal of Optical Communication and Networking*, vol. 4, no. 6, pp. 480–489, 2012.

7. H. Nishihara, M. Haruna, and T. Suhara, *Optical Integrated Circuits*, McGraw-Hill, New York, 1989.

8. A. K. Das and P. P. Sahu, "Compact integrated optical devices using high index contrast waveguides," *IEEE Wireless and Optical Communication*, Bangalore, India, IEEE Xplore Digital No. 01666673, pp. 1–5, 2006.

9. N. Takato, T. Kominato, A. Sugita, K. Jinguji, H. Toba, and M. Kawachi, "Silica based integrated optic Mach Zehnder multi/demultiplexer family with channel spacing of 0.01-250 nm," *IEEE Selected Areas Communications*, vol. 8, no. 6, pp. 1120–1127, 1990.

10. A. Neyer, "Integrated optical multichannel wavelength multiplexer for monomode systems," *IEE Electronics Letter*, vol. 20, no. 18, pp. 744–746, 1984.

11. T. Y. Tsai, Z. C. Lee, J. R. Chen, C. C. Chen, Y. C. Fang, and M. H. Cha. "A novel ultra compact two mode interference wavelength division multiplexerfor 1.5 μm operation," *IEEE Journal of Quantum Electronics*, vol. 41, no. 5, pp. 741–746, 2005.

12. B. J. Offrein, G. L. Bona, F. Horst, W. M. Salemink, R. Beyeler, and R. Germann. "Wavelength tunable optical add after drop filter with flat pass band for WDM networks," *IEEE Photonics Technology Letters*, vol. 11, no. 2, pp. 239–241, 1999.

13. M. Okuno, "Highly integrated PLC type optical switches for OADM and OxC systems," *IEEE Optical Fiber Conference (OFC)*, vol. 1, pp. 169–170, 2003.

14. P. P. Sahu, "Tunable optical add/drop multiplexers using cascaded Mach Zehnder coupler," *Fiber and Integrated Optics* (Taylor and Francis), vol. 27, no. 1, pp. 24–34, 2008.

15. P. P. Sahu, "Polarization insensitive thermally tunable Add/Drop multiplexer using cascaded Mach Zehnder coupler," *Applied Physics: Lasers and Optics* (Springer), vol. B92, pp. 247–252, 2008.

16. M. Kuznetsov, "Cascaded coupler Mach-Zehnder channel dropping filters for wavelength division multiplexed optical system," *IEEE Journal of Lightwave Technology*, vol. 12, no. 2, p. 225, 1994.

17. C. Kostrzewa, R. Moosburger, G. Fisehbech, B. Schuppert, and K. Petermann, "Bandwidth optimization of optical add/drop multiplexers using cascaded couplers and Mach-Zehnder sections," *IEEE Photonics Technology Letters*, vol. 7, no. 8, p. 902, 1995.

18. R. Kashahara et al., "New structures of silica-based planar light wave circuits for low power thermooptic switch and its application to 8×8 optical matrix switch," *Journal of Lightwave Technology*, vol. 20, no. 6, pp. 993–1000, June 2002.

19. S. Sohma, "Low switching power silica based super high delta thermo optic switch with heat insulating grooves," *IEE Electronics Letters*, vol. 38, no. 3, pp. 127–128, 2002.

20. A. K. Das and P. P. Sahu, "Minimization of heating power for thermooptic waveguide type devices," *Journal of Optics*, vol. 32, no. 3, pp. 151–167, July–September 2003.

21. M. Yagi, S. Nagai, H. Inayoshi, and K. Utaka, "Versatile multimodes interference photonic switches with partial index modulation regions," *IEE Electronics Letter*, vol. 36, no. 6, pp. 533–534, 2000.

22. R. Krahenbuhl, M. M. Howerton, J. Dubinger, and A. S. Greenblatt. "Performance and modeling of advanced Ti:LiNbO$_3$ digital optical switches," *IEEE Journal of Lightwave Technology*, vol. 20, no. 1, pp. 92–99, 2002.

23. Y. Hida, H. Onose, and S. Imamura, "Polymer waveguide thermooptic switch with low electric power consumption at 1.3μm," *IEEE Photonics Technology Letters*, vol. 5, no. 7, pp. 782–784, 1993.

24. L. Tancevski, A. Bononi, and L. A. Rusch, "Output power and SNR swings in cascades of EDFAs for circuit packet switched optical networks'," *IEEE Journal of Lightwave Technology*, vol. 17, pp. 733–742, 1999.

25. G. Keiser, *Optical Fiber Communications*, McGraw-Hill Inc., New York, 1999.

26. M. Karasek and J. A. Valles, "Analysis of channel addition / removal response in all optical gain-controlled cascade of Erbium doped fiber amplifier," *IEEE Journal of Lightwave Technology*, vol. 16, pp. 1795–1803, 1998.

27. P. P. Sahu, "Thermally tunable EDFA gain equalizer using point symmetric cascaded Mach-Zehnder Filter," *Optics Communications*, vol. 281, no. 4, pp. 573–579, 2008.

28. P. P. Sahu, "Polarization independent thermally tunable EDFA gain equalizer using cascade Mach-Zehnder coupler," *Applied Optics*, vol. 47, no. 5, pp. 718–724, 2008.

29. C. R. Doerr et al., "Dynamic wavelength equalizer in silica using the single filtered arm interferometer," *IEEE Photonics Technology Letters*, vol. 11, pp. 581–583, 1999.

30. A. C. Baishya, S. K. Srivastav, and P. P. Sahu, "Cascaded Mach Zehnder coupler for dynamic EDFA gain equalization applications," *Journal of Optics* (Springer), vol. 39, no. 1, pp. 42–47, 2010.

31. F. Halsall, *Digital Communications, Computer Networks and Open Systems*, Addison-Wesley, Reading, MA, 1996.

32. B. Jain and A. Agarwala, *Open System Interconnection*, McGraw-Hill Inc., New York, 1993.

33. W. Stallings, *Data and Computer Communications*, Prentice-Hall, Upper Saddle River, NJ, 1999.

34. E. Murphy, S. Hayes, and M. Enders, *TCP/IP: Tutorial and Technical Overview*, Prentice-Hall, 1992.

35. D. Bertsekas and R. Gallager, *Data Networks*, Prentice-Hall, Upper Saddle River, NJ, 1992.

36. J. Spragins, J. Hammond, andK. Pawlikowski, *Telecommunications Protocols and Design*, Prentice-Hall, Upper Saddle River, NJ, 1985.

37. P. Smith, *Frame Relay: Principles and Applications*, Addison-Wesley, Reading, MA, 1993.

38. R. Onvural, *Asynchronous Transfer Mode Networks: Performances Issues*, Artech House, London, 1994.

39. P. P. Sahu, "Effect of altitude and minimum elevation angle on polar LEO satellite constellation for Global continuous coverage," *Proceedings of International Forum Cum Conference on IT and Communication at the New Millennium*, Thailand, vol. 2, pp. 405–409, 2000.

40. S. R. Pratt, R. A. Raines, C. E. Fossa, and M. A. Temple, "An operational and performance overview of the IRIDIUM low earth orbit satellite system," *IEEE Communication Surveys*, http://www.comsoc.org/pubs/surveys Second Quarter 1999.

41. H. Helgert, *Integrated Services Digital Networks: Architectures, Protocols and Standards*, Addison-Wesley, Reading, MA, 1991.

42. J. Hecht, *Understanding Fiber Optics*, 3rd ed., Prentice-Hall, 1999.

43. L. Kazovsky, S. Benedetto, andA. E. Willner, *Optical Fiber Communication Systems*, Artech House, Boston, MA, 1996.

44. E. Ciaramella, G. Contestabile, A. D. Errico, C. Loiacono, and M. Presi, "High-power widely tunable 40-GHz pulse source for 160-gb/s OTDM systems based on nonlinear fiber effects," *IEEE Photonics Technology Letters*, vol. 6, no. 3, pp. 753–755, 2004.

45. J. P. Powers, *An Introduction to Fiber Optic Systems*, Irwin, Homewood, IL, 1993.

46. A. R. Chraplyvy, "Lirnits on lightwave communications imposed by optical-fiber nonlinearities," *IEEE/OSA Journal of Lightwave Technology*, vol. 8, pp. 154, 1557, October 1990.

47. F. Forghieri, R. W. Tkach, A. R. Chraplyvy, and D. Marcuse, "Reduction of four-wave mixing crosstalk in WDM systems using unequally spaced channels," *IEEE Photonics Technology Letters*, vol. 6, no. 6, pp. 754–756, 1994.

48. A. R. Chraply, "Optical power limits in multi-channel wavelength-division-multiplexed systems due to stimulated Raman cattering," *IEEE Electronics Letters*, vol. 20, no. 2, pp. 58–59, 1984.

49. A. K. Ghatak and K. Thyagragan, Laser: Fundamentals and Applications, Sringer, New York, 2010.

50. M. Gotoda, T. Nishirnura, and Y. Tokuda, "Widely tunable SOA integrated DBR laser with combination of sampled-grating and superstructure grating," *Proceedings, 19th IEEE International Semiconductor Laser Conference*, Matsue, Japan, pp. 147–148, 2004.

51. J. Hecht, *Understanding Lasers: An Entry-Level Guide*, 2nd ed., Wiley, 2004.

52. L. A. Johansson, Z. Hu, D. J. Blumenthal, L. A. Coldren, Y. A. Akulova, and G. A. Fish, "40-GHz dual-mode-locked widely tunable sampled-grating DBR laser," *IEEE Photonics Technology Letters*, vol. 17, no. 2, pp. 285–287, February 2005.

53. M. Azizoglu, *Phase Noise in Coherent Optical Communications*, Ph.D. Dissertation, Department of Electrical Engineering and Computer Science, Massachusetts Institute of Technology, 1991.

54. P. E. Green, *Fiber Optic Networks*, Prentice-Hall, Englewood Cliffs, NJ, 1993.

55. J. Kani and K. Iwatsuki, "A wavelength-tunable optical transmitter using semiconductor optical amplifiers and an optical tunable filter for metro/access DWDM applications," *IEEE/OSA Journal of Lightwave Technology*, vol. 23, no. 3, pp. 1164–1169, March 2005.

56. H. Kobrinski and K.-W. Cheung, "Wavelength-tunable optical filters: Applications and technologies," *IEEE Communications Magazine*, vol. 27, p. 5343, October 1989.

57. A. Inoue, M. Shigehara, M. Ito, M. Inai, Y. Hattori, and T. Mizunami, "Fabrication and application of fiber Bragg grating – A review," *Optoelectronics - Devices and Technologies*, vol. 10, pp. 119–130, March 1995.

58. B. Mukherjee, *Optical Communication Networks*, McGraw-Hill, 2006.

2 Different Optical Network Node

One of the most essential parts in wavelength division multiplexing (WDM) network is the hardware part of the node having a workstation interconnected to optical fibers and transceivers potentially accessing any of the available wavelength channels in each fiber. In designing the network interface of the node one should choose the number and the type of transmitters and receivers (whether fixed-tuned or tunable) at each node. These are determined on the protocol, degree of access and connectivity desired in the network as well as on practicality and cost considerations. A WDM network protocol is either a single-hop protocol (in which communication takes place directly between two nodes without being routed through intermediate nodes) or a multihop system [1–17] (in which information from a source node to a destination node may be routed through the electronics at intermediate nodes in the network). In multihop networks, the transmitter or receiver is less tuning than single-hop networks. Each node has at least one transmitter and one receiver. When both the transmitters and the receivers are fixed-tuned to certain wavelength channels, and there is more than one channel, then a static multihop topology must be established over the passive-star coupler.

Since a node in the optical backbone plays a key part, flexible operations of the functions such as routing, restoration and reconfiguration are provided. These flexible operations are based on node architectures where WDM, optical matrix switches and add/drop multiplexing devices are the key hardware devices. The architecture of the node is based on the topology and operation of the network, which is either a regular or mesh architecture. Two types of node architecture are used in optical networks–non-reconfigurable node and reconfigurable node.

2.1 NON-RECONFIGURABLE NODE

In this section, non-reconfigurable node is discussed. In this type of node, there is only fixed but no flexible operation as per the conditions of optical network.

2.1.1 Non-Reconfigurable Wavelength Router Node

A wavelength router transmits the signals from different input to different output port fibers on the basis of the destination address of the signals. In general, a wavelength-routing device has demultiplexing of the different wavelengths from each input port, and then multiplexing signals at each output port as shown in Figure 2.1. Because of its non-reconfigurable nature, there is no switching stage between the demultiplexers and the multiplexers, and the routes for different signals arriving at any input port are fixed (these devices are referred as routers rather than switches), or reconfigurable, in which case the routing function of the switch is controlled electronically.

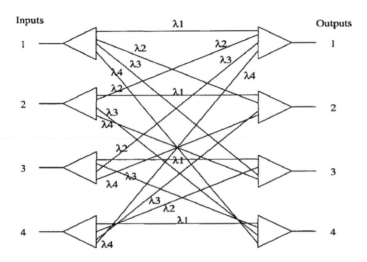

FIGURE 2.1   (a) A 4 × 4 non-reconfigurable wavelength router.

(*Continued*)

An on-reconfigurable wavelength router comprises a first stage having demultiplexers which separate each of the wavelengths on an incoming fiber and a second stage having multiplexers which recombine wavelengths channels to a single output. The outputs of the demultiplexers are connected to the inputs of the multiplexers. The router has more than one incoming fibers and outgoing fibers. On each incoming fiber, there are M wavelength channels. A 4 × 4 non-reconfigurable wavelength router with $M = 4$ is shown in Figure 2.1a. The router is non-reconfigurable because the path of a given wavelength channel, after it enters the router on a particular input fiber, is fixed [12]. The wavelengths one ach incoming fiber are separated using a grating demultiplexer. Finally, information from multiple WDM channels are multiplexed before launching them back on to an output fiber. In between the demultiplexers and multiplexers, there are direct connections from each demultiplexer output to each multiplexer input. The wavelength on which the input port gets routed to which output port depends on an outing matrix.

2.1.2   ARRAYED WAVEGUIDE GRATING-BASED NODE

Figure 2.1b shows a non-reconfigurable node having an arrayed waveguide grating (AWG) which uses a fixed routing of an optical signal from a given input port to an output port based on the wavelength of the signal. Signals with different wavelengths arriving to an input port are routed to a different output port by using AWG. Also, different signals using the same wavelength are incident simultaneously to different input ports, and still do not interfere with each other at the output ports. In comparison to a passive-star coupler using a given wavelength on a single input port, the AWG with N input and N output ports is enabled to do the routing at a maximum of N connections, as opposed to a maximum of N connections in the passive-star coupler. Also, the AWG can be easily fabricated at low cost using an integrated optic concept. The disadvantage of AWG is that it acts as a device with a fixed routing matrix which cannot be reconfigured [18,19].

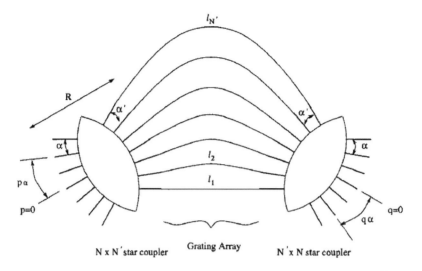

FIGURE 2.1 (CONTINUED)   (b) An array waveguide grating-based node.

(*Continued*)

Here AWG works as a fixed router for a tunable optical transmitter or receiver. Figure 2.1b shows an AWG having two passive-star couplers connected by a grating array. The first star coupler has N inputs and N' outputs (where $N < N'$), while the second one has N' inputs and N outputs. The inputs to the first star are separated by an angular distance of α, and their outputs are separated by an angular distance α'. The grating array consists of N' waveguides, with lengths, $l_1, l_2, \ldots l_{N'}$ where $l_1 < l_2 < \ldots < l_{N'}$. The length difference between any two adjacent waveguides is a constant A_1. In the first star coupler, a signal on a given wavelength entering from any of the input ports is separated and transmitted to its N' outputs, which are also the N' inputs of the second star of grating array. The signal is sent through the paths of grating array, experiencing a different phase shift in each waveguide for the length of the waveguides and the wavelength of the signal transmitted.

Figure 2.1c represents an AWG-based node used in a Fiber-To-The-Curb (FTTC) network [12], which makes a single broadband access infrastructure providing many application services to different service providers of end users. The WDM demultiplexer (WDDM) is linked to the broader Internet via a metro ring network, a wide-area LAN or a long-haul optical network. The ONUs are written as Local Access Points (LAPS) which are distribution centers for bandwidth to end users. This is known as an *open access network*. This access network is maintained by an *access-network operator* (ANO) [12].

2.1.3 Node Architecture of a Passive-Star WDM Network

Figure 2.2 shows the simplest and most popular interconnection node for passive-star WDM network acting on a broadcast communication provider. The broadcast capability of the star coupler with multiple WDM channels contributes to a wide

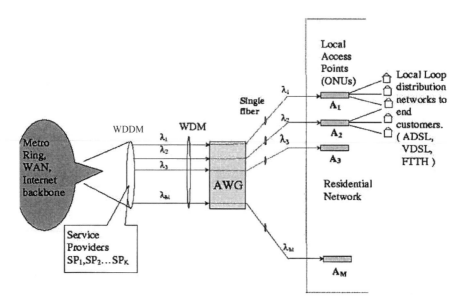

FIGURE 2.1 (CONTINUED)   (c) AWG is used as node in an optical network (FTTC – Fiber To The Curb) [12].

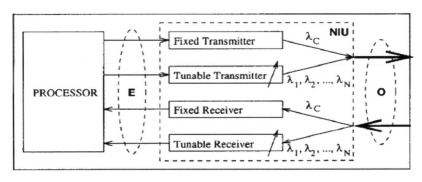

PSC -- Passive Star Coupler   O -- Optical inferface

NIU -- Network Interface Unit   E -- Electronic interface

FIGURE 2.2   Node architecture in a passive-star WDM network [12].

range of possible media access protocols [20]. With its advantage of reliability, the drawback of having a passive network medium is the requirement of handling of additional processing and hardware at the node for routing. The broadcast capability of the star coupler also prevents the reuse of wavelengths to create more simultaneous connections. The node in a network requires a workstation connected to the network medium via optical fiber, and the node accesses any of the available wavelength channels on each fiber. The design of the network interface for the node requires to know the number of transmitters and receivers as well as the type of transmitters and receivers – fixed-tuned or tunable – to place at each node. These selections usually depend on the protocol, degree of access and connectivity desired in the network, as well as on practicality and cost considerations.

2.2 RECONFIGURABLE WAVELENGTH-ROUTING NODE

A reconfigurable wavelength-routing switch (WRS) node, also known as a wavelength-selective cross connect (WSC), requires photonic switches inside the routing element. Figure 2.3 are presents $N \times N$ reconfigurable WRS node having N number of incoming fibers and N number of outgoing fibers. On each incoming fiber, there are M number of wavelength channels. The wave lengths on each incoming fiber are separated by a grating demultiplexer. The outputs of the demultiplexers are directed to an array of M number of 2×2 optical switches between the demultiplexer and the multiplexer stages. All signals on a given wavelength are sent to the same switch. The switched signals are then directed to the outputs of the multiplexers. Finally, information streams from multiple WDM channels are multiplexed before launching them back into an output fiber.

Space-division optical-routing switches are made with multiple 2×2 optical crosspoint elements arranged in a banyan-based structure [21]. The space-division switches (which is one per wavelength [22–24]) transmit signals from any input to output using a given wavelength. Such switches based on relational devices [9] are capable of switching very high-capacity signals. The 2×2 crosspoint elements are taken for making slowly tunable space-divisions witches and are reconfigured to be adaptive with the changing traffic requirements [23–25].

Figure 2.3b represents an electrical switch-based node architecture [12]. In this node, the WDM signals arrived at an input fiber, first encounter a WDDM [26,27], and different wavelengths are separated. The bits of each wavelength entering to a receiver node are converted into electrical signals, and are again converted back to optical signals on a wavelength and then multiplexed with WDM.

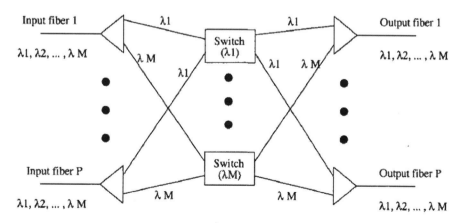

FIGURE 2.3 (a) Reconfigurable photonic switch-based node.

(*Continued*)

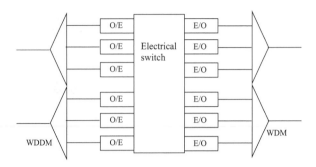

FIGURE 2.3 (CONTINUED)   (b) Electrical switch-based node architecture [12].

2.2.1 ADD/DROP MULTIPLEXER-BASED RECONFIGURABLE NODE IN A RING WDM NETWORK

In a traditional SONET ring network [28], there is a requirement of add–drop multiplexer (ADM) for each wavelength at every node to get traffic add/drop of that particular wavelength. With the development of WDM technology, few hundred wavelengths can now be accommodated simultaneously by a single fiber. With the emerging of optical ADMs (OADM [29,30]), it is possible for a node to avoid dropping of most of the wavelength channels optically and only drop the wavelengths having the traffic destined to the node. Figure 2.4a represents the architecture of a typical node in a SONET/WDM ring network [12,28]. For some wavelengths (say λ_1), since there is no need to add or drop any of its timeslots, these can be optically passed through the node without dropping. For other wavelengths ($\lambda_2 \ldots \lambda_s$) where only one timeslot is added or dropped, an electronic ADM is used.

In this structure, one of the key components of node is the OADM [29,30]. Figure 2.4b shows an OADM-based reconfigurable node which is normally applied in a single WDM ring network. It is also upgraded to operate over multiple wavelengths by using more OADMs and appropriate terminal equipment at the network nodes, i.e., the same fiber cable is also reused in a WDM ring network. In this W-wavelength ring network, an OADM drops information from and adds the wavelength having traffic to the ring to be set to operate at the node. OADMs reduce the network cost and make it easy by allowing traffic to bypass intermediate nodes without expensive 0–E–O conversion [12]. An OADM made to operate permanently on a fixed wavelength is known as a fixed OADM (FOADM) [15]; OADMs tuned with

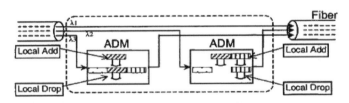

FIGURE 2.4   (a) A WDM ring network node architecture [12].

(*Continued*)

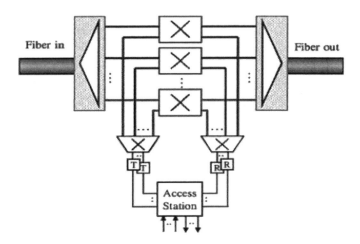

FIGURE 2.4 (CONTINUED)   (b) OADM based on reconfigurable node in a WDM ring network [12].

some control mechanism via external energy is called as a reconfigurable OADM (ROADM) [29,30]. ROADMs add/drop traffic on to/from different wavelengths with control mechanism giving desirable flexibility and enabling fast provisioning of dynamic traffic, with lower cost. To set up a connection, ROADMs at the source and destination nodes are tuned to the same free wavelength [12].

In a SONET-based [31] WDM ring, there are two types of nodes – non-inter section nodes and intersection nodes. In a non-intersection node, two interfaces are connected to it with its two neighbors, and a local interface is required for adding or dropping traffic. Figure 2.5 shows the architecture of a non-intersection node in a unidirectional ring. A bidirectional ring consists of two unidirectional rings requiring extra hardware. Most architectures are directional rings and bidirectional SONETADMs (SADMs) based (back-to-back double SADMs for both directions assembled as one unit) [12]. The architecture of the intersection node depends on the availability of the hardware. The OADM-and SADM-based architectures are popular [12]. In this architecture, a digital cross connect (DXC) connects low-speed streams between the two rings.

Two rings interconnect at either one or multiple points. Usually, two physical intersections are considered due to the fault-recovery concern (when a node failure occurs at one intersection node, the rest of the nodes should still be connected so that the auto recovery mechanism is used to continue the traffic flow in a ring network). SONET-ring-based protection mechanism is verified by time, and we are not proposing any changes. Figure 2.6 shows an architecture that is interfaced with a double-ring network at SONET level [31], where traffic, either going to local ports or to another ring, is dropped by OADMs and SADMs is then relayed by optical cross-connect (OXC) to their desired destination. Figure 2.7a and b shows double-ring network node architectures connecting two rings at wavelength level. Figure 2.7c shows another double-ring network node architecture providing a mixed connection with wavelength and SONET level which contributes to maximum flexibility. The nodes in Figure 2.7a and b do not crossconnect two rings together if they are used alone.

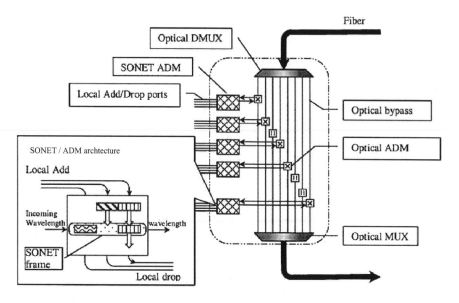

FIGURE 2.5 Modified node architecture of a WDM ring network for accommodating SONET signal [12].

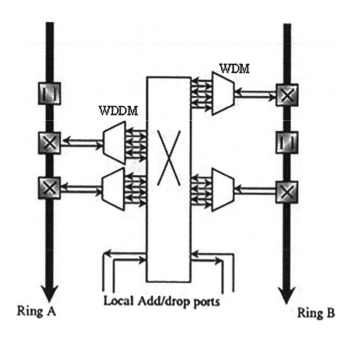

FIGURE 2.6 Node architecture interface with SONET for double-ring network [12] (WDDM – wavelength division demultiplexing, WDM – wavelength division multiplexing.

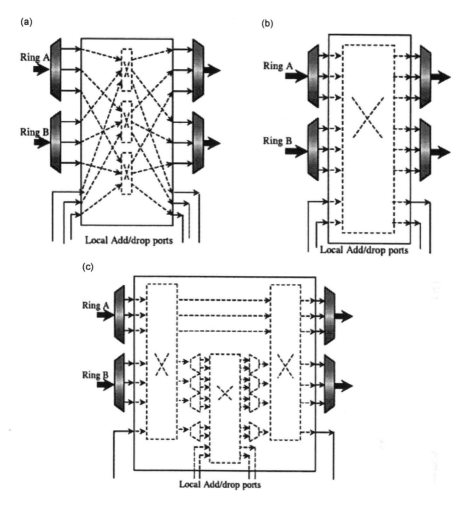

FIGURE 2.7 Node architecture interface with WDM for a double-ring network [12]. (a) Multiple switch block, (b) single switch block and (c) mixed architectures with SONET and WDM.

2.2.2 WAVELENGTH CONVERTIBLE NODE ARCHITECTURE

Figure 2.8 shows a wavelength convertible reconfigurable 2×2 node architecture in which wavelength is converted if it is not available to get assignment of a connection request. There are two types of wavelength convertible node architectures [32,33]. Figure 2.8a shows a node architecture in which wavelengths are shared inside the node, whereas Figure 2.8b shows a node architecture in which wavelengths are shared per fiber optic link. In the case of wavelength convertible architecture (a), the wavelengths are switched to an output fiber as per the routing path decided by a wavelength router, whereas architecture (b) switches the wavelength channel inside

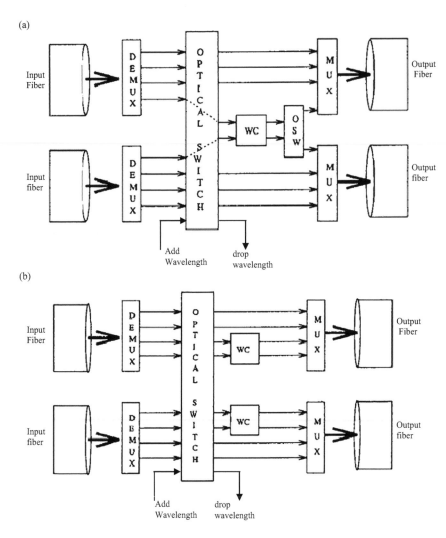

FIGURE 2.8 Wavelength convertible node architecture: (a) Shared per node [12] and (b) shared per link (WC – wavelength converter; OSW – optical switch) [12].

the same fiber link. This type of architecture is used normally in double WDM ring network. It is also employed in WDM mesh optical network, but a number of input/ output fibers are related to the number of links connected to the node.

2.2.3 RECONFIGURABLE NODE ARCHITECTURE IN WDM-BASED MESH OPTICAL NETWORK

Nationwide optical network does not follow a regular topology but is based on mesh topology. In this type of WDM mesh optical network, the number of bidirectional fiber link depends on the number of links connected to the node.

2.2.3.1 Wavelength-Router–Based Reconfigurable Node

The wavelength-router–based reconfigurable WDM mesh optical node architecture consists of a wavelength multiplexer/demultiplexer [26,27], optical matrix switches [24,25] and wavelength router [12,32,33]. Figure 2.9 shows a schematic block diagram of a reconfigurable node. In the figure, there are N number of input/output fibers and each fiber has M number of wavelengths. At first, the M number of wavelengths in node are multiplexed by using W-MUX and then switching of each wavelength to a particular output fiber is carried out by thermooptic integrated optical devices via thermooptic effect, depending on the wavelength-routing function of the node.

2.2.3.2 Fully Wavelength Convertible Node
Architecture of a WDM Mesh Network

Figure 2.10 shows the fully wavelength convertible $N \times N$ node architecture of a WDM-based mesh optical network [12,32–34]. In this type of node, all the wavelengths convertible and the wavelengths in any input fiber are switched to any output fiber as per the routing. So it is fully flexible as per the routing.

2.2.4 SONET over WDM Node Architecture
for a Mesh Optical Network

Figure 2.11a shows SONET over a WDM reconfigurable node architecture for a WDM mesh optical network. There are two sections in this architecture – optical domain section (ODS) and access section (AS). The ODS does wavelength routing

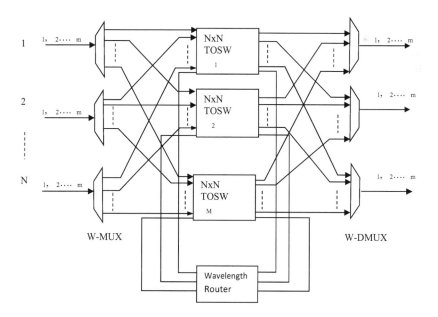

FIGURE 2.9 A schematic block diagram of a reconfigurable node consisting of $N \times N$ TOSW, W-MUX and W-DMUX.

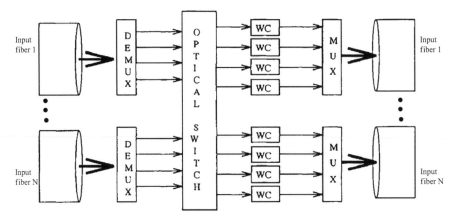

FIGURE 2.10   WC-based N × N node architecture of a WDM mesh optical network [12].

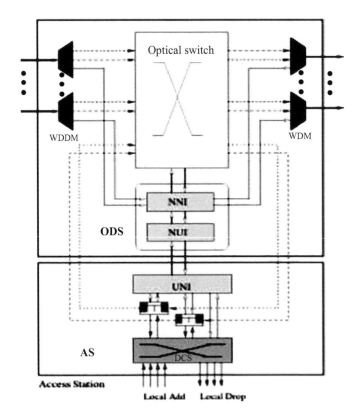

FIGURE 2.11   (a) Traffic grooming node architecture (DXC – digital crossconnect, UNI – user-to-network interface. NUI – network-to-user interface and NNI – network-to-network interface, ODS – optical domain section, AS – access section [12].)

(*Continued*)

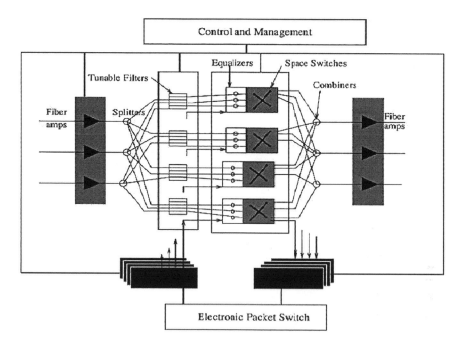

FIGURE 2.11 (CONTINUED)   (b) Transport node of an optical network [12].

and wavelength multiplexing/demultiplexing, whereas the AS carries out local traffic adding/dropping and low-speed traffic grooming functionalities. The ODS has an OXC, network control and management unit (NC&M), and an optical multiplexer/ demultiplexer. In the NC&M, network-to-network interfaces (NNI) interchange control messages between the nodes by configuring OXC on a dedicated wavelength channel. The network-to-user interface (NUI) transmits NNI and exchanges control information to the user-to-network interface (UNI) and the control component of the access station. In the figure, each access station has several SONET ADMs [12,31]. Each SONET ADM splits a high-rate SONET signal into lower rate components. In order to transmit or receive traffic on a wavelength in a node, the wavelength is added or dropped at the node through a SONET ADM. In the figure each SONET ADM has a fixed transceiver transmitting only on one wavelength. The DXC interlinks with the low-speed traffic streams between the access station and the ADMs. A low-speed traffic stream on one wavelength is either dropped to the local client (IP router, ATM switch, etc.) or switched to another ADM and transmits another wavelength. SONET components (ADM, DCS, etc.) and SONET framing schemes have TDM-based fast multiplexing/demulitplexing capability, in comparison to the software-based scheme. The disadvantage of this approach is expensive due to having SONET components such as ADM and DCS. Both kinds of access stations are used together for the connection with an OXC to get a multiservice platform for accessing an OXC in the network.

2.2.5 Transport Node of a WDM Optical Network

Resource budgeting in the network has a direct impact on the cost of setting up the network. The WRS is based on the prototype used in transport node of an optical network used in a RACE project [35]. Also, the prototype considers that OXCs and transmission equipment at a node are integrated together to form the corresponding WRS.

2.2.6 IP over WDM Network Node Architecture

Figure 2.12 represents IP over WDM reconfigurable node architecture in a mesh optical network. It has an ODS and an AS. The ODS comprises an OXC) [12], NC&M, and optical multiplexer/demultiplexer [26,27]. In the figure, each access station consists of transmitters and receivers (transceivers). Traffic originated from access station is transmitted as an optical signal on one wavelength channel by a transmitter. Traffic destined to access node is transformed from an optical signal to electronic data in a receiver. Both tunable and fixed transceivers are needed in the access station. A tunable transceiver is tuned to different wavelengths so that an optical signal is transmitted on any free wavelength in its tuning range. A set of fixed transceivers, one per wavelength, make a group together to form a transceiver array. The size of a fixed transceiver array is equal or smaller than the number of wavelengths on a fiber, but the number of transceiver arrays is equal to or smaller than the number of fibers connecting a node. The AS provides a flexible software-based bandwidth-provisioning capability to the network. Multiplexing low-speed connections to high-capacity lightpaths are made by the multiprotocol level shifting/Internet protocol (MPLS/IP) router with a software-based queuing scheme.

2.2.7 Node Architecture for Multicasting Optical Network

Figure 2.13 represents a hybrid approach, in which the incoming optical bit streams are converted to electronic data, the data is switched using an electronic crossconnect and then the electronic bit streams are again converted back to the optical domain. The signal in a channel arriving on the input fiber link D is replicated into three copies in the electronic domain – one copy is dropped locally at the node and the remaining two are switched to different channels on outgoing fiber links 1 and **2**. (Along with the light tree, the switch is also used to establish lightpaths from a source to a destination by a unicast connection from input fiber link **2** to output fiber link D.) This is also called as "opaque" switch architecture [12], which is nowadays focused for its use in optical network due to the existence of a mature technology to design high-bandwidth multichannel non-blocking electronic crossconnect fabrics at low cost.

Figure 2.14 represents a multicast-capable all-optical switch which makes cross-connection of the optical channels in an optical network. Here, the switch operation is made by bit encoding schemes, as conflicted to a switch with 0–E–O conversion in Figure 2.13. For multicasting in all-optical switches, optical splitters are used to replicate an incoming bit stream to two or more outputs in the figure. A signal arriving

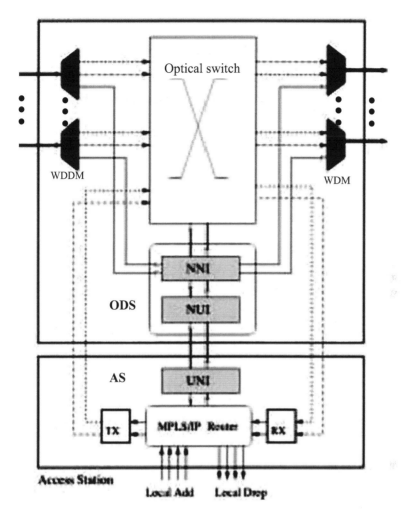

FIGURE 2.12 Node architecture for IP over a WDM network (UNI – user-to-network interface. NUI – network-to-user interface, NNI – network-to-network interface and MPLS/IP – Multiprotocol level shifting/Internet protocol [12].)

on wavelength λ_b from input fiber link D is sent to the optical splitter X for separating it into three identical copies – one is dropped locally at the node, whereas the other two are sent to output fiber links 1 and 2. The signal arriving on wavelength λ, from input fiber link 2 bypasses to the node.

Figure 2.15 represents an architecture with a "transparent" switch. Here, as a replacement of two optical switches, one optical switch is used for switching additional signals from the splitters among a larger number of ports. An advantage of this architecture is that the fan-out of a signal is not restricted by the splitting ration of a splitter. By using a proper switch configuration, output from one splitter is fed as an input of another splitter, thus obtaining a higher splitting ratio of a signal.

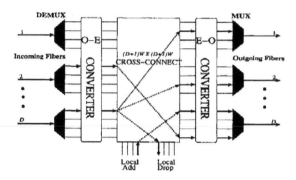

FIGURE 2.13   Opaque switch-based node architecture for multicasting optical network using O–E and E–O converter and electrical crossconnect [12].

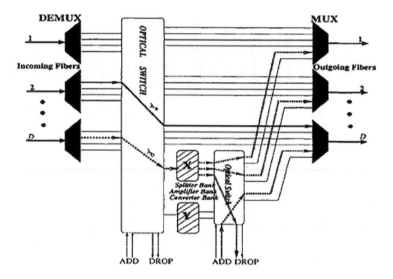

FIGURE 2.14   Node architecture for multicasting optical network using optical crossconnect, X splitter and Y optical splitter [12].

In the figure, an incoming signal from an input fiber D is split by a three-way splitter X – One is dropped locally and the other two are switched to output fiber links 1 and 2. Again, the signal arriving from input fiber link 2 on wavelength goes around the node without any local drop [12].

2.2.8   TRAFFIC GROOMING NODE ARCHITECTURE
FOR AN OPTICAL MESH NETWORK

Traffic grooming concept [36–38] is worked in an optical network for efficient use of wavelength channels. Traffic grooming node based on SONET technology is discussed in this section. Figure 2.16 represents a partial traffic grooming WDM mesh

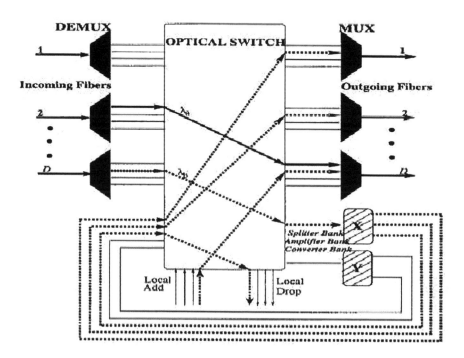

FIGURE 2.15 Node architecture for supporting multicasting with all optical crossconnect [12].

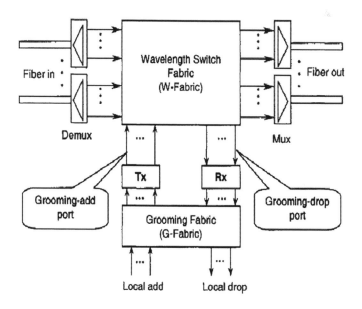

FIGURE 2.16 Partial traffic grooming WDM mesh network node architecture [12].

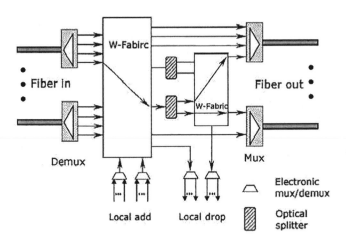

FIGURE 2.17   Node architecture for source grooming [12].

network node architecture having wavelength switch fabric (W-Fabric) and grooming fabric (G-Fabric). For provisioning a connection request, there are two types of resource constraints – wavelengths and grooming ports [38]. Typically, the more the number of wavelengths the network has, the less the number of grooming ports, node needs and vice versa.

Figure 2.17 represents the source grooming node architecture used normally in a tree topology-based optical network [12,34]. In the figure, duplication is made in the optical domain using an optical splitter by splitting the power of an optical signal from one input port to multiple output ports. For an OXC using opaque technology, traffic duplication is made by copying the electronic bit stream from one input port to multiple output ports.

2.2.9   Node Architecture of Optical Packet-Switched Network

Two categories of optical packet-switched networks are slotted (synchronous) and unslotted (asynchronous) networks [12]. At the input ports of each node, packets arrive at different times. Since the switch fabric changes its state incrementally by establishing one input–output connection at an arbitrary time or jointly establishing multiple input–output connections together simultaneously, it makes switching of multiple time-aligned packets together or switching each packet individually "on the fly." In both cases, bit-level synchronization and fast clock recovery are required for packet-header identification and packet demarcation. Figure 2.17 represents a node architecture in a slotted optical packet-switched network. The packets are considered to be of same size in a slotted network in a variation of slotted network and even if the packets are of variable length, each packet's length is an integral multiple of a slot. A fixed-size time slot contains both the payload and the header. The time slot has a longer duration than the guard time provided for the whole packet. All the input packets arriving at the input ports are required to be aligned in phase with one another before entering the switch fabric. To successfully synchronize all the incoming packets, it is required to analyze what types of delay variation a packet experiences.

Since delay variations are relatively little, it is compensated statistically as a replacement for a dynamically packet-by-packet basis. The time delay of each packet inside a node depends on the switch fabric and contention-resolution scheme used in the node. Depending on the operation of a switch fabric, a packet uses different paths with unequal lengths within a switch fabric [12]. The fast time jitter that arises due to the dispersion between different wavelengths and unequal optical paths varies from packet to packet at the output of the switch; so a fast output synchronization interface might be required. Thermal effects within a node are very little because it varies more slowly and is easily controlled.

Figure 2.18 represents a functional diagram of a node architecture of a slotted network. A passive tap separates out a small amount of power from the incoming signal (or packet) for header reading. The header-processing circuit identifies a preamble at the beginning of the packet and header information. It also passes the timing information of the incoming packet to the control unit to configure the synchronization stages and switch fabric. The input synchronization stage aligns the packets before they enter the switch fabric. The output synchronization stage shown in Figure 2.17 compensates for the fast time jitter that occurs inside the node.

Figure 2.19 represents the node architecture based on packet transmission for unslotted networks. The fixed length of fiber delay lines grasps the packet when header processing and switch reconfiguration are taking place. There is no packet-alignment stage, and all the packets experience the same amount of delay with the same relative position in which they arrived, provided there is no contention. The unslotted network avoids the requirement of synchronization stages. For a traffic load, its network throughput is lower than that of the slotted networks because contention is more likely to occur.

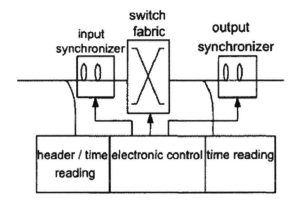

FIGURE 2.18 Functional block for synchronization of packets in a slotted optical packet-switched network node.

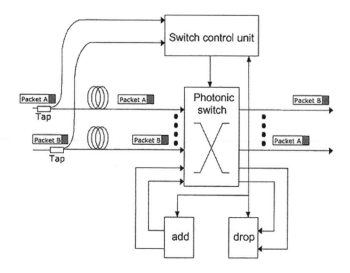

FIGURE 2.19   An unslotted optical packet-switched network node architecture [12].

2.3 NETWORK NODE BASED ON DELIVERY AND COUPLING SWITCH

Figure 2.20 shows an $n \times m$ network node architecture based on delivery and coupling switch (DC-SW). It consists of $n \times m$ number of DC-SW switches in which signal of any input fiber is to be switched to any output fiber through DC-SW as per destination. The main advantage of this architecture is that it can be extendable. The figure shows the architecture of DC-SW used in fiber optic networks. Here, each input optical signal is sent to its destination output port by using a $1 \times n$ switch unit. Each $1 \times n$ switch has arrayed 1×2 switch elements. The switched optical signals transmitted to the same output port are provoked by an optical coupler. For equalization of the output optical signal power level, each 1×2 switch element has a variable optical attenuator (VOA) [39] along with the 1×2 switch. PLC-TO-SW is taken as a DC-SW I in this type of the node [40,41]. The performances of the DC-SW architecture are enhanced due to low loss, high reliability, ease of fabricating arrayed switch modules and employability for a VOA in PLC-TO-SW.

2.4 MULTIHOP NETWORK NODE ARCHITECTURE

Figure 2.21 shows a multihop grooming node architecture consisting of wavelength-switch fabric (W-Fabric) [43] and an electronic-switch fabric, which makes switching operation of low-speed traffic streams. This node architecture provides both partial and full grooming multihop operation of the node. The electronic-switch fabric works as a G-Fabric [11,12,42,43]. With this hierarchical switching and multiplexing architecture, the OXC makes the switching of the low-speed traffic streams from one wavelength channel to other wavelength channels and groom them with other low-speed streams without using any extra network element.

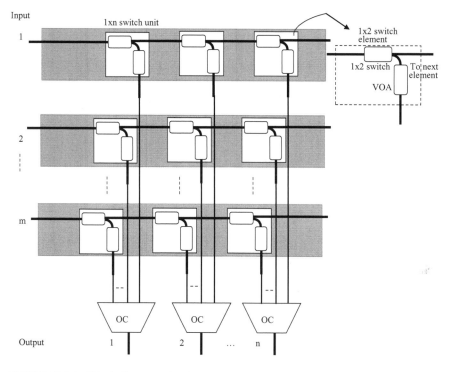

FIGURE 2.20 Block diagram of DC-SW consisting of variable optical attenuator (VOA), optical coupler (OC) and 1 × 2 switch.

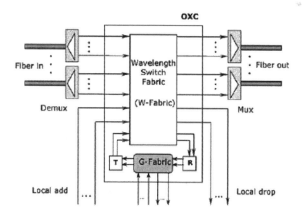

FIGURE 2.21 Multihop grooming node architecture [12].

The wavelength capacity is OC–N and the lowest input port speed of the electronic switch fabric is OC–M ($N > M$), and the ratio between N and M is a grooming ratio. In this architecture, only a few wavelength channels are sent to the G-Fabric for switching of finer granularity. The number of ports connecting to the W-Fabric and

G-Fabric finds how much multihop grooming capability is necessary. In full grooming, every OC–N wavelength signal reaching at the OXC is demultiplexed into its constituent OC–M streams before it comes to the switch fabric. The switch fabric makes the switching of these OC–M traffic streams in a non-blocking manner [11]. Then, the switched streams are multiplexed back onto different wavelength channels. The node with full grooming functionality is made using the opaque approach.

SUMMARY

Since optical network mainly has nodes, in this chapter, we have tried to discuss node architectures used in different optical backbones. We have mentioned both non-reconfigurable and reconfigurable node architectures. This chapter provides single-hop and multihop traffic grooming node architectures which are used in traffic grooming optical backbone discussed later in this book. We have also mentioned node architecture $v = $ based wavelength converters. We have also discussed interfaces such as IP over WDM and SONET over WDM used in an optical backbone.

EXERCISES

2.1. Construct an 8×8 Banyan interconnect node by using a 2×2 OXC.

2.2. Construct a 4×4 node of optical node having four wavelength channels with WDM, optical switches wavelength demultiplexer, ADM and wavelength router.

2.3. Construct a 3×3 wavelength convertible node of optical node having four wavelength channels with WDM, optical switches wavelength demultiplexer, ADM and wavelength router.

2.4. What are the uses of wavelength converter in a node architecture. How does it enhance the performance of the network using a wavelength converter.

2.5. How can you introduce traffic grooming concept in the node to enhance the performance of optical network.

2.6. In a WDM network node, if two signals of the same wavelength arriving from different ports need to go to the same output, then a conflict occurs. How can you resolve the conflict.

2.7. What are the advantages and disadvantages of network node using DC-SW.

2.8. Design a 4×4 node of optical packet switch node with optical switches, ADM and wavelength router.

2.9. What are the advantages and disadvantages of array waveguide grating in an optical node.

2.10. Show the design of an 8×8 star coupler that has three stages of 2×2 couplers with four couplers in each stage. If each node transmits with optical power P, what is the power received by another node. Show that the overall power splitting loss is 10 log 8 dB.

2.11. Draw a 2×2 slotted packet switch node architecture.

2.12. Draw a 3×3 unslotted packet switch node architecture.

2.13. Draw a 4×4 node architecture with source grooming.

2.14. Draw a 4×4 node architecture with grooming and wavelength conversion capability.

2.15. Design a 4×4 node architecture with source grooming, wavelength conversion and wavelength adding/dropping capability.

REFERENCES

1. J.-F. P. Labourdette and A. S. Acarnpora, "Partially reconfigurable multihop lightwave networks," *Proceedings, IEEE Globecom '90*, San Diego, CA, pp. 34–40, December 1990.

2. J.-F. P. Labourdette and A. S. Acampora, "Logically rearrangeable multihop lightwave networks," *IEEE Transactions on Communications*, vol. 39, pp. 1223–1230, August 1991.

3. D. Banerjee, B. Mukherjee and S. Ramamurthy, "The multidimensional torus: Analysis of average hop distance and application as a multihop lightwave network," *Proceedings, IEEE International Conference on Communications (ICC '94)*, New Orleans, LA, pp. 1675–1680, May 1994.

4. C. Chen and S. Banerjee, "Optical switch configuration and lightpath assignment in wavelength routing multihop lightwave networks," *Proceedings, IEEE INFOCOM '95*, Boston, MA, pp. 1300–1307, June 1995.

5. M. Eisenberg and N. Mehravari, "Performance of the multichannel multihop lightwave network under nonuniform traffic," *IEEE Journal on Selected Areas in Communications*, vol. 6, pp. 1063–1078, August 1988.

6. R. Gidron and A. Temple, "Teranet: A multihop multi-channel ATM lightwave network," *Proceedings, IEEE International Conference on Communications (ICC) '91*, Denver, CO, pp. 602–608, June 1991.

7. J. Iness, S. Banerjee and B. Mukherjee, "GEMNET: A generalized, shuftle-exchange-based, regular, scalable, and modular multihop network based on WDM lightwave technology," *IEEE/ACM Transactions on Networking*, vol. 3, no. 4, pp. 470–476, August 1995.

8. M. J. Karol and S. Z. Shaikh, "A simple adaptive routing scheme for congestion control in ShuffleNet multihop lightwave networks," *IEEE Journal on Selected Areas in Communications*, vol. 9, pp. 1040–1051, September 1991.

9. R. Libeskind-H'adas and R. Melhem, "Multicast routing and wavelength assignment in multihop optical networks," *IEEE/ACM Transactions on Networking*, vol. 10, no. 5, pp. 621–629, October 2002.

10. S. Chen, K. Nahrstedt and Y. Shavitt, "A QoS-Aware Multicast Routing Protocol," *IEEE Journal on Selected areas in communications*, vol. 18, no. 12, pp. 2580–2590, 2000.

11. B. Mukherjee, "WDM-based local lightwave networks – Part 11: Multihop systems," *IEEE Network Magazine*, vol. 6, no. 4, pp. 20–32, July 1992.

12. B. Mukherjee, *Optical Communication Networks*, McGraw-Hill, New York, 2006.

13. G. Panchapakesan and A. Sengupta, "On multihop optical network topology using Kautz digraph," *Proceedings, IEEE INFOCOM '95*, Boston, MA, pp. 675–682, April 1995.

14. G. N. Rouskas and M. H. Ammar, "Dynamic reconfiguration in multihop WDM networks," *Journal of High Speed Networks*, vol. 4, no. 3, pp. 221–238, 1995.

15. K. Sivarajan and R. Ramaswami, "Multihop networks based on de bruijn graphs," *Proceedings, IEEE INFOCOM '91*, Bal Harbour, FL, pp. 1001–1011, April 1991.

16. K. W. Tang, "CayletNet: A multihop WDM-based lightwave network," *Proceedings, IEEE INFOCOM '94*, Toronto, Canada, vol. 3, pp. 1260–1267, June 1994.

17. S. B. Tridandapani and J. S. Meditch, "Supporting multipoint connections in multihop WDM optical networks," *Journal of High-Speed Networks*, vol. 4, no. 2, pp. 169–188, 1995.

18. K. Okamoto, K. Moriwaki and S. Suzuki, "Fabrication of 64 × 64 arrayed-waveguide grating multiplexer on silicon," *Electronic Letters*, vol. 31, pp. 184–186, February 1995.

19. K. Okamoto and A. Sugita, "Flat spectral response arrayed waveguide grating multiplexer with parabolic waveguide horns," *Electronic Letters*, vol. 32, pp. 1661–1662, August 1996.

20. M.-S. Chen, N. R. Dono, and R. Ramaswcami, "A media access protocol for packet-switched wavelength division multi-access metropolitan area networks," *IEEE Journal on Selected Areas in Communications*, vol. 8, pp. 1048–1057, August 1990.

21. J. Sharony, K. Cheung, and T. E. Stern, "The wavelength dilation concept in lightwave networks: Implementation and system considerations," *IEEE/OSA Journal of Lightwave Technology*, vol. 11, pp. 900–907, 1993.

22. S. Sohma, "Low switching power silica based super high delta thermo optic switch with heat insulating grooves," *IEE Electronics Letters*, vol. 38, no. 3, pp. 127–128, 2002.

23. R. Kashahara, M. Yanagisawa, T. Goh, A. Sugita, A. Himeno, M. Yasu, and S. Matsui, "New structures of silica-based planar light wave circuits for low power thermooptic switch and its application to 8 × 8 optical matrix switch," *Journal of Lightwave Technology*, vol. 20, no. 6, pp. 993–1000, June 2002.

24. H. S. Hinton, "Photonic switching fabrics," *IEEE Communications Magazine*, vol. 28, pp. 71–89, April 1990.

25. P. P. Sahu, "Photonic switching using KDP based mechanically controlled directional coupler and its fiber optic networks," *Proceedings of ICOT-2004*, Jalgao, India, pp. 568–575, 2004.

26. P. P. Sahu and A. K. Das, "Compact integrated optical devices using high index contrast waveguides," *IEEE Wireless and Optical Communication*, ieeexplore Digital No-01666673, pp. 1–5, 2006.

27. P. P. Sahu, "Compact optical multiplexer using silicon nano-waveguide," *IEEE Journal of Selected Topics in Quantum Electronics*, vol. 15, no. 5, pp. 1537–1541, 2009.

28. P. P. Sahu, "WDM hierarchical fiber optic ring networks," *Journal of Optical Communication*, vol. 27, pp. 1–8, 2007.

29. P. P. Sahu, "Tunable optical Add/Drop multiplexers using cascaded Mach Zehnder coupler," *Fiber and Integrated Optics* (Taylor and Francis), vol. 27, no. 1, p. 24, 2008.

30. P. P. Sahu, "Polarization insensitive thermally tunable Add/Drop multiplexer using cascaded Mach Zehnder coupler," *Applied Physics: Lasers and Optics* (Springer), vol. B92, pp. 247–252, 2008.

31. G. Keiser, *Optical Fiber Communications*, McGraw Hill Inc., New York, 1999.

32. P. P. Sahu and R. Pradhan, "Blocking probability analysis for shared protected optical network with wavelength converter," *Journal of Optical Communication*, vol. 28, pp. 1–4, 2007.

33. P. P. Sahu and R. Pradhan, "Reduction of blocking probability in restricted shared protected optical network," *Proceedings of XXXIII OSI Symposium on Optics and Optoelectronics*, pp. 11–14, 2008.

34. P. P. Sahu, "A new shared protection scheme for optical networks," *Current Science Journal*, vol. 91, no. 9, pp. 1176–1184, 2006.

35. G. Hill et al., "A transport network layer based on optical network elements," *IEEE/OSA Journal of Lightwave Technology*, vol. 11, pp. 667–679, May–June 1993.

36. K. Zhu and B. Mukherjee, "Traffic grooming in an optical WDM mesh network," *IEEE Journal on Selected Areas in Communications*, vol. 20, no. 1, pp. 122–133, January 2002.

37. K. Zhu and B. Mukherjee, "A review of traffic grooming in WDM optical networks: Architectures and challenges," *SPIE Optical Networks Magazine*, vol. 4, no. 2, pp. 55–64, March/April 2003.

38. P. P. Sahu, "New traffic grooming approaches in optical networks under restricted shared protection," *Photonics Communication Networks*, vol. 16, pp. 223–238, 2008.

39. P. P. Sahu, "Variable optical attenuator using thermo optic two mode interference with fast response time," *Applied Optics*, vol. 48, no. 21, pp. 4213–4218, 2009.

40. C. R. Giles and M. Spector, "The wavelength add/drop multiplexer for lightwave communication networks," *Bell Labs Technical Journal*, vol. 4, no. 1, pp. 207–229, 1999.

41. R. V. Sclirnidt and R. C. Alferness, "Directional coupler switches, modulators, and filters using alternating 6P techniques," *Photonic Switching*, pp. 71–80, IEEE Press, New York, 1990.

42. K. Zhu and B. Mukherjee, "On-line approaches for provisioning connections of different bandwidth granularities in WDM mesh networks," Proceedings, OFC '02, Anaheim, CA, pp. 549–551, March 2002.

43. H. Zhu and B. Mukherjee, "Online connection provisioning in metro optical WDM networks using reconfigurable OADMs (ROADMs)," *IEEE/OSA Journal of Lightwave Technology*, vol. 23, no. 10, pp. 2893–2901, December 2005.

3 Devices in Optical Network Node

Recently, Fiber-optic networks have become essential to fulfill the skyrocketing demands of bandwidth in present day's communication networks. In these networks, flexible operations such as routing, restoration and reconfiguration are provided by the nodes, where optical matrix switches [1–4], wavelength division multiplexing (WDM) [5,6] and add/drop multiplexing (ADM) devices [7–10] are the key devices. The basic design of these optical devices has not changed for hundreds of years. They had bulky and heavy components requiring careful alignment, protection against vibration, moisture and temperature drift. In the early 1970s, in order to make them more compatible with modern technology, integrated optics concept has emerged [11]. At that time, the availability of low-loss optical fibers together with the invention of the laser caused an increasing interest in compact optical systems, in which conventional integrated circuit (IC) processing is used to miniaturize optical ICs (OIC) or photonic IC (PIC), and the wires and radio links are replaced by optical waveguides in the backbone of networks. OICs [11] would have a number of advantages compared with other bulk optical system – enhanced reliability, protection against vibration and electromagnetic interference, low loss propagation, small size, light weight, large bandwidth (multiplexing capability), low power consumption and mass-scale fabrication economy. Other than optical communication, OIC is also used for sensor technology. There are mainly three basic passive components and two basic active components in constructing the above devices. The passive components are directional coupler (DC) [11–16], multimode interference (MMI) coupler [12], [16–20], and two-mode interference (TMI) coupler [16,21–23]. The active components are Mach–Zehnder (MZ) device with phase controller and delay line coupler with phase controller (MZ with unequal arms). Apart from these, there is an array of waveguide grating components.

3.1 BASIC COMPONENTS OF INTEGRATED WAVEGUIDE DEVICES

As discussed earlier, integrated waveguide devices are based on two types of basic components passive components and active components. Here, we tried to discuss basic passive components like DC, TMI and MMI couplers and array waveguide grating components.

3.1.1 DIRECTIONAL COUPLER

Figure 3.1 shows a three-dimensional (3D) view of a typical asymmetric directional waveguide coupler consisting of two rectangular waveguides – waveguide-1 of width w_1 and thickness t_1 and waveguide-2 of width w_2 and thickness t_2, where β_1 and β_2 are the propagation constants in wave guides 1 and 2 before coupling, respectively. The refractive indices of spacing in the coupling region, core-1, core-2 and their surroundings are n_3, n_2, n_4 and n_1, respectively. The gap between two waveguides in the coupling region is h. The input powers P_1 and P_2 are incident in waveguide-1 and waveguide-2, respectively, when the output powers P_3 and P_4 are found in waveguide-1 and waveguide-2, respectively, after coupling. The coupling takes place in the $0 < z < L$ region in which the even and odd modes can propagate with propagation constants βe and βo. The phase shift between the even and odd modes becomes π when the propagation distance L_π is given by [14]

$$L_\pi = \pi/(\beta_e - \beta_o) \tag{3.1a}$$

In a symmetrical DC where $t_1 = t_2$, $n_2 = n_4$ and $w_1 = w_2$, i.e., $\beta_1 = \beta_2$, considerable coupling occurs in the $h < 8\,\mu m$ range [18]. On the other hand, in an asymmetrical DC where $t_1 \neq t_2$, $n_2 \neq n_4$ and $w_1 \neq w_2$, and hence, $\beta_1 \neq \beta_2$, the coupling is not noticeable unless h is less than $5\,\mu m$. The power transfer due to mode coupling is generally characterized by a phase mismatch $(\beta_1 - \beta_2)$ between the two waveguides, and the coupling coefficient is determined by [14]

$$k = \tfrac{1}{2}(\beta_e - \beta_o) \tag{3.1b}$$

To study mathematical analysis of DC, it is required to know coupled mode theory, which is discussed in the next section.

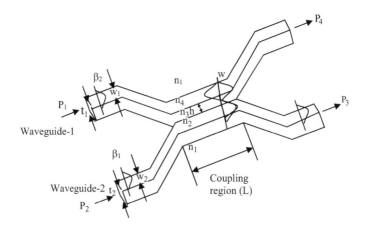

FIGURE 3.1 A 3D view of an asymmetric DC of coupling length L consisting waveguide-1 and -2.

3.1.1.1 Coupled Mode Theory

Coupled mode theory is a powerful tool for studying the optical waveguide coupling behavior. The concept of coupled mode theory is based on two-mode coupling theory. It is seen that when the energy is incident on one of the waveguides, then there is a periodic exchange between two waveguide-1 and 2. To explain the coupling behavior, we should know the coupled mode equations, which describe the variation of amplitude of modes propagating in each individual waveguide of the coupler.

The coupled mode equations may be written as [23]

$$\frac{da}{dz} = -j\beta_1 a(z) - jk_{12}b(z) \tag{3.2}$$

$$\frac{db}{dz} = -j\beta_2 b(z) - jk_{21}a(z) \tag{3.3}$$

The k_{12} and k_{21} represent the strength of coupling between two modes and are also called as coupling coefficients. In the absence of coupling, $k_{12} = k_{21} = 0$. The coupling coefficients depend on the waveguide parameters, separation between the waveguides in coupling region h and wavelength.

3.1.1.2 Power Transferred between Two Waveguides Due to Coupling

In order to solve coupled mode equations, we have considered the trial solutions of equations (3.1) and (3.2) as follows [14]:

$$\left.\begin{array}{c} a(z) = a_0 e^{-j\beta_1 z} \\[2mm] b(z) = b_0 e^{-j\beta_2 z} \end{array}\right\} \tag{3.4}$$

Substituting $a(z)$ and $b(z)$ in equations (3.11) and (3.12), we get

$$a_0(\beta - \beta_1) - k_{12}b_0 = 0 \tag{3.5}$$

$$b_0(\beta - \beta_2) - k_{21}a_0 = 0 \tag{3.6}$$

So, we can write from equations (3.14) and (3.15),

$$\beta^2 - \beta(\beta_1 + \beta_2) + (\beta_1\beta_2 - k^2) = 0 \tag{3.7}$$

Thus,

$$\beta_{e,o} = \frac{1}{2}(\beta_1 + \beta_2) \pm \left[\frac{1}{4}(\beta_1 - \beta_2)^2 + k^2\right]^{1/2} \tag{3.8}$$

where

$$k = \sqrt{k_{12}k_{21}} \tag{3.9}$$

In the coupling region, there are two independent modes called as even and odd modes propagating with propagation constants β_e and β_o, respectively. The suffixes e and o represent even and odd modes, respectively. The general solutions are written as [14]

$$a(z) = a_e e^{-j\beta_s z} + a_o e^{-j\beta_a z} \tag{3.10}$$

$$b(z) = \left\{ (\beta_e - \beta_1)/k_{12} \right\} a_e e^{-j\beta_e z} + \left\{ (\beta_o - \beta_1)/k_{12} \right\} a_o e^{-j\beta_o z} \tag{3.11}$$

where a_e and a_o are the amplitudes of even and odd modes, respectively. Equations (3.10) and (3.11) are coupled wave fields in waveguide-1 and 2, respectively. The behavior of the coupled waves can be determined by obtaining propagation constants. Since the waves in the two waveguides are propagated in the same direction in the case of DC, the propagation constants are $\beta_1 > 0$ and $\beta_2 > 0$, respectively. The solutions of the coupled mode equations are rewritten as

$$a(z) = \left(a_e e^{-j\sqrt{k^2 + \delta\beta^2}\, z} + a_o e^{j\sqrt{k^2 + \delta\beta^2}\, z} \right) e^{j\beta_{av} z}$$

$$b(z) = \left[\left\{ (\beta_e - \beta_1)/k_{12} \right\} a_e e^{-j\sqrt{k^2 + \delta\beta^2}\, z} + \left\{ (\beta_o - \beta_1)/k_{12} \right\} a_o e^{-j\sqrt{k^2 + \delta\beta^2}\, z} \right]$$

$$a_a e^{-j\beta_{av} z} \tag{3.12}$$

where $2\delta\beta = \beta_1 - \beta_2$ and $2\beta_{av} = \beta_1 + \beta_2$. The constants a_e and a_o for even and odd modes are determined by boundary conditions. We assume that at $z = 0$, the mode is launched in waveguide-1 with unit power and there is no power in waveguide-2. By considering boundary conditions, the power flows in waveguide-2 and 1 are given by

$$P_4/P_1 = |A(z)|^2 = 1 - \frac{k^2}{k^2 + \delta\beta^2} \sin^2\left[\left(k^2 + \delta\beta^2 \right)^{1/2} z \right] \tag{3.13}$$

$$P_3/P_1 = |B(z)|^2 = \frac{k^2}{k^2 + \delta\beta^2} \sin^2\left[\left(k^2 + \delta\beta^2 \right)^{1/2} z \right] \tag{3.14}$$

where $k = \sqrt{k_{12} k_{21}}$

The powers of waves propagating along two guides vary periodically. The maximum power transfer occurring at a distance L_π is obtained as [24,25]

$$P_{4,\max}/P_1 = \frac{1}{1 + \left(\delta\beta/k \right)^2}. \tag{3.15}$$

where $L_\pi = \dfrac{\pi}{2\sqrt{k^2 + \delta\beta^2}}$

As $\delta\beta \to 0$, the maximum power transfer increases. At $\delta\beta = 0$ there is a complete power transfer between two waveguides. This is called as synchronous or symmetric DC ($\beta_1 = \beta_2$).

3.1.1.3 Coupling Coefficient

The coupling coefficient of asymmetric DC with gap h between the coupling waveguides (2D model) derived by Marcuse [24] is written as

$$k = |k| = \frac{2K_2 K_4 \gamma_3 e^{-h\gamma_3}}{k_0^2 \beta \left\{ \left(n_2^2 - n_3^2\right)\left(n_4^2 - n_3^2\right)\left(w_1 + 1/\gamma_1 + 1/\gamma_3\right)\left(w_2 + 1/\gamma_1 + 1/\gamma_3\right) \right\}^{1/2}} \quad (3.16)$$

where

$$K_2 = \sqrt{n_2^2 k_0^2 - \beta^2}, \; K_4 = \sqrt{n_4^2 k_0^2 - \beta^2}$$

$$\gamma_3 = \sqrt{\beta^2 - n_3^2 k_0^2}$$

$$k_0 = 2\pi/\lambda$$

The propagation constants for even and odd modes are given by

$$\left.\begin{array}{l} \beta_e = \beta + k \\ \beta_0 = \beta - k \end{array}\right\} \quad (3.17)$$

3.1.2 MMI COUPLER

Figure 3.2a shows the schematic diagram of $M \times M$ MMI device in which the central structure is a multimode waveguide designed to support a large number of modes (typically ≥ 3). In order to launch light into and recover light from the multimode waveguide, a number of access waveguides (usually single-mode waveguides) are placed at its beginning and end of the central structure of width w_{mmi} and thickness t. Such devices are generally called as $M \times M$ MMI couplers, where M is the number

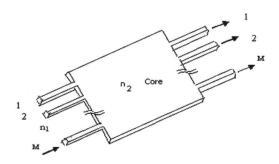

FIGURE 3.2   (a) A 3D MMI coupler with M number of input and M number of output access waveguides.

(Continued)

of input/output access waveguides. The refractive indices of MMI core and cladding are n_2 and n_1 respectively.

The principle of operation of MMI is based on self-imaging by which the input field is replicated in single or multiple images periodically along the propagation direction of the waveguide. There are a number of methods to describe the self-imaging phenomena – ray-optics approach [26], hybrid methods [27], guided mode propagation analysis [28], etc. The guided mode propagation analysis is probably the most comprehensive method to analyze self-imaging in multimode waveguide, because it not only supplies the basis for numerical modeling and design but also explains the mechanism of MMI.

In MMI waveguide for wide width, the electric field is present along the Y direction in *TE* mode, and for *TM* mode the electric field is present along the X direction. This follows the field distribution of *TE* and *TM* modes in Figure 3.2b and Figure 3.3, respectively.

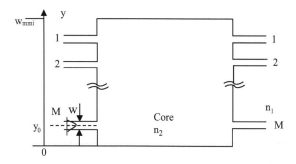

FIGURE 3.2 (CONTINUED)   (b) A 2D representation of an $M \times M$ MMI coupler.

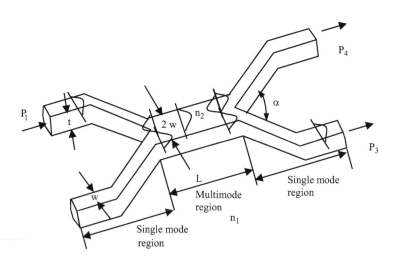

FIGURE 3.3   A schematic diagram of a TMI coupler of coupling length L.

3.1.2.1   Guided Mode Propagation Analysis

The self-imaging phenomena should be analyzed by 3D $M \times M$ multimode structures as shown in Figure 3.2b. As the lateral dimensions are much larger than the transverse dimensions, it is justified to assume that the modes have the same transverse behavior everywhere in the waveguide. So, the problem can be analyzed using two-dimensional (2D) (lateral and longitudinal) structures, as shown in Figure 3.2b. The analysis based on 2D representation of the multimode waveguide can be obtained from the actual 3D physical multimode waveguide by effective index method [11].

The input field profile $H(y, 0)$ incident on MMI coupler is summation of mode field distribution of all modes in 2D approximation as follows [29],

$$H(y,0) = \sum_i b_i H_i(y) \qquad (3.18)$$

where b_i is the mode field excitation coefficient which can be estimated using overlap integrals based on the field orthogonality relations and $H_i(y)$ = mode field distribution of the ith mode.

The composite mode field profile at a distances inside a multimode coupler can be represented in 2D representation as a super position of all guided modes [29]:

$$H(y,z) = \sum_{i=0}^{m-1} b_i H_i(y) \exp\left[j(\beta_0 - \beta_i)z \right] \qquad (3.19)$$

where m is the total number of guided modes and β_i is the propagation constant of the ith mode. For high index contrast, it is approximately written as [28]

$$\beta_i \approx k_0 n_r - \frac{(i+1)^2 \pi \lambda}{4 n_r w_e^2} \qquad (3.20)$$

where

$$w_e = w_{mmi} + w_p = \text{equivalent width or effective width} \qquad (3.21)$$

w_{mmi} = physical width of MMI coupler

$$w_p = \frac{\lambda}{\pi} \left(\frac{n_1}{n_r} \right)^{2\sigma} \left(n_r^2 - n_1^2 \right)^{-1/2}$$

$$= \text{lateral penetration depth related to Goos-Hahnchen shift} \qquad (3.22)$$

n_r is the effective index of the MMI core, w_{mmi} is the width of the multimode waveguide, n_1 is the refractive index of multimode wave guide cladding is the wavelength and $k_0 = 2\pi/\lambda$. $\sigma = 0$ for *TE* mode and $\sigma = 1$ for *TM* mode. Defining $L\pi$ as the beat length of the two lowest order modes, it is given in Ref. [28] as

$$L_\pi = \frac{\pi}{\beta_0 - \beta_1} \approx \frac{4 n_r w_e^2}{3\lambda} \qquad (3.23)$$

where β_0 = propagation constant of fundamental mode and β_1 = propagation constant of first-order mode.

3.1.2.2 Power Transferred to the Output Waveguides

At the end of the MMI section, optical power is either transferred to the output wave-guide or lost out at the end of the multimode waveguide. Again, the mode field at the access waveguide of same width, w is assumed to be mode 0. Each mode of the MMI coupler contributes to mode 0 at the output access waveguide. The mode field of the output waveguide is the sum of the contribution of all the modes guided in the MMI section. So, the mode field at the Mth waveguide can be written as [29,30]

$$H_M(y,L) = \sum_{i=0}^{M-1} c_{M,i} H_i(y) \exp\left[j \frac{i(i+2)\pi L}{3L_\pi} \right] \tag{3.24}$$

where $c_{M,i}$ = measure of field contribution of ith mode to Mth output waveguide. The $c_{M,i}$ is evaluated from a simple sinusoidal mode analysis [30].

In MMI coupler, there are two types of interference – general interference and restricted interference. In the case of general interference, the self-imaging mechanism is independent of modal excitation and the single image is formed at a distance [28]

$$L = p(3L_\pi) \tag{3.25}$$

where p = even for direct image and p = odd for mirror image. The multiple images are formed at

$$L = \frac{p}{2}(3L_\pi) \tag{3.26}$$

In the case of restricted interference, there is a restriction of excitation of some selected modes. There are two types of restricted interference – paired and symmetric. In the case of paired interference [28], and N-fold images are formed at a distance, $L = \frac{p}{N}(L_\pi)$ where $p \geq 0$ and $N \geq 1$ are integers having no common divisor. In the case of symmetric interference and N-fold images are formed at a distance $L = \frac{p}{N}(3L_\pi/4)$, where $p \geq 0$ and $N \geq 1$ are integers having no common divisor. The N images are formed with an equal spacing of w_{mmi}/N. The N-way splitter can be realized using this principle [31]. The transition from DC to MMI structure with Ridge structure by reducing etch depths in between two coupling waveguides of DC is reported by Darmawan et al. [15].

3.1.3 TMI Coupler

Figure 3.3 shows the schematic diagram of a TMI coupler consisting of two single-mode entrances of core width w and thickness and exit waveguides of same size and a TMI core of width $2w$ and length L. The operating principle of the TMI coupler is based on TMI in the coupling region. When light is incident on one of the input waveguides, only fundamental and first-order mode with propagation constants β_{00} and β_{01}, respectively, are excited in the coupling region [16,22]. These two modes interfere with each other while propagating along the direction of propagation. Depending on the relative phase differences $\Delta\varphi$ at the end of the coupling region, the light powers are coupled into two output waveguides.

3.1.3.1 Power Transferred to Output Waveguides

Like DC, in the case of TMI DC, we have to use the same coupled mode equations for the calculation of power transfer to the output waveguides. So, the powers coupled into two single-mode identical waveguides of TMI coupler are approximately given by [16]

$$\frac{P_3}{P_1} = \sin^2\left(\Delta\varphi/2\right) \tag{3.28}$$

$$\frac{P_4}{P_1} = \cos^2\left(\Delta\varphi/2\right) \tag{3.29}$$

where

$$\Delta\varphi = \Delta\beta.L, L = \text{length of multimode region and } \Delta\beta = \beta_{00} - \beta_{01} \tag{3.30}$$

The coupling length for getting maximum power transfer from waveguide-1 to -2 is found to be

$$L_{co} = \pi n/\Delta\beta = nL_\pi \tag{3.31}$$

where is an odd integer and

$$L_\pi = \pi/\Delta\beta \tag{3.32}$$

3.1.4 ARRAY WAVEGUIDE GRATING

Figure 3.4 shows a schematic structure of Array waveguide grating (AWG) [32,33] having $N \times N$ input star coupler and $N \times N$ output star coupler. It has two passive-star couplers connected to each other by a grating array. The first star coupler consists of N inputs and N' outputs, (where $N \ll N'$), whereas the second one has N' inputs and N outputs. The inputs to the first star are alienated by an angular distance of a and their outputs are estranged by an angular distance. The grating array has N' waveguides, with lengths $l_1, l_2, \ldots l_N$, where, $l_1 < l_2 < \ldots < l_{N'}$. The difference in length between any two adjacent waveguides is a constant Δl. In the first star coupler, a signal on a given wavelength entering from any of the input ports is split and transmitted to its N' outputs which are also N' inputs of the second star coupler. The signal transmitted through the grating array obtaining a different phase shift in each waveguide depends on the length of the waveguides and the wavelength of the signal.

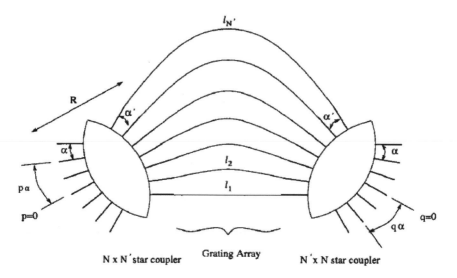

FIGURE 3.4   Array waveguide grating.

The constant difference in the lengths of the waveguides makes a phase difference of $\beta \, \Delta l$ in adjacent waveguides, where $\beta = 2\pi n_{eff}/\lambda$ is the propagation constant in the waveguide, n_{eff} is the effective refractive index of the waveguide and λ is the wavelength of the light. At the input of the second star coupler, the phase difference in the signal shows that the signal will constructively recombine only at a single output port. Signals of different wavelengths coming into an input port will each be separated to a different output port. Also, different signals using the same wavelength is simultaneously incident on different input ports, and still not interfere with each other at the output ports.

Two signals of same wavelength coming from two different input ports do not interfere with each other in the grating because there is an additional phase difference obtained by the distance between the two input ports. The two signals will be joint in the grating but is separated out again in the second star coupler and directed to different outputs. This phase difference is given by $kR(p - q)\alpha\alpha'$, where k is a propagation constant which is not a function of wavelength, where R is the constant distance between the two foci of the optical star, p is the input port number of the router and q is the output port number of the router. The total phase difference is given by

$$\phi = \frac{2\pi \Delta l}{\lambda} + kR(p - q)\alpha\alpha' \tag{3.33}$$

The transmission power from a particular input port p to a particular output port q is maximum when the phase difference is equal to an integral multiple of 2π. Thus, for only one wavelength λ, ϕ is satisfied with an integral multiple of 2π, and this λ is transmitted from input port p to output port q. Alternately, for a given input port and a given wavelength, the signal is transmitted to the output port.

3.1.5 MZ Active Device

Figure 3.5 shows a 2×2 MZ active device [34–36] consisting of an MZ section of equal arm length with phase controllers and two 3-dB DCs of coupling lengths L_0 and L_1. The phase controller is a device that changes the phase of the wave using external power P. The input power P_1 is incident in waveguide-1 when the output powers P_3 and P_4 are obtained as cross and bar states, respectively. The 3-dB coupler consists of two waveguides having a small gap h between them. The core width of waveguide is w. The refractive index of core and cladding are n_2 and n_1 respectively.

The coupling section of DC can be described with the coupled mode equations of DC with a small gap as follows:

$$\frac{dA}{dz} = -jKB \text{ and } \frac{dB}{dz} = -jKA \qquad (3.34)$$

where A and B are the normalized electric fields in the upper and lower waveguides, K = coupling coefficient of DC. There are two orthogonal polarization modes propagating in the planar waveguide of thermooptic MZ (TOMZ) device – *TE* and *TM* polarization modes.

3.1.5.1 *TE* Polarization

In the case of *TE* polarization modes, analytical solution of equation (3.63) following equations (3.21) and (3.22) for each individual (kth) coupler of the length L_k ($k = 0, 1$) is given by

$$A(L_k) = A(0)\cos\left(K^{TE}L_k\right) - jB(0)\sin\left(K^{TE}L_k\right) \qquad (3.35)$$

$$B(L_k) = B(0)\cos\left(K^{TE}L_k\right) - jA(0)\sin\left(K^{TE}L_k\right) \qquad (3.36)$$

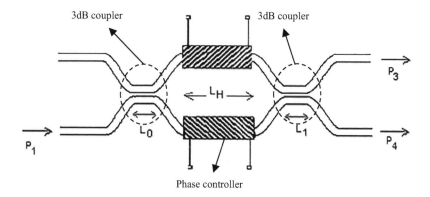

FIGURE 3.5 A schematic diagram of a planar waveguide-type TOMZ switching unit with 3-dB DC and a heater of length L_H.

where $A(L_k)$ and $B(L_k)$ are amplitudes of coupling waveguide-1 and 2, respectively, with length L_k. K^{TE} is the coupling coefficient of TE mode for DC with a small coupling gap. In calculating K^{TE} by using Marcuse's equation [30], the propagation constant is determined from dispersion equations for TE mode [18]. Equations (3.35) and (3.36) represent the coupled electric fields in the upper and lower waveguides after coupling in the coupling region of length L_k. In matrix form, equations (3.35) and (3.36) can be written as

$$
\begin{pmatrix} A(L_k) \\ B(L_k) \end{pmatrix} = T_k \begin{pmatrix} A(0) \\ B(0) \end{pmatrix}
$$

$$
= \begin{pmatrix} C_k^{TE} & -jS_k^{TE} \\ -jS_k^{TE} & C_k^{TE} \end{pmatrix} \begin{pmatrix} A(0) \\ B(0) \end{pmatrix} \tag{3.37}
$$

where $S_k^{TE} = \sin\left(K^{TE}L_k\right)$

$$
C_k^{TE} = \cos\left(K^{TE}L_k\right) \text{ and } T_k = \begin{pmatrix} C_k^{TE} & -jS_k^{TE} \\ -jS_k^{TE} & C_k^{TE} \end{pmatrix} \tag{3.38}
$$

The MZ section is a phase shifter in which phase changes with heating power P applied on the device via a thin film heater. In the case of TE mode, this phase change occurs mainly due to thermooptic effect with application of heating power [10]. The electric fields in the upper and lower waveguides are written as

$$
A(Z) = A(0)\exp\left(-j\Delta\phi(P/2)\right)
$$

$$
B(Z) = B(0)\exp\left(j\Delta\phi(P)\right) \tag{3.39}
$$

In matrix form, we can write [37]

$$
\begin{pmatrix} A(Z) \\ B(Z) \end{pmatrix} = T_{MZ}^{TE} \begin{pmatrix} A(0) \\ B(0) \end{pmatrix}
$$

$$
= \begin{pmatrix} \exp\left(-j\Delta\varphi(P)/2\right) & 0 \\ 0 & \exp\left(j\Delta\varphi(P)/2\right) \end{pmatrix} \begin{pmatrix} A(0) \\ B(0) \end{pmatrix} \tag{3.40}
$$

where

$$
T_{MZ}^{TE} = \begin{pmatrix} \exp\left(-j\Delta\varphi(P)/2\right) & 0 \\ 0 & \exp\left(j\Delta\varphi(P)/2\right) \end{pmatrix} \tag{3.41}
$$

$\Delta\phi(P)$ = thermooptic phase change obtained with the application of heating power P for TE mode

$$= \frac{2\pi}{\lambda}\frac{dn}{dT}\Delta T_c L_H \tag{3.42}$$

L_H = heater length,

$\dfrac{dn}{dT}$ = thermooptic refractive index coefficient

λ = wavelength and ΔT_c = temperature difference between two cores.

The transfer matrix of MZ coupler for TE mode is written as

$$T = T_1^{TE} T_{MZ}^{TE} T_0^{TE} \tag{3.43}$$

The output electric field A_{out}^{TE} and B_{out}^{TE} for upper and lower waveguide scan be expressed as

$$\begin{pmatrix} A_{out}^{TE} \\ B_{out}^{TE} \end{pmatrix} = T_1^{TE} T_{MZ}^{TE} T_0^{TE} \begin{pmatrix} A_{in}^{TE} \\ B_{in}^{TE} \end{pmatrix} \tag{3.44}$$

$$= \begin{pmatrix} T_{21}^{TE} & -T_{22}^{TE*} \\ T_{22}^{*} & T_{21}^{TE*} \end{pmatrix} \begin{pmatrix} A_{in}^{TE} \\ B_{in}^{TE} \end{pmatrix} \tag{3.45}$$

where A_{in}^{TE} and B_{in}^{TE} are the input fields of TE mode in upper and lower waveguides, respectively. T_{21}^{TE} and T_{22}^{TE} are the matrix elements with relation $\left|T_{21}^{TE}\right|^2 + \left|T_{22}^{TE}\right|^2 = 1$. From equation (3.73), we can write

$$A_{out}^{TE} = T_{21}^{TE} A_{in}^{TE} - T_{22}^{TE*} B_{in}^{TE} \tag{3.46}$$

$$B_{out}^{TE} = T_{22}^{TE} A_{in}^{TE} + T_{21}^{TE*} B_{in}^{TE} \tag{3.47}$$

Considering input field B_{in}^{TE} in lower waveguide only, we can write equations (3.46) and (3.47) as follows:

$$A_{out}^{TE} = -T_{22}^{TE*} B_{in}^{TE} \tag{3.48}$$

$$B_{out}^{TE} = T_{21}^{TE*} B_{in}^{TE} \tag{3.49}$$

The cross-state transmitted power function for TE mode is written as

$$\left(P_3/P_1\right)_{TE} = \left|T_{22}^{TE}\right|^2 = \left|a_0^{TE}\right|^2 + \left|a_1^{TE}\right|^2 + 2a_0^{TE} a_1^{TE}\cos\left[\Delta\phi(P)\right] \tag{3.50}$$

where $a_0^{TE} = C_0^{TE} S_0^{TE}$ and $a_1^{TE} = S_0^{TE} C_1^{TE}$. Considering 3-dB couplers of same coupling length ($L_0 = L_1$) in both sides of the MZ section, we get $a_0^{TE} = a_1^{TE} = 0.5$. The cross- and bar-state transmitted powers of TE mode can be written as

$$\left(P_3/P_1\right)_{TE} \sim \cos^2\left[\left(\Delta\phi(P)/2\right)\right]$$ (3.51)

$$\left(P_4/P_1\right)_{TE} \sim \sin^2\left[\left(\Delta\phi(P)/2\right)\right]$$ (3.52)

Similarly, the cross- and bar-state powers of TM mode are written as

$$\left(P_3/P_1\right)_{TM} \approx \cos^2\left[\left(\Delta\phi(P)_{TM}/2\right)\right]$$ (3.53)

$$\left(P_4/P_1\right)_{TM} \approx \sin^2\left[\left(\Delta\phi(P)_{TM}/2\right)\right]$$ (3.54)

The phase change is obtained due to the application of external power P. Two types of external power can be applied – thermooptic and electrooptic powers. Due to these powers, there are two types of MZ devices – TOMZ device and electrooptic MZ (EOMZ) device. The details of these devices are discussed later in this chapter.

3.2   WAVELENGTH DIVISION MULTIPLEXER/ DEMULTIPLEXER-BASED WAVEGUIDE COUPLER

The wavelength multiplexer/demultiplexer can be developed using basic waveguide coupler components such as DC), TMI coupler and MMI coupler. The multiplexing is achieved by cross coupling of one wavelength and bar coupling of the other wavelength, and in the case of demultiplexing, it is vice versa. There are different types of wavelength multiplexer/demultiplexer – DC-based, TMI-based and MMI-based multiplexer/demultiplexer [5,6,14]. Figure 3.6a shows the basic structure of a four-channel wavelength multiplexer/demultiplexer based on a waveguide coupler in which there are two waveguide couplers of L_π for wavelength λ_1 in the first level and one coupler with beat length $2L_\pi$. The figure also shows the working of multiplexing wavelengths λ_1, λ_2, λ_3 and λ_4. It is seen that wavelength multiplexing/demultiplexing made by using TMI coupler is more compact than that using DC and MMI coupler.

3.2.1   WDM-BASED TMI COUPLER

For TMI coupler-based multiplexing/demultiplexing of two wavelengths λ_1 and λ_2, we have to consider four guided propagation constants $\beta_{00}^{\lambda_1}$, $\beta_{01}^{\lambda_1}$, $\beta_{00}^{\lambda_2}$ and $\beta_{01}^{\lambda_2}$, and the required coupling length can be approximately written as [6]

$$L_c = \frac{\pi}{\Delta\beta|_{\lambda_1} - \Delta\beta|_{\lambda_2}}$$ (3.55)

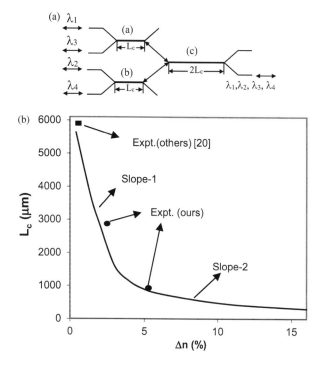

FIGURE 3.6 (a) Four-channel cascaded multiplexer/demultiplexer, (b) variation of coupling length with Δn ($n_1 = 1.447$, $V = 2.4$).

where

$$\Delta\beta\big|_{\lambda 1} = \left(\beta_{00}^{\lambda_1} - \beta_{01}^{\lambda_1}\right), \; \Delta\beta\big|_{\lambda 2} = \left(\beta_{00}^{\lambda_2} - \beta_{01}^{\lambda_2}\right) \qquad (3.56)$$

These propagation constants are determined by using an effective index method [18] where $\beta_{00}^{\lambda_1}$, $\beta_{01}^{\lambda_1}$, $\beta_{00}^{\lambda_2}$ and $\beta_{01}^{\lambda_2}$ are ~6.0556, 5.9807, 5.974 and 5.9023 $(\mu m)^{-1}$, respectively, for the wavelengths $\lambda_1 = 1.52\,\mu m$ and $\lambda_2 = 1.54\,\mu m$, of the coupler with $\Delta n = 5\%$. L_c is calculated as 980 μm, which is about six times less than that of the TMI coupler with $\Delta n = 0.6\%$, using Ti:LiNbO$_3$ [20].

Keeping the normalized frequency ~2.4 for single-mode waveguide access of TMI coupler, we have determined the variation of coupling length of multiplexer/demultiplexer with Δn for wavelengths 1.52 and 1.56 μm using equation (3.17) as shown in Figure 3.6b. The experimental result of TMI multiplexer using Ti:LiNbO$_3$ reported in Ref. [31] is represented by a black rectangle showing the almost agreement with theoretical value. The black circle represents L_c obtained experimentally by us with Δn ~2% and 5%, showing almost close to theoretical value [14]. The curve in the figure has two slopes – slope-1 which represents the compact TMI multiplexer region where $L_c > 980$ μm and its corresponding $\Delta n < 5\%$ and slope-2 which represents the ultra compact TMI multiplexer region, where $L_c < 980$ μm and its corresponding

$\Delta n > 5\%$. It is evident from the figure that L_c in slope-2 decreases slowly with Δn in comparison with slope-1.

Figure 3.6a shows the block diagram of a four-channel multiplexer/demultiplexer consisting three TMI couplers – having two couplers of the same coupling length L_c and other one of $2L_c$. For the four-channel multiplexer/demultiplexer with $\Delta n = 5\%$, considering $\lambda_1 = 1.52$, $\lambda_2 = 1.54$, $\lambda_3 = 1.56$ and $\lambda_4 = 1.58\,\mu$m, the device length can be obtained approximately as ~5 mm.

3.3 OPTICAL SWITCHING

Optical switching is required to change the optical signal path from one input fiber to the other fiber or from one direction to the other. The concept to switch in origi-nated from electronics field. In case of switching, two basic types are *circuit switch-ing* and *cells witching* [38]. In an optical field, circuit switching provides wavelength routing, and cell switching gives optical packet switching and optical burst switch-ing. For the transparency of signals considered here, there are two types of switching: opaque and transparent. The switching devices are of two types: logic and relational switching.

Logic switching is carried out by a device where the data (or the information-carrying signal) launched into the device makes the control over the state of the device in such a way that some Boolean function, or a combination of Boolean functions, is carried out on the inputs. In a logic device, form at and rate of data are changed or converted in intermediate nodes; thus, logic switching provides opaque switching. Further, some of its components perform the change of states or it switches as fast as or faster than the signal bit rate. Based on the logic device and ideal performance in electronic field, logic switching is used in an electronic field. But, traditional. optical–electronic–optical (o–e–o) conversion in optical networks is still widely applied due to having lack of proper logic devices operated in the optical domain. The most current optical networks use electronic processing and consider the optical fiber only as a transmission medium. Switching and processing of data are carried out by converting an optical signal back to its "native" electronic form. Such a network relies on electronic switches, i.e., logic devices. It shows a high degree of flexibility in terms of switching and routing functions for optical networks; however, the speed of electronics is not able to deal with the high band width of an optical fiber. Also, an electro optic conversion at an intermediate node in the network produces extra delay and cost. These factors make motivated toward the develop-ment of all-optical networks where optical switching components switch high-band-width optical data stream switch out electro optic conversion. *Relational switching* is used to set up a relation between inputs and outputs. The relation function depends on the control signals applied to it and is independent of the contents of the signal or data inputs. In switching devices, the control of the switching function is performed electronically, with the optical stream being transparently routed from a given input of the switch to a given output. Such transparent switching permits the switch to be independent of the data rate and format of the optical signals. Thus, the strength of a relational device permits signals at high bit rates to pass through it. Due to the limits of optical hardware, various kinds of optical switching devices basically use *relational switching*, which provides more advantages for optical networks in terms of optical hardware limits.

There are different optical/photonic switches – TOMZ switch [36], TMI switch [39], X-junction type switch [40], MMI switch [41,42], etc.

3.3.1 MZ SWITCH

MZ switch consists of a 3-dB coupler and an MZ section of equal arms. The 3-dB coupler divides signal equally into two output access waveguides when power is incident into one of the input waveguides. Due to 3-dB effect, without phase change, within MZ section signal will be transferred to cross-state output waveguide. Through thermooptic or electrooptic effects, the path difference between two arms of MZ section can be changed. While the phase difference between signal of two arms is π, then the signal will be transferred into the bar-state output waveguide. There are different types of MZ switch – TOMZ switch and EOMZ switch. In TOMZ switch, the phase change in MZ arms is controlled by thermooptic effect via a thin film heater, whereas in EOMZ switch, the phase change in MZ arms is controlled by electro-ooptic effect via an electrode placed on one of the MZ arms. The 3-dB couplers are made of either DC or MMI coupler or TMI coupler.

3.3.1.1 TOMZ Switch-Based DC

Figure 3.7 shows a TOMZ switch consisting of an MZ section of equal arm length with thermooptic phase controllers and two 3-dB DCs of coupling lengths L_0 and L_1. The thermooptic phase controller is a thin film heater, which changes the phase of the wave via thermooptic effect. The input power P_1 is launched into waveguide-1, and the output powers P_3 and P_4 are obtained as cross and bar states, respectively. The 3-dB coupler consists of two waveguides having a small gap h between them. The core width of the waveguide is w. The refractive index of core and cladding are n_2 and n_1 respectively.

There are two orthogonal polarization modes propagating in the planar wave-guide of TOMZ device – *TE* and *TM* polarization modes.

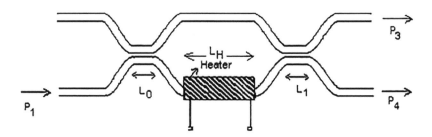

FIGURE 3.7 A schematic diagram of a planar waveguide-type TOMZ switching unit with 3-dB DC and a heater of length L_H.

3.3.1.2 *TE* Polarization

The cross- and bar-state transmitted powers of *TE* mode obtained from equations (3.51) and (3.52) are rewritten as

$$\left(P_3/P_1\right)_{TE} \sim \cos^2\left[\left(\Delta\phi(P)_{TE}/2\right)\right]$$

$$\left(P_4/P_1\right)_{TE} \sim \sin^2\left[\left(\Delta\phi(P)_{TE}/2\right)\right]$$

where $\Delta\phi(P)_{TE}$ = thermooptic phase change due to application of heating power P for *TE* mode

$$= \frac{2\pi}{\lambda}\frac{dn}{dT}\Delta T_c L_H \tag{3.57}$$

L_H = heater length,

$\dfrac{dn}{dT}$ = thermooptic refractive index coefficient

λ = wavelength and ΔT_c = temperature difference between two cores.

Similarly, the cross- and bar-state powers of *TM* mode obtained from equations (3.53) and (3.54) are rewritten as

$$\left(P_3/P_1\right)_{TM} \approx \cos^2\left[\left(\Delta\phi(P)_{TE}/2 + \Delta\phi_s/2\right)\right]$$

$$\left(P_4/P_1\right)_{TM} \approx \sin^2\left[\left(\Delta\phi(P)_{TE}/2 + \Delta\phi_s/2\right)\right]$$

where $\left[\Delta\phi(P)_{TM} = \Delta\phi(P)_{TE} + \Delta\phi_s\right]$. The phase change with applying heating power arises not only due to isotropic thermooptic effect but also by an anisotropic stress optic effect when the waveguide is heated by a thin film heater locally [39,41]. This is called as secondary stress optical effect. When the waveguide is heated locally via the heater, the glass (SiO_2) can expand freely to the Si substrate. But it cannot expand freely in the parallel direction, because it is surrounded by other glass (SiO_2). So, a compressive stress occurs only in the parallel direction, and it mainly induces a refractive index increase in the *TM* mode. The refractive index increase due to stress optic effect provides an extra phase change $\Delta\phi_s$ in *TM* mode apart from the thermooptic phase change with application of heating. The extra stress optic phase change in *TM* mode is given by

$$\Delta\phi_S = \frac{2\pi}{\lambda}\frac{d(n_{TM} - n_{TE})}{dT}\Delta T_c L_H \tag{3.58}$$

where $\dfrac{d(n_{TM} - n_{TE})}{dT}$ = temperature rate of birefringence produced by heater.

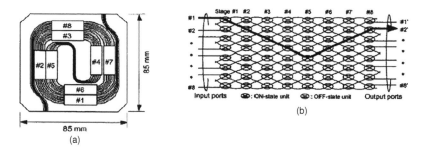

FIGURE 3.8   (a) Layout of an 8×8 optical matrix switch demonstrated by Kasahara et al. [2] using SiO_2/SiO_2-GeO_2 waveguide. (b) Arrangement of eight switching units in each stage.

The cross-and bar-state powers are obtained as a function of heating power applied on the MZ section of the device. Since both sides of MZ section 3-dB couplers are used, the TOMZ device shows a cross state where signal will be fully transferred to the other waveguide. As the heating power applied on MZ section increases, the power transferred to cross state decreases, and at a particular heating power, the power will remain in parallel state waveguide and the transferred power to cross state becomes almost zero. The state is called as bar state. The heating power required to obtain bar state is called as bar state power, and at bar sate, the thermooptic phase $\Delta\phi(P)$ is π. It is seen that due to anisotropic thermooptic effect, the stress optic phase is included in *TM* mode apart from the thermooptic phase as mentioned earlier. At bar state, a slight amount of power is transferred to cross state via *TM* mode. This provides a crosstalk of the switch. To reduce crosstalk, it is required to release the stress developed in the silica waveguide due to anisotropic thermooptic effect. Figure 3.8b shows an 8×8 optical matrix switch consisting of a 64 TOMZ device (Figure 3.8c) which has a switching power and response time of ~ 360 mW and 4.9 ms, respectively.

3.3.1.3   EOMZ-Based DC

The balanced bridge interferometer switch (Figure 3.9) comprises an input 3-dB coupler having two input waveguides and at middle electrodes to allow changing the effective path length over the two arms, and a final 3-dB coupler [11]. Light incident on the upper waveguide is divided into half by the first coupler. With no voltage applied to the electrodes, the optical path length of the two arms enters the second

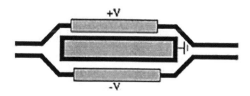

FIGURE 3.9   An EOMZ-based 3-dB DC.

coupler in phase. The second coupler acts like the continuation of the first, and all the light are crossed over to the second waveguide to provide the cross state. To achieve the bar state, voltage is applied to an electrode, placed over one of the interferometer arms to electrooptically produce a 180° phase difference between the two arms. In this case, the two inputs from the arms of the interferometer combine at the second 3-dB coupler out of phase, with the result that light remains in the upper waveguide.

3.3.1.4 MMI Coupler-Based MZ Switch

Figure 3.10 shows a 4×4 MMI coupler-based 4×4 optical switch [43] having five TOMZ structures. The switch-based SOI waveguide has length of 50 mm, response time of 30 μs and heating power of 330 mW. Each switching element uses two 3-dB MMI coupler on both sides of MZ section with a thermooptic phase changer. There is no heating in the MZ section, the switch is in cross state and when the heating power is applied to get thermooptic phase of π, then the switch is in cross state.

3.3.1.5 TMI Coupler-Based MZ Switch

Figure 3.11 shows single MZ optical switching element [39] having two 3-dB TMI coupler on both sides of the MZ section with a thermooptic phase changer. There is no heating in the MZ section, the switch is in cross state and when the heating power is applied to get thermooptic phase of π, then the switch is in cross state. In cross state of the switch, $P_4 \sim 0$ and in bar state, $P_3 \sim 0$.

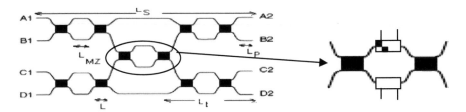

FIGURE 3.10 Architecture of an SOI 4×4 optical matrix switch demonstrated by Wang et al. [43] (L_P = length of input/output waveguide, L = length of 3-dB coupler, L_{MZ} = MZ section length = L_H = heater length, L_S = length of 4×4 optical matrix switch).

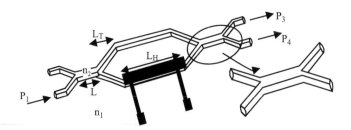

FIGURE 3.11 A schematic diagram of a TOMZ switch with a thin film heater of length L_H, transition region of length L_T and 3-dB TMI couplers of length L.

3.3.2   X-JUNCTION SWITCH

Figure 3.12a shows an X-junction switch structure [40] in which for a small intersection angle θ the symmetric X junction can be treated as a zero-gap DC with branches 1 and 4 forming the top waveguide and branches 2 and 3 the bottom waveguides. The actual pattern of the X junction is approximated by a staircase configuration along the direction of propagation. In the symmetric X junction, there is no mode conversion, and therefore there is no phase shift at the steps. For each lateral input mode two modes of equal amplitude (even and odd) existed in the symmetric junction. For the sake of simplicity, we restrict ourselves to the two fundamental TE modes of the waveguide system. Under these conditions, coupled mode equations are easily shown to be

$$A_{i1} = c_i A_{i0} \tag{3.59}$$

where A_{i1}, A_{i0} are mode amplitudes before and after the step, respectively ($i = 1$ for even and $i = 0$ for odd mode). The coupling coefficient is given by

$$c_i = \frac{2\sqrt{\beta_{i0}\beta_{i1}}}{\beta_{i0} + \beta_{i1}} \frac{I_{i0,i1}}{\sqrt{(I_{i0,i0} \cdot I_{i1,i1})}} \tag{3.60}$$

β_{i0} and β_{i1} are the local normal mode propagation constants before and after the steps. Overlap integrals are defined as

$$I_{im,in} = \int_{-\infty}^{\infty} E_{im}(x) \cdot E_{in}(x) \cdot dx \tag{3.61}$$

where $m, n = 0, 1$. For simple field distributions, $E_{im}(x)$ and $E_{in}(x)$ are given by sinusoidal and exponential functions and overlap integrals are analytically obtainable. This prevents the need for the time-consuming numerical simulation.

The propagation of modes from one step to the other adds a phase factor $\Delta z\, \beta_{im}$. There is also radiation at the steps. This radiation is modeling loss caused by the waveguide taper. The steps are appropriate approximations to the photomasks made by electron beam stair step raster scanning process. Since the propagation constants of even β_e and odd modes β_o are different (the difference $\Delta\beta\,(z) = \beta_e - \beta_o$), their

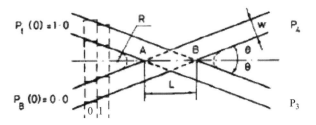

FIGURE 3.12   An X-junction switch architecture.

interference provides an optical power at each step. For a small intersection angle θ the symmetric Junction can be treated as a zero-gap DC with branches 1 and 4 forming the top waveguide and branches 2 and 3 the bottom waveguide. We get top and bottom waveguide optical powers as follows:

$$P_t(z) = (A_e - A_o)^2 + 2A_e A_o \cos^2\left[\Delta\beta(z)z/2\right] \tag{3.62}$$

$$P_b(z) = (A_e - A_o)^2 + 2A_e A_o \sin^2\left[\Delta\beta(z)z/2\right] \tag{3.63}$$

where z is the distance from the input plane $z = \Sigma \Delta z$, P_t and P_b are the optical powers in the top and bottom waveguides, respectively. Repeating this procedure, the power distribution is tracked along the device. For multimode operation each input mode must be considered separately. Corresponding even and odd modes give power distribution for this input mode. The total power distribution is the superposition of all the modes at each step. Rapid changes at the intersecting points are caused by coupling between converging guides and radiation losses. As a result, the power coupled to the top and bottom waveguides of output section of X-junction depends on an intersecting angle θ. The X junction is taken as an intersecting waveguide switch shown in Figure 3.13. The properly fabricated electrode is shown in the figure, where both cross and bar states can be electrooptically achieved with good crosstalk.

In this X-junction switch, $\Delta\beta(z)$ depends on the electric field E applied on the electrode. The X-junction device can be used as a thermooptic switch in which waveguide material is thermooptic materials and thermooptic heaters are used instead of electrodes [40].

3.3.3 DC-BASED ELECTROOPTIC SWITCH

Figure 3.14a shows a DC-based electrooptic switch [11,44] consisting of a pair of optical channel waveguides that are parallel and in close proximity over a finite interaction length. Light incident on one of the waveguides is transferred to the second waveguide through evanescent coupling. The coupling strength depends on the interwaveguide separation, and the waveguide mode size also depends on the optical wavelength and confinement factor of the waveguide. If the two waveguides are indistinguishable, complete coupling between the two waveguides is obtained at a beat length which is related to the coupling strength. However, by placing electrodes

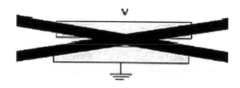

FIGURE 3.13   An X-junction electrooptic switch.

over the waveguides, the difference in propagation constants of the waveguides is sufficiently increased so that no light couples between the two waveguides. Therefore, the cross state is obtained with the application of no voltage, and the bar state is obtained with the application of a switching voltage. Unfortunately, the interaction length is required to be accurate for good isolation, and these couplers are wavelength specific.

Switch fabrication tolerance, as well as the ability to achieve good switching for a relatively wide range of wavelengths, is overcome by using the so-called reversed delta-beta coupler (Figure 3.14b). In this device, the electrode is split into at least two sections. The cross state is obtained by applying equal and opposite voltages to the two electrodes.

Other types of switches include the mechanical fiber-optic switch and the thermooptic switch. These devices show slow switching (about milliseconds) and is used in circuit-switched networks. One mechanical switch has two ferrules, each with polished end faces that can rotate to switch the light appropriately. Thermooptic waveguide switches, on the other hand, are fabricated on a glass substrate and are operated by the use of the thermooptic effect. One such device uses a zero-gap DC configuration with a heater electrode to increase the waveguide index of refraction.

3.3.4 GATE SWITCHES

In the $N \times N$ gate switch-based amplifier gates [45], each input signal first passes through a $1 \times N$ splitter. The signals then pass through an array of N_2 gate elements and are then recombined in $N \times$ combiners and sent to the N outputs. The gate elements can be implemented using optical amplifiers that can either be turned on or off to pass only selected signals to the outputs. The amplifier gains can compensate for coupling losses and losses incurred at the splitters and combiners. A 2×2 amplifier gate switch is shown in Figure 3.15. A disadvantage of the gate switch is that the splitting and combining losses limit the size of the switch.

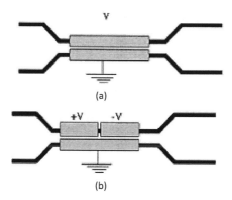

(a)

(b)

FIGURE 3.14 A DC-based electrooptic switch.

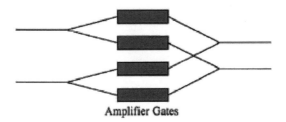

Amplifier Gates

FIGURE 3.15   Switches based on amplifier gates.

3.4   OPTICAL CROSSCONNECT (OXC)

An optical crossconnect (OXC) makes switching operation of wavelength having optical signals from input to output ports with rout specified for destination [43]. It is based on an optical matrix switch. As per input and output ports in OXC, the number of inputs and outputs of optical matrix switch is selected. The optical matrix switch is based on basic switch elements, and for $N \times N$ optical matrix switch, the number of switch elements/units is $2N-3$. These elements are usually considered to be wavelength insensitive, i.e., incapable of demultiplexing different wavelength signals on a given input fiber. A basic crossconnect element is a 2×2 crosspoint element which is shown in Figure 3.16. There are two states of 2×2 crosspoint element – cross and bar states. In cross state, the signal from the upper input port is routed to the lower output port, and the signal from the lower input port is routed to the upper output port. In the bar state, the signal from the upper input port is routed to the upper output port.

There are two types of OXC architectures demonstrated using two types of technologies:

1. the generic directive structure where light is physically transmitted to one of two different outputs.
2. the gate switch structure in which optical amplifier gates are made to select and filter input signals to specific output ports.

Different types of switching elements are already discussed earlier. These switching elements are based on DC, X-branching structure, MMI coupler, TMI coupler, MZ structure, etc.

Cross state   Bar state

FIGURE 3.16   A 2×2 crossconnect element.

3.4.1 Architecture-Based Crossconnect

Figure 3.17 shows OXC crossconnect based on Clos architecture [46]. It is used for building multistage TDM switching systems. The advantage is that it implements the fewest switching crosspoints for providing a large range of scalability that provides strict or rearrangeably non-blocking traffic paths.

In the figure, the number of second-stage switches is dependent on blocking: in fully non-blocking, $k \geq 2n - 1$; in rearrangeably non-blocking, $k \geq n$ [47]. Crossconnect Switch Architecture up to 2048×2048 ports and 10 Gbps per port are also reported.

3.4.2 Micro Electro Mechanical Systems (MEMS)

Recently, micro electro mechanical systems (MEMS) is one of the most promising approaches for large-scale OXCs. Optical MEMS-based switches are either mirrors and membranes based or planar moving waveguides based [46,48]. MEMS-based switches follow two major approaches – 2D and 3D approaches. The 3D optical MEMS based on mirror is more preferred for compact, large-scale switching fabrics. This type of switch has high application flexibility in network design due to low insertion loss and low wavelength dependency under various operating conditions. Furthermore, this switch shows minimal degradation of optical signal-to-noise ratio caused by crosstalk, polarization-dependent loss (PDL), and chromatic and polarization mode dispersions.

Figure 3.18 shows the basic unit of a 3D MEMS optical switch. The optical signals passing through the optical fibers at the input port are switched independently by the gimbal-mounted MEMS mirrors with two- axis tilt control and are then focused onto the optical fibers at the output ports. In the switch, any connection between input and output fibers is accommodated by controlling the tilt angle of each mirror. As a result, the switch can handle several channels of optical signals directly without costly optical–electrical or electrical–optical conversion. The 3D MEMS-based 0–0–0 switch is built in sizes ranging from 256×256 to 1000×1000 bidirectional port machines [48]. In addition, 8000×8000 ports may be fabricated

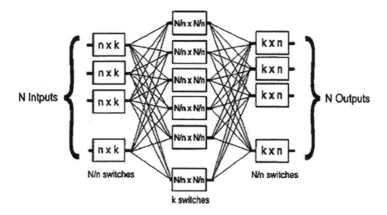

FIGURE 3.17 A 3-stage Clos architecture.

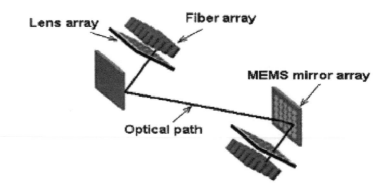

FIGURE 3.18   A schematic diagram of a 3D MEMS optical switch.

within the foreseeable future. The port count is only one dimension to the scalability of a 0–0–0 switch. All-optical switch based on this type is bit-rate and protocol independent. The combination of thousand ports and bit-rate independence may provide unlimited scalability.

Optical MEMS approach provides miniature devices with optical, electrical, and mechanical functionalities at the same time, fabricated using batch process techniques as derived from microelectronic fabrication. Optical MEMS provides intrinsic characteristics for very low crosstalk, wavelength insensitivity, polarization in sensitivity and scalability.

3.5   OPTICAL ADM (OADM)

Optical ADM (OADMs) provide capability to add and drop wavelength traffic in the network like synchronous optical network (SONET) ADMs. Figure 3.19 shows a generic ADM placed at network nodes connecting one or two (bidirectional) fiber pairs and making a number of wavelength channels to be dropped and added. This reduces the number of unnecessary optoelectronic conversions, without affecting the traffic that is transmitted transparently through the node.

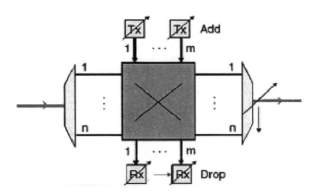

FIGURE 3.19   WSOADM architectures.

An OADM is employed in both linear and ring network architectures operating in either fixed or reconfigurable mode [9,10]. In fixed OADMs, the add/drop and through channels are predetermined, and the adding and dropping of wavelength channels are not tuned by external arrangement. In reconfigurable OADMs, the channels that are added/dropped pass through the node with dynamically reconfigured external arrangement as required by the network. Thus the reconfigurable OADMs are more complex but more flexible as they provide on-demand provisioning without manual intervention.

Reconfigurable OADMs are classified into two categories – partly reconfigurable and fully reconfigurable architectures [9,10,49]. In partly reconfigurable architectures, there is a capability to select the predetermined channels to be added/dropped, with a predetermined connectivity matrix between add/drop and through ports, restricting the wavelength-assignment function. Fully-reconfigurable OADMs provide the ability to select all the channels to be added/dropped, but they also offer connectivity between add/drop and through ports, which enables flexible wavelength assignment with the use of tunable transmitters and receivers. Reconfigurable OADMs have two main generations. The first is mainly applied in linear network configurations and support no optical path protection, while the second provides optical layer protection.

Two types of fully reconfigurable OADMs are – wavelength-selective (WS) and broadcast-selective (BS) architectures [43], which are shown in Figure 3.20. The WS architecture has wavelength demultiplexing/multiplexing and a switch fabric interconnecting express and add/drop ports, whereas the BS has passive splitters/couplers and tunable filters. The overall loss obtained by the through path of the BS is noticeably lower than that of the WS approach, significantly improving the optical signal-to-noise ratio (OSNR) of the node, and therefore its performance is better in a transmission link or ring. In addition, the BS provides superior filter concatenation performance, features such as drop and continue, and good scalability in terms of add/drop percentage.

The theory and applications of cascaded MZ (CMZ) filters consisting of delay lines (MZ coupler with unequal arms) are already reported by different authors [9,10,49,50], in which Y symmetry CMZ filter is chosen for add/drop filter application, because of lower pass band in comparison to point symmetry CMZ couplers [9]. Figure 3.21 illustrates 2×2 N-stage Y symmetry CMZ coupler having N number of delay line section with arm lengths L_A and L_B (where path difference between two arms $\Delta L = L_A - L_B$), thin film heater of length $L_H (L_H \approx L_A)$ and width W_H on the

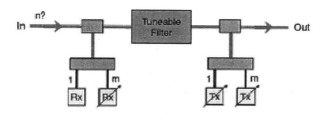

FIGURE 3.20 Fully reconfigurable BS OADM architectures.

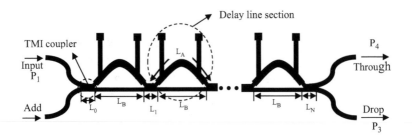

FIGURE 3.21 N-stage Y symmetric CMZ circuit using TMI coupler and thin film.

curved arm of MZ section and $N + 1$ number of TMI) couplers of width $2w$ (where w = width of single-mode access waveguide). The core and cladding are chosen to be SiON and SiO_2 respectively, due to availability of wide index contrast, compatibility with conventional silicon-based IC processing, high stability, etc. In the figure, the couplers of the device are considered to act as one long coupler with total coupling length L distributed in different ways over all individual couplers of the circuit, where $L = \sum_{k=0}^{N} L_i$ and L_i is the length of the ith coupler ($i = 0, 2, 3, \ldots N$). The coupling length distribution which controls transmission characteristics of the filter is discussed later. Each TMI coupler consists of a two-mode coupling region in which only fundamental and first-order mode with propagation constants β_{00} and β_{01}, respectively, are excited in the coupling region [11], and coupling coefficient (k_T) of TMI coupler is represented by $(\beta_{00} - \beta_{01})/2$. From the geometry of the figure length of each delay line section is obtained as $L_B \approx H^2/\Delta L$, where H is the height of the delay line section. The refractive indices of the core and its cladding are n_1 and n_2 respectively. The input power P_1 is launched into lowermost waveguide and the output powers P_3 and P_4 are cross-and bar-state powers, respectively. The normalized cross-state power of N-stage Y symmetric CMZ coupler is derived as [10]

$$\frac{P_3}{P_1} = \left\{ \sum_{k=0}^{N} |a_k|^2 + 2 \sum_{i=0}^{N} a_i \sum_{k=i+1}^{N} a_k \cos\left[(k-i)\Delta\phi(\lambda, \Delta L, P) \right] \right\} e^{-N\alpha L_B} \qquad (3.64)$$

where α is bending loss coefficient function of bending radius [10,18]. The coefficients a_k of normalized cross-state power are estimated from the coupling coefficient k_T of TMI couplers [10]. $\Delta\phi(\lambda, \Delta L, P)$ is the phase difference for the length difference between two arms of delay line section plus the phase shift obtained by heating the curved arm with heating power P and is written as[10]

$$\Delta\varphi(\lambda, \Delta L, P) = 2\pi \left[\varphi(\Delta L) + \varphi(P) \right] \left\{ 1 - \frac{\lambda - \lambda_{ref}}{\lambda_{ref}} \right\} \qquad (3.64)$$

where $\varphi(\Delta L) = \dfrac{n_{eff} \Delta L}{\lambda}$, n_{eff} is the effective index at wavelength λ and λ_{ref} is the reference wavelength.

When the waveguide is heated through the thin film heater, the glass (SiO$_2$) can expand freely to the Si substrate. But it cannot expand freely in the parallel direction, because it is surrounded by other glass (SiO$_2$). So, a compressive stress is developed in the parallel direction, and it mainly induces a refractive index increase in the *TM* mode. In the case of *TM* mode $\varphi(P)$, phase change due to application of heating power P occurs not only due to isotropic thermooptic phase $\varphi_T(P)$, $\varphi_T(P) = \frac{dn}{dT} \Delta T_c(P) \frac{L_H}{\lambda}$, where $\frac{dn}{dT}$ = thermooptic index coefficient, $\Delta T_c(P)$ is the temperature difference between two cores) but also by an anisotropic stress optic phase $\varphi_S(P)$. The anisotropic stress optic phase change for the temperature difference $\Delta T_c(P)$ between two cores obtained by heating via heater is written as [10]

$$\varphi_S(P) = \frac{d(n_{TM} - n_{TE})}{dT} \Delta T_c(P) \frac{L_H}{\lambda}$$

where $\frac{d(n_{TM} - n_{TE})}{dT}$ = temperature rate of increase of birefringence depending on stress optical coefficient $\frac{\delta(n_{TM} - n_{TE})}{\delta S}$ and Young's modulus $\frac{\delta S}{\delta V}$, and thermal expansion coefficient $\frac{\delta V}{\delta T}$ is expressed as [10]

$$\frac{d(n_{TM} - n_{TE})}{dT} = \frac{\partial(n_{TM} - n_{TE})}{\partial S} \cdot \frac{\partial S}{\partial V} \frac{\partial V}{\partial T} \qquad (3.65)$$

In the case of *TE* mode, the phase change with the application of heating power $\phi(P)$ is only an isotropic phase change $\phi_T(P)s$.

3.5.1 THERMOOPTIC DELAY LINE STRUCTURE

Considering the above polarization-dependent characteristics and reduction structure of polarization dependence, a thermooptic delay line structure is shown in Figure 3.22. The structure consists of four sides or boundaries – top surface (side-A), bottom surface (side-B), left surface (side-C) and right surface (side-D). The thermal analysis of conventional and low-power thermooptic device structure with silicon trench at the bottom surface are already studied [9,10] and optimized the parameters such as heater width, total cladding width, and trench width. Like these structures, the delay line structure has two waveguides with the addition of a groove of width W_G and depth H_G made in between two waveguide cores at its top surface for release of stress anisotropy, inducing mainly a refractive index increase in the *TM* polarization [10]. It has an air medium in which the temperature is taken to be ambient temperature of air medium which is close to room temperature T_I. The bottom and right-side surfaces of the silicon trench, made just below the heater in the proposed structure, are attached to the substrate, whereas left-side surface of the trench is attached to the heat insulator. The position of waveguide cores and their sizes, heater

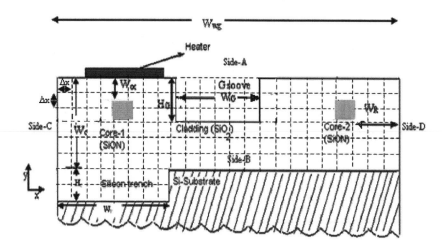

FIGURE 3.22 Cross-sectional view of the proposed thermooptic delay line structure consisting of single stress releasing grooves of depth H_G and width W_G and a silicon trench of trench height H_T and width W_T (Cladding width $= W_c$, upper cladding width $= W_{oc}$, and device width $= W_{wg}$).

size and its position, upper cladding thickness W_{oc} and total cladding thickness W_c, and trench width and thickness are the same as those of the conventional structure. Both side surfaces of the proposed structure are taken as heat insulator for suppressing lateral heat diffusion. The temperature gradient obtained from the temperature profiles by using implicit finite difference temperature equations [10] is an important factor for study of stress release groove in which, for more magnitude of these values, the stress release is also more.

The implicit temperature equations are made at discrete points. The first step in this method is to find these points. The temperature distribution of the waveguide region is made with a heat flux of q_0 via heater and is divided into several small regions of same width, same length and height of Δx and assigning to each reference point that is at its center. This reference point is termed as a nodal point or node. Two types of nodes – interior nodes, which are situated inside the thermooptic structure, and surface nodes/exterior nodes, which are situated on the surface or boundary of the thermooptic structure as shown in Figure 3.22. For computation, these equations of the nodes in short form are written using implicit temperature equations as [10]

$$a_{i,i}T_i^{p+1} + \sum a_{i,j}T_j^{p+1} = b_i \tag{3.66}$$

Where superscript p indicates the time t ($t = p\Delta t$, where $\Delta t = (\Delta x)^2/4\alpha$). The first subscript i of coefficients $a_{i,l}$ shows the equation number, and second subscript i states the node number. Similarly, the first subscript i of coefficients $a_{i,j}$ indicates the equation number, and second subscript j denotes the neighboring node number of the ith node. T_i^{p+1} and T_j^{p+1} are denoted as the temperatures of ith node and its neighboring node j, respectively. The coefficients of the temperature equations for all interior and surface nodes are derived easily from implicit temperture equations.

temperature equations [10] and are as follows [10,11]:

$a_{i,i} = 8$ for all interior nodes, silicon trench nodesand air-exposed top surface nodes,

= 1, for side surface nodes (side-C and -D) andbottom surface nodes attached to the substrate and stress releasing groove nodes,

= 3, for heater-exposed top surface nodes.

$a_{i,j} = -1$, for all interior nodes, side surface nodes ($j = i + 1$ for side-C and $j = i - 1$ for side-D), heater exposed to top surface nodes ($j = i + m$, $m =$ total number of nodes in a row of the device) and air-exposed top surface nodes ($j = i \pm 1$), silicon trench nodes and stress releasing groove nodes.

= −2, for air-exposed top surface nodes ($j = i + m$)

= 0, otherwise.

$b_i = 4T_i^p$, for all interior nodes.

= $2T_i^p + q_0\Delta x/k$, for heater-exposed top surface nodes.

= 0, for side surface nodes (side-C and -D).

= T_l, for bottom surface nodes.

= T_a, for stress relieving groove nodes.

= $4T_i^p + 2BiT_a$, for air-exposed top surface node and stress relieving groove nodes.

k = thermal conductivity of waveguide medium.

Bi = Biot's number.

After keeping the initial temperature of all nodes at room temperature, the heat flux q_0 of the heater is set at a value, and the old temperatures of all the nodes are updated with new tempertures by putting $p = p + 1$ till t is equal to time to get the required temperature difference between the cores.

Figure 3.23 represents the polarization-independent tunable transmission characteristics of 5-stage Y symmetric CMZ filter based on the proposed thermooptic delay line structure with $\Delta n = 5\%$, $\lambda_{ref} = 1.55$ μm, cladding index = 1.447, waveguide core width $w = 1.5$ μm, $\Delta L = 20.5$ μm and $L_B = 462$ μm. The difference between transmission characteristics of TE and TM polarization for the structure having groove is lesser than that of the conventional structure having no groove. The shift of resonant wavelength due to heating the curved arm of delay line section by heater is the same in both TE and TM polarization because anisotropic stress developed by the temperature difference between two cores (showing additional phase difference in TM polarization) is relieved by the groove. The reduction of peak normalized cross-state power is obtained due to bending loss which is approximately 0.1 dB per MZ section. The heating power (H) needed per delay line section to obtain $\Delta T_c(P)$ of 6°C and 12°C for tuning of ADM based on CMZ coupler with conventional structure to wavelengths 1.56 and 1.57 μm from 1.55 μm is estimated by using the equation $H = q_0 \cdot W_H \cdot L_H$ (where $q_0 =$ heat flux to achieve these temperature difference, $W_H =$ heater width and $L_H =$ heater length) as 84.2 and 178 mW per delay line section, respectively, whereas those needed for the structure having grooves to tune to the same wavelengths are 53 and 108.4 mW, respectively.

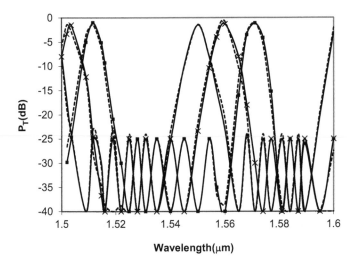

FIGURE 3.23 Polarization-independent tunable transmission characteristics of 5-stage Y symmetric CMZ filter based on a proposed thermooptic delay line structure with $\Delta n = 5\%$, $\lambda_{ref} = 1.55$ μm, cladding index = 1.447, waveguide core width $w = 1.5$ μm, $\Delta L = 20.5$ μm and $L_B = 462$ μm.

3.6 SONET/SDH

With the development of WDM optical network, it is required to increase the transmission capacity in each individual wavelength. It is seen that for accommodation of a connection request/service, we do not need this much of bandwidth to allocate a dedicated wavelength to this connection. We should have hierarchical digital time multiplexing to accommodate more number of channels for a wavelength channel. In this direction, there is a standard signal format known as SONET in North America and synchronous digital hierarchy (SDH) in other parts of the world. This section mentions the basic concepts of SONET/SDH, its optical interfaces and fundamental network implementations.

3.6.1 TRANSMISSION FORMATS AND SPEEDS OF SONET

Several vendors throughout the world started developing standards for formats of SONET frame to interconnect different connections and services for a wavelength channel for fiber-optic communication. There is a need for the development of a common standard. In this direction, ANSI T1.105 standards are developed for SONET in North America [51] and ITU-G. 957 standards for SDH in other parts of the world [IEEE 802.17]. In fact, there is a slight difference for implementation of these standards. Figure 3.24 shows the structure of a basic synchronous transport signal (STS)-1 frame of SONET having a 2D structure consisting of 90 columns by 9 rows of bytes. There are three overloads – section overload and line overload at the beginning of the frame and path overload in the middle of the frame. Section overload connects adjacent pieces of equipment, whereas line overload connects two

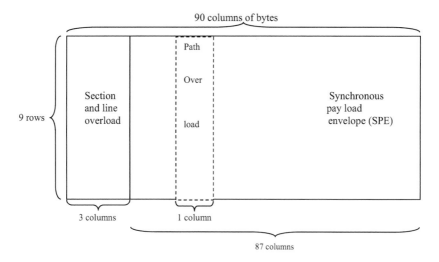

FIGURE 3.24   An STS-1 frame structure.

SONET devices [52–54]. Path overload provides complete end-to-end connection. The fundamental SONET frame has a 125 μs duration.

$$\text{Overload per frame of STS-1} = (4\,\text{bytes/row}) \times (9\,\text{rows/frame}) \times (8\,\text{bits/byte})$$

$$= 288\,\text{bits}$$

$$\text{Information bits per frame of STS-1} = (86\,\text{bytes/row}) \times (9\,\text{rows/frame}) \times (8\,\text{bits/byte})$$

$$= 6192\,\text{bits}$$

$$\text{Total number bits per frame of STS-1} = 6192\,\text{bits} + 288\,\text{bits}$$

$$= 6480\,\text{bits}$$

Since the frame length is 125 μs, the transmission bit rate of the basic SONET signal is given by

$$\text{STS-1} = 6480\,\text{bits}/125\mu s = 51.84\,\text{Mbps}$$

$$\text{STS-1} = (90\,\text{bytes/row}) \times (9\,\text{rows/frame}) \times (8\,\text{bits/byte})/(125\mu s/\text{frame})$$

$$= 51.84\,\text{Mbps}$$

This is called an STS-1 signal where STS represents a synchronous transport signal. All other SONET signals are integral multiples of this bit rate. Figure 3.25 shows STS-N signals in which the transmitted bit rate is $N \times 51.84$ Mbps. Each frame of STS-N has $N \times 90$ column bytes and same 9 rows within 125 μs duration. When an STS-N signal is used to modulate an optical source, the logical STS-N signal is scrambled to avoid log strings of ones and zeros and to allow easier clock recovery at

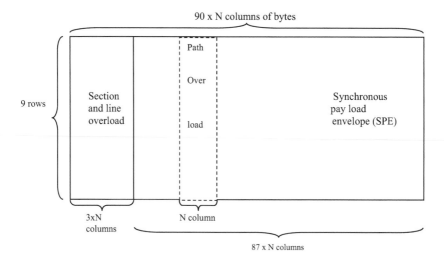

FIGURE 3.25   An STS-N frame structure.

the receiver. After undergoing electrical to optical conversion, the resultant physical layer optical signal is called OC-N, where OC represented an optical carrier. The value can have range 1-192 but ANSI T1.105 standard recognizes the value of $N = 1$, 3, 12, 24, 48 and 192.

In SDH, the basic rate is equivalent to STS-3 or 155.52 Mbps. This is called as synchronous transport module – STM-1. Higher rates can be written as an integral multiple of STM-1 $\times M$ or STM-M where $M = 1, 2, \ldots, 64$. The values of M supported by ITU-T recommendations are $M = 1, 4, 16$ and 64. These are equivalent to SONET OC-N signals where $N = 3M$. This shows compatibility between SONET and SDH. Table 3.1 shows commonly listed values of OC-N and STS-M. Figure 3.26 shows SONET STS-192 digital transmission hierarchy and its SDH equivalent. In the figure, lower-level time division multiplexer is STS-1 for SONET, whereas that for SDH is STM-1. There are five levels of hierarchy in SONET, whereas there are four levels for SDH.

TABLE 3.1

STS/OC Specifications

Electrical Level	SONET Level	SDH Equivalent	Line Rate (Mbps)	Overload Rate (Mbps)
STS-1	OC-1	–	51.84	2.304
STS-3	OC-3	STM-1	155.52	6.912
STS-12	OC-12	STM-4	622.08	27.648
STS-48	OC-48	STM-16	2488.32	110.592
STS-192	OC-192	STM-64	9953.28	442.368

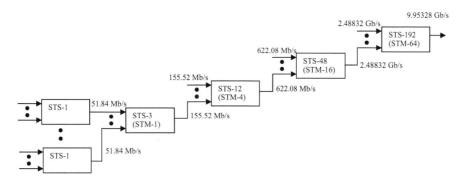

FIGURE 3.26 A SONET STS-192 or SDH STM-64 digital transmission hierarchy multiplexer.

3.6.2 SONET/SDH RINGS

Normally, SONET/SDH technologies are configured as a ring architecture. This is done to create a loop diversity for uninterrupted service protection purposes in case of link or equivalent failures. This SONET/SDH rings are commonly called self-healing rings, since the traffic flowing along a certain path can automatically be switched to an alternative or backup path while failure or degradation of the link segment occurs.

There are three main features yielding eight possible combinations of ring types. First, there can be either two or four fibers running between the nodes on a ring.

3.7 OPTICAL REGENERATOR

As discussed in Figure 1.6, there are three optical windows of low propagation loss for optical fiber transmission characteristics – first window centered at 0.85 μm and with propagation loss ~0.82 dB/km, second window centered at 1.30 μm and with propagation loss ~0.3 dB/km and a third window centered at 1.55 μm and with propagation loss ~0.2 dB/km. Attenuation in optical fiber is due to the impurity content in glass (water vapor) and Rayleigh scattering which is caused by fluctuation in the refractive index. As lower attenuation is obtained in the third window, optical network uses this window for signal transmission. But still for long-distance communication (more than 100 km) it requires an optical regenerator/repeater. Optical regenerator mainly amplifies the signal so that it can compensate signal power loss and normally placed it at an interval of 40 km [55]. Since it has mainly optical amplifiers, it is discussed in the next section.

3.7.1 OPTICAL AMPLIFIERS

An optical signal transmits a long distance typically 80 km at a stretch in current deployment before it needs amplification. Optical networks cover a wide area, but these networks having a diameter of covering area (specially nationwide network) beyond 87 km need all-optical amplifiers for long-distance links. All-optical amplification is different in which before amplification it needs optoelectronic conversion and after amplification electrooptical conversion. The optical amplifier acts only to amplify the power of a signal, but not to restore the shape or timing of the signal. There are three types of optical amplification – inline optical amplifier, preamplifier and power amplifier [56].

Inline amplification: There is only inline amplification without getting reshaping and retiming of the signals to compensate only the transmission loss.

Preamplification: This type of amplification is used as a front-end amplification for optical receivers. The week signal is amplified before detection, so that signal-to-noise degradation arises due to thermal noise in the receiver which can be suppressed.

Power amplification: It is placed just after the transmitter to boost the transmitted power and to increase the transmission distance without amplification. This boosting technique is used in undersea optical fiber communication where the transmission distance is 200–250 km. It is also used for compensation of coupler insertion loss and power splitting loss.

In communication networks using SONET and SDH, the optical fiber is only required as a transmission medium, the optical signals are amplified by first converting the information stream into an electronic data signal, and then retransmitting the signal optically. Such a process is referred to as 3R (reamplification, reshaping and retiming).

The reshaping of the signal regenerates the original pulse shape, eliminating noise/distortion. Reshaping applies mainly to digitally modulated signals, but in some cases, it is also used for analog signals. The retiming of the signal synchronizes the signal to its original bit timing pattern and bit rate. Retiming applies only to digitally modulated signals.

Also, in a WDM system having optoelectronic regeneration, each wavelength is to be separated before being amplified electronically, and then recombined before being retransmission. Thus, in order to replace optical multiplexers and demultiplexers in optoelectronics amplifiers, optical amplifiers must boost the strength of optical signals without first converting them to electrical signals. The main problem is that optical noise is amplified with the signal. Also, the amplifier includes spontaneous emission noise, since optical amplification normally uses the principle of stimulated emission, similar to the approach used in a laser. Optical amplifiers are classified into two basic classes: optical fiber amplifiers (OFA) and semiconductor optical amplifiers (SOAs), which is mentioned in detail in the following section. In Table 3.2, comparison between OFAs and SOAs is presented. Besides, there is a new kind of optical amplifier called Raman amplifier, which is explained in detail in the following sections.

3.7.2 OPTICAL AMPLIFIER CHARACTERISTICS

The performance parameters and characteristics of an optical amplifier are gain, gain bandwidth, gain saturation, polarization sensitivity and noise amplification [52].

Gain is a ratio of the output power of a signal to its input power. The performance of amplifiers are represented by gain efficiency as a function of pump power in dB/mW, where pump power is the energy required for amplification. The gain

TABLE 3.2
Difference of Characteristics of OFAs and SOAs

Features	OFA	SOA
Maximum internal gain	25–30	20–25
Insertion loss (dB)	0.1–2	6–10
Polarization sensitivity	Negligible	<2 dB
Saturation output power (dBm)	13–23	5–20
Noise figure (dB)	4.6	7–12

bandwidth of an amplifier defines as a range of frequencies or wavelengths over which the amplifier amplifies effectively. In a network, the gain bandwidth provides the number of wavelength channels obtained for a given channel spacing. The gain saturation point of an amplifier states that when the input power is increased beyond a certain value, the carriers (electrons) in the amplifier are unable to output any additional light energy. The saturation power is typically defined as the output power at which there is a 3-dB reduction in the ratio of output power to input power. Polarization sensitivity refers to the dependence of the gain on the polarization of the signal. The sensitivity is measured in dB and refers to the gain difference between the *TE* and *TM* polarizations.

In optical amplifiers, the dominant source of noise is amplified spontaneous emission (ASE) arising from the spontaneous emission of photons in the active region of the amplifier. The amount of noise produced by the amplifier is a function of factors such as the amplifier gain spectrum, the noise bandwidth and the population inversion parameter which represents the degree of population inversion that has been achieved between two energy levels [52].

3.7.3 SEMICONDUCTOR LASER AMPLIFIER

A semiconductor laser amplifier (Figure 3.27) is a modified semiconductor laser, which has different facet reflectivity and different device length [57]. A weak signal is sent through the active region of the semiconductor, in which stimulated emission makes it a stronger signal emitted from the semiconductor.

The basic structures of semiconductor laser amplifiers are the Fabry–Perot amplifier (FPA) having the Fabry–Perot cavity with partially reflective facets and

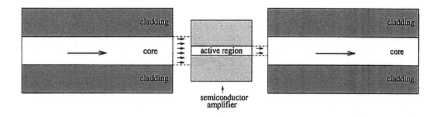

FIGURE 3.27 Semiconductor laser amplifier.

traveling-wave amplifier (TWA) having non-resonant cavity with less reflective facets. The primary difference between the two is in the reflectivity of the end mirrors. FPAs have a reflectivity around 30%, whereas TWAs have a reflectivity of around 0.01%. In order to prevent lasing in the FPA, the bias current is operated below the lasing threshold current. When an optical signal enters the FPA cavity, it gets amplified as it reflects back and forth between two mirror facets until it emits at higher intensity. The amplifier gain is written as [57]

$$G = \frac{P_{out}}{P_{in}} \tag{3.67}$$

where P_{out} and P_{in} are the output and input powers of FPA. The higher reflections in the FPA cause Fabry–Perot resonances in the amplifier, resulting in narrow passbands of around 5 GHz. By decreasing the reflectivity, the amplification is performed in a single pass so that no resonance occurs. So the performance of TWAs is better than that of FPAs in the case of WDM networks. In TWA, the input signal gets amplified only once with the principle of semiconductor laser action during a single pass through TWA. The amplifier gain is derived as

$$G = \exp\left[\Gamma\left(g_m - \alpha\right)L\right] = \exp\left[g(z)L\right] \tag{3.68}$$

where Γ = optical confinement factor, g_m = material gain coefficient, α = absorption coefficient, L = amplifier length and $g(z)$ = overall gain per unit length. Semiconductor amplifiers based on multiple quantum wells (MQW) are studied by many authors. These amplifiers have higher bandwidth and gain saturation than bulk devices and also fast on–off switching times. The disadvantage is a higher polarization sensitivity. Currently, SOAs attract more interest in both research and industry fields because its advantage of semiconductor amplifiers is the ability to integrate them with other components. By turning a drive current on and off, the amplifier can act as a gate, blocking or amplifying the signal devices.

3.7.4 Doped Fiber Amplifier

Doped fiber amplifier (DFA) consists of a length of fiber lightly doped with an element [58] (rare earth elements such as Erbium (Er), ytterbium (Yb), neodymium (Nd) and praseodymium (Pr)) are used to amplify light (Figure 3.28). The active medium length doped with rare earth element is nominally ~ 10–30 m. The doping concentration is ~1000 parts per million weight. The most common doping element is erbium contributing gain for wavelengths between 1525 and 1560 nm. Sometimes Yb is added with Er++ to increase pumping efficiency and amplifier gain. For amplification of 1300 nm window, the doping elements used are Pr and Nd. At the ending of the length of the fiber amplifier, a laser emits a strong signal at a lower wavelength (taken to be the pump wavelength) to backup the fiber. Figure 3.29 shows a simplified energy diagram Er^{3+}-doped silica fiber in which metastable states and pump energy levels are $^4I_{13/2}$ and $^4I_{11/2}$, respectively. The metastable states have slightly longer

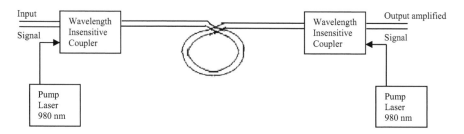

FIGURE 3.28   Erbium-doped fiber amplifier.

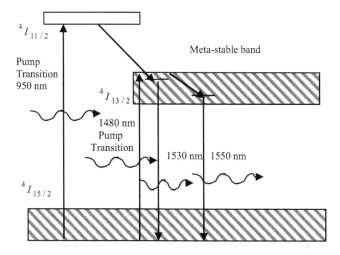

FIGURE 3.29   Energy level diagram of various transition processes of Er^{3+}-doped silica.

lifetimes in comparison with that of pump energy levels. The pump energy levels, metastable energy levels and ground energy levels are actually bands of closely spaced energy levels that form manifold due to the effect known as stark splitting by broadening with thermal effects. The gap between the top of $^4I_{15/2}$ and the bottom of the ground state band $^4I_{13/2}$ is around 0.841 eV, which corresponds to the wavelength 1480 nm, and the energy gap between the bottom of $^4I_{15/2}$ and the bottom of meta-stable band $^4I_{13/2}$ is ~0.814 eV corresponding to the wavelength 1527 nm. This pump signal excites the doped atoms into a higher energy level. This allows the data signal to stimulate the excited atoms to release photons. Most erbium-DFAs (EDFAs) are pumped by lasers with a wavelength of either 980 or 1480 nm. The input signal power (P_{in}) and output

amplified signal (P_{out}) of an EDFA can be expressed in terms of the principle of energy conversion [52]

$$P_{out} \leq P_{in} + \frac{\lambda_p}{\lambda_s} \cdot P_p \qquad (3.70)$$

where λ_p and λ_s is pump and signal wavelengths, respectively, and P_p = pump signal power. The maximum output amplified signal power depends on the ratio λ_p/λ_s. For pumping, it is required to have $\lambda_p < \lambda_s$, and for getting gain, it is also necessary to have $P_{in} \ll P_p$. The power conversion efficiency (PCE) is written as

$$\text{PCE} = \frac{P_{out} - P_{in}}{P_p} \tag{3.71}$$

The maximum value of PCE is $\sim \lambda_p/\lambda_s$. The quantum conversion efficiency (QCE) is defined as $\text{QCE} = \dfrac{\lambda_s}{\lambda_p} \text{PCE}$. The maximum value of QCE is unity in which all the pump photons are converted into signal photons. The amplifier gain (G) is written as

$$G = \frac{P_{out}}{P_{in}} \le 1 + \frac{\lambda_p P_p}{\lambda_s P_{in}} \tag{3.72}$$

To achieve the gain G, the input power signal cannot exceed a value given by

$$P_{in} \le \frac{(\lambda_p/\lambda_s)P_p}{G-1}$$

The gain of fiber-doped amplifier not only depends on pump power but also on fiber length signal emission cross-section and rare earth element concentration. The maximum gain G_{max} is written as

$$G_{max} = \exp(\rho \sigma L) \tag{3.73}$$

The maximum possible gain of DFA is written as [52]

$$G \le \min\left\{\exp(\rho \sigma L), 1 + \frac{\lambda_p}{\lambda_s} \cdot \frac{P_p}{P_{in}}\right\} \tag{3.74}$$

The maximum possible output power is expressed with min of two possible values as [52]

$$P_{out} \le \min\left\{P_{in}\exp(\rho \sigma L), P_{in} + \frac{\lambda_p}{\lambda_s} \cdot P_p\right\} \tag{3.75}$$

The 980-nm pump wavelength provides gain efficiencies of around 10 dB/mW and 1480-nm pump wavelength gives efficiencies of around 5 dB/mW. Typical gains are of ~25 dB. Experimentally, EDFAs have been shown to achieve gains of up to 51 dB with the maximum gain limited by internal Rayleigh backscattering (RBS), and a portion of scattered light is back reflected towards the launch end within the optical waveguide. The 3-dB gain bandwidth for the EDFA is around 35 nm and the saturation power is ~20 dBm. A wide-band EDFA with 25 dB of flat gain over 77 nm

(1528–1605 nm) and dynamic gain clamping over 13 dB of input range (25–483 W) are established experimentally.

Apart from Erbium-doped fiber, there are praseodymium-doped fluoride fiber amplifier (PDFFA) and Yb-DFA. A limitation to optical amplification is unequal gain spectrum of optical amplifiers. The gain spectrum DFA with non-uniform gain at different wavelengths is shown in Figure 3.30a, and Figure 3.30b shows the EDFA gain spectrum with different inversion levels (where the inversion levels are function of concentration of rare earth doped elements). While an optical amplifier contributes gain across a range of wavelengths, it requires to amplify all wavelengths equally. Apart from this disadvantage, optical amplifiers amplify noise with signal and also the active region of the amplifier spontaneously transmits photons providing the noise and hence degrading the performance of optical amplifiers. The spontaneous noise in the EDFA is mainly due to the spontaneous recombination of electrons and holes in the active medium of EDFA. This noise is modeled as a stream of infinitely random short pulses distributed along the amplifying medium. Such a random process is characterized by a noise power spectrum. The power spectral density of spontaneous nose [52] is

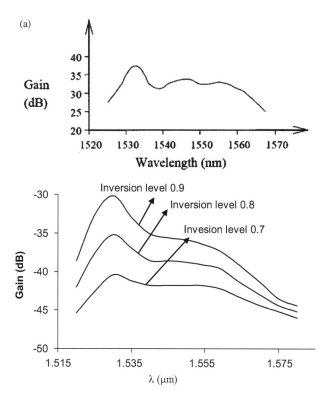

FIGURE 3.30 (a) An EDFA gain spectrum. (b) Gain spectrum for different inversion levels.

$$S_{SPN}(f) = h\nu n_{sp}[G(f)-1] = P_{SPN}/\Delta\nu_{opt} \tag{3.76}$$

where P_{SPN} = the spontaneous noise power in optical frequency band $\Delta\nu_{opt}$. and n_{sp} denotes population inversion between two energy levels – metastable and lower energy levels.

$$n_{sp} = \frac{n_2}{n_2 - n_1}$$

n_2 and n_1 are populations of electrons of metastable state and lower energy levels. The presence or absence of spontaneous noise depends on whether co-directional or counter-directional pumping is used. The signal-to-noise ratio depends on input signal power P_{in}, gain G, population inversion n_{sp}, front-end receiver electrical bandwidth B and quantum efficiency η.

Non-uniform gain of EDFA gain equalizer makes power imbalance in an optical network. There are a number of approaches for equalizing the non-uniform gain of an EDFA. There is a notch filter (a filter that attenuates the signal at a selected frequency) centered at around 1530 nm is used to suppress the peak in the EDFA gain. If multiple EDFAs are cascaded, another peak occurs around the 1560-nm wavelength. In the network, it is required to have dynamic EDFA gain equalizer. Using this approach we can develop these dynamic gain equalizers (DGEs) that are discussed in the next section.

3.7.5 RAMAN AMPLIFIER

Raman amplifiers based on Raman's effect is used for optical amplification in long-haul and ultra-long-haul fiber-optical transmission systems. It is one of the first widely commercialized nonlinear optical devices in telecommunications [59,60]. The schematic diagram of a Raman amplifier is presented in Figure 3.31. When an optical field is incident on a molecule, the bound electrons oscillate at an optical frequency. This induced oscillating dipole moment generates optical radiation at the same frequency, with a phase shift due to the medium's refractive index's time-ously, the molecular structure is oscillate data the frequencies of various molecular vibrations. Therefore, the induced oscillating dipole moment also comprises the difference frequency terms between the optical and vibration frequencies. These terms provide Raman scattered light in the reradiated field. In a solid-state quantum-mechanical description, optical photons are in elastically scattered by quantized molecular vibrations represented by optical phonons. Photon energy is lost (the molecular lattice is heated) or gained (the lattice is cooled), shifting the frequency of the light. The components of scattered light that shifted to lower frequencies are Stokes lines, while those shifted to higher frequencies are anti-Stokes lines. The frequency shift is due to oscillation frequency of the lattice phonon that is generate do annihilated. (The anti-Stokes process is not mentioned as it is typically orders of magnitude weaker than the Stokes process in the context of optical communications, making it irrelevant.) Raman scattering occurs in all materials, but in silica glass the domain an Raman lines are due to the bending motion of the Si–0–Si bond (see bond

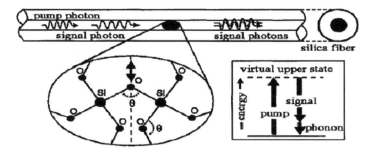

FIGURE 3.31   A schematic demonstration of Raman amplification using simulated Raman scattering with Stoke lines concept.

angle in Figure 3.31). Raman scattering is stimulated by signal light at an appropriate frequency shift from a pump, making stimulated Raman scattering (SRS). In this process, pump and signal light are coherently coupled by the Raman process. In a quantum-mechanical description, shown in the energy-level diagram in Figure 3.31, a pump photon is converted into a second signal photon that is an exact replica of the first one, and the remaining energy produces an optical photon. The initial signal photon, therefore, has been amplified.

There is a disadvantage – requirement of more pump power, roughly tens of milliwatts per dB of gain, whereas the tenth of a milliwatt per dB is required for EDFAs for small signal powers. There are advantages that Raman amplifiers are made along the transmission fiber itself. Now, Raman amplification is an accepted technique for enhancing system performance. Raman amplification has enabled dramatic increases in the reach and capacity of lightwave systems. Novel Raman pumping schemes have recently been reported.

3.8   CHANNEL EQUALIZERS

Optical networks are dynamic in nature due to non-uniform and unpredictable traffics in the networks. It leads the requirements for dynamic compensation of signal power with optical amplifiers. As discussed in the previous section regarding non-uniform gain spectrum of optical amplifiers, compensation is required to get flatness of amplifier gain profiles. The gain of EDFA spectrum depends on pumping power exciting Er^{3+} ions to the metastable state. The number of excited Er^{3+} ions to metastable state is increased with an increase of pumping power. It is indicated in terms of inversion level as the ratio between the number of Er^{3+} ions exited to the metastable state and total number of Er^{3+} ions in EDFA. The variation of input powers and changes of the temperature also contribute to the variation of power profile of EDFA [61]. In order to overcome the power imbalance problems, the imbalance power profile of all wavelength channels is equalized. Due to imperfections in gain-flattening filters, changes in the amplifier operating conditions and changes in channel loading, a DGE is required with a smooth, low-ripple spectral attenuation profile that is the negative of the deviations of the amplifier gain profile from the desired profile. To compensate for unequal channel powers, resulting from dropping and adding channels, one needs a dynamic channel-power equalizer (DCE), which provides a flat attenuation profile across the full bandwidth of each channel, but the attenuation for each channel can be adjusted accordingly. There are different ways of making DGE. Every channel is equalized individually by using a channel equalizer or demux/mux combination with variable optical attenuation [61]. The use of channel separation and recombination by demux and mux become lossy and expensive for a large number of channels. Single gain equalizing device is used to flatten a large number of wavelength channels simultaneously [51,62–63]. Another way of flattening is that using many EDFAs in cascade, one can flatten WDM channels. The use of many EDFAs in cascade is costly to equalize non-uniform amplified power for a large number of channels. A single gain equalizing device is preferred due to having less lossy property. One of such EDFA gain equalizer filter based on CMZ coupler [62,63] is very simple and more cost effective. The CMZ filters made up of silica on silicon waveguide flatten the bandwidth of 30 nm (from 1.53 to 1.56 micron) but gain equalization is not dynamically adjusted. As discussed earlier, in optical network, dynamic gain equalization is required because the reconfiguration and automatic wavelengths switching in optical network change the power profile dynamically. Moreover, the changes in the network load require the dynamic adjustment of equalization of EDFA spectrum. This dynamic adjustment of EDFA gain equalization is performed with tuning of gain equalizer. The tuning is obtained by building heaters into the waveguides to vary the coupling strength and phase shift in delay line coupler via thermooptic effect. The problems of these gain equalizers are large bending loss and device length due to presence of more number of stages. Figure 3.32 shows thermally tunable CMZ filter used as dynamic EDFA gain equalizer. It has an N-stage point symmetry CMZ coupler consisting of N number of delay line sections with arm lengths L_A and L_B (where path difference between two arms, $\Delta L = L_A - L_B$), thin film heater of length $L_H (L_H \approx L_A)$ and width W_H made on a curved arm of MZ section and $N+1$ number of TMI couplers of width $2w$.

The cross-sectional view xx' is a thermooptic delay line structure that has a silicon trench with optimized device parameters such as heater width, total cladding width and trench width. Like these structures, the delay line structure has two waveguides with a groove of width W_G and depth H_G made in between two waveguide cores for release of stress anisotropy, which enhances mainly a refractive index increase in the *TM* polarization [63]. It has an air medium where the temperature is taken to be ambient temperature of air medium which is slightly more than the room temperature T_I. In the proposed structure, bottom and right-side surfaces of the trench portion are attached to the substrate, whereas left-side surface of the trench is attached to the heat insulator. The position of waveguide cores and their sizes, heater size and its position, upper cladding thickness W_{oc} and total cladding thickness W_c, and trench width and thickness are kept the same as those of the conventional structure. Both side surfaces of the structure are assumed as heat insulator for suppressing lateral heat diffusion.

The couplers of the device act as one long coupler with total coupling length L, which is represented with different ways of distribution over all individual couplers of the circuit as $L = \sum_{k=0}^{N} L_i$ and L_i is the length of the ith coupler ($i = 0, 2, 3, ..., N$).

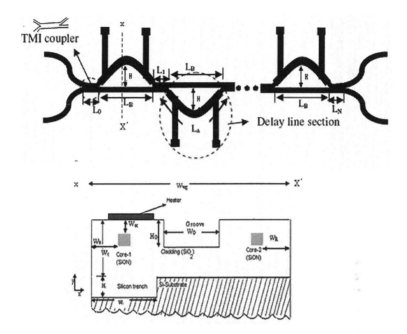

FIGURE 3.32 Schematic diagram of an N-stage 2×2 point symmetric MZ filter consisting of N number of delay line sections with thin film heaters and $(N + 1)$ number.

The coupling length distribution controls transmission characteristics of the filter. In each TMI coupler only fundamental and first-order mode with propagation constants β_{00} and β_{01} respectively are excited in the coupling region [63], and the coupling coefficient (k_T) of TMI coupler is represented by $(\beta_{00} - \beta_{01})/2$. Expanding $k_T(\lambda)$ to the first-order approximation to the wavelength λ is derived in terms of the reference wavelength λ_{ref} as

$$k_T(\lambda) = k_T(\lambda_{ref}) \left\{ 1 - \frac{\lambda - \lambda_{ref}}{\lambda_{ref}} \right\} \qquad (3.77)$$

From the geometry, horizontal length of each delay line section is obtained as $L_B \approx H^2/\Delta L$, where H is the height of delay line section. The total length of the device is written as [63]

$$L_T = N \cdot L_B + L + L_I \qquad (3.78)$$

where L_I = length of input and output portion of the device. The refractive indices of the core and its surrounding cladding are n_1 and n_2 respectively. The input power P_1 is incident on the lowermost waveguide when the output powers P_3 and P_4 are obtained as cross and bar states, respectively. The normalized cross power of N-stage points symmetric CMZ coupler is given by [63]

$$\frac{P_3}{P_1} = \left\{ \left[\sum_{k=0}^{N} |a_k|^2 + 2\sum_{i=0}^{N} a_i \sum_{k=i+1}^{N} a_k \cos\left[(k-i)\Delta\varphi(\lambda,\Delta L,P)\right] \right] e^{-N\alpha L_B} \right\} \quad (3.78)$$

where α is the bending loss coefficient depending on bending radius [17], and the coefficients a_ks are determined from the coupling coefficient $k_T(\lambda)$ of TMI couplers (as expressed in equation (1)). $\Delta\varphi(\lambda,\Delta L,P)$ is the phase difference obtained due to path difference between two arms of delay line section plus the phase shift introduced by heating the curved arm. Expanding to first order in wavelength, $\Delta\varphi(\lambda,\Delta L,P)$ is derived as [66,65]

$$\Delta\varphi(\lambda,\Delta L,P) = 2\pi\left[\varphi(\Delta L) + \varphi(P)\right]\left\{1 - \frac{\lambda - \lambda_{ref}}{\lambda_{ref}}\right\} \quad (3.79)$$

where $\varphi(\Delta L) = \dfrac{n_{eff}\Delta L}{\lambda}$, n_{eff} is the effective index at wavelength λ and λ_{ref} is the reference wavelength. When the waveguide is heated locally via the heater, the glass (SiO$_2$) can enlarge freely to the Si substrate. But it is not enlarged freely in the parallel direction, because it is surrounded by other glass (SiO$_2$). As a result, compressive stress is built only in the parallel direction, and it mainly induces a refractive index increase in the TM mode. So, the refractive index increase due to stress optic effect gives rise to an extra phase change $\phi_S(P)$ in TM mode apart from thermooptic phase change $\phi_T(P)$ with the application of heating. The anisotropic stress optic phase change for the temperature difference $\Delta T_c(P)$ between the two cores obtained by heating via heater is expressed as [63]

$$\phi_S(P) = \frac{d(n_{TM} - n_{TE})}{dT}\Delta T_c(P)\frac{L_H}{\lambda} \quad (3.80)$$

where $\dfrac{d(n_{TM} - n_{TE})}{dT}$ is the temperature rate of birefringence depending on stress optical coefficient $\dfrac{\delta(n_{TM} - n_{TE})}{\delta S}$, Young's modulus $\dfrac{\delta S}{\delta V}$ and thermal expansion coefficient $\dfrac{\delta V}{\delta T}$ is expressed as [63]

$$\frac{d(n_{TM} - n_{TE})}{dT} = \frac{\partial(n_{TM} - n_{TE})}{\partial S} \cdot \frac{\partial S}{\partial V}\frac{\partial V}{\partial T} \quad (3.81)$$

So in the case of TM mode $\varphi(P)$, phase change due to application of heating power P occurs not only due to isotropic thermooptic phase $\varphi_T(P)$ ($\varphi_T(P) = \dfrac{dn}{dT}\Delta T_c(P)\dfrac{L_H}{\lambda}$, where $\dfrac{dn}{dT}$ = thermooptic index coefficient, $\Delta T_c(P)$ is the temperature difference between two cores) but also by an anisotropic stress optic phase $\varphi_S(P)$. In the case of

TE mode, the phase change due to the application of heating power $\varphi(P)$ is only an isotropic phase change $\varphi_T(P)$.

Figure 3.33 illustrates the equalization of EDFA gain spectrum of different inversion levels, obtained by using 2-stage point symmetric CMZ filter-based proposed structure. The solid line of the figure shows an EDFA gain spectrum of inversion levels 0.7, 0.8 and 0.9 [63], whereas the dashed and dotted lines show gain equalized spectrum of *TE* and *TM* polarization, respectively, and EDFA gain spectrum. In the figure, EDFA spectrum of inversion level 0.7 is equalized with a 2- stage point CMZ coupler by considering that central wavelength 1.55 μm and equalization of EDFA spectrum of inversion levels 0.8 and 0.9 are obtained by shifting the central wavelength to 1.555 and 1.56 μm, respectively, by making the temperature difference $\Delta T_c(P)$ of 3°C and 6°C between two cores with heater on core-1. The equalized gain spectrums of *TE* and *TM* polarization of the device structure almost overlapped. So, the gain spectrums of the proposed structure are almost polarization independent. The heating power required per delay line section to obtain $\Delta T_c(P)$ of 3°C and 6°C for tuning of EDFA gain equalization with conventional structure and proposed structure are estimated by using implicit finite difference method and shown in Table 3.3. In the table, the reduction of the heating power of the low-power thermooptic structure is obtained by ~1.6 times with respect to that of the conventional structure.

Further, for large-range EDFA gain spectrum, its equalization is made by shifting the central wavelength via thermal tuning.

In some EDFA gain equalizers, either DGE or DCE functionality is made, but the basic design parameters for these two cases are different for different device designs required to accomplish the two different functionalities. In particular, the DGE makes an important role in WDM networks because of its ability to control the power profile of the wavelength channels, hence maintaining a high quality of service (QoS) and providing more flexibility in transmission management. The key requirements of future dynamic WDM signal equalizers are low insertion loss, wide

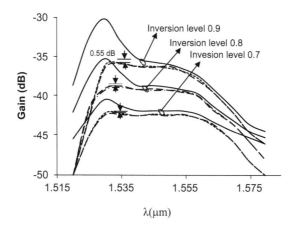

FIGURE 3.33 Equalization of an EDFA gain spectrum of inversion levels 0.7, 0.8 and 0.9 by using 2-PB CMZ circuit with $\Delta T_c(P) = 0$, 3°C and 6°C.

TABLE 3.3

Heating Power of Conventional and Proposed Structures, Thickness of Waveguide (Wc) ~ 15 μm, Thickness of Upper Cladding Layer (Woc) ~ 3 μm, Heater Length L_H ~ 482 μm, Heater Width W_H~ 6 μm and Wavelength $\lambda_{ref} = 1.55$ μm

$\Delta T_c(P)$, Temperature Difference between Two Waveguides Cores of Delay Line Section (°C)	Heating Power (mW) of Conventional Structure (Per Delay Line Section) (mW)	Heating Power (mW) of Proposed Structure (Per Delay Line Section) (Trench Height H_T ~ 12 μm, Trench Width $W_T = 18$ μm, Groove Depth $H_G = 9$ μm Groove Width, $W_G = 9$ μm)
~3	42.1	26.5
~6	89	54.2

bandwidth, fast equalization speed, small size and low cost. There are dynamic WDM equalizer structures including MEMS filters, MZ interferometer (MZI) filters, acoustooptic filters, digital holographic filters and liquid-crystal modulators. These structures use few cascaded (or parallel) optical filters, whose weights are dynamically optimized to obtain a smooth spectral equalization. It is difficult to obtain channel-by-channel equalization unless the number of optical amplifiers used in an equalizer subsystem is equal to the number of WDM channels.

3.9 WAVELENGTH CONVERSION

In a wavelength-routed network we require the same wavelength to be allocated on all the links in the path between the source and destination station [66–68]. This requirement is known as the wavelength-continuity constraint [66]. This constraint makes the wavelength-routed network blocking calls only when there is no availability of same wavelength for all the links in the path between the source and destination. Thus, wavelength-continuity network may suffer higher blocking than a circuit-switched network.

It eliminates the wavelength-continuity constraint problem for removal of blocked connections by converting the wavelength to another wavelength available in outgoing link in the intermediate node. This technique is known as wavelength conversion in the intermediate node.

Figure 3.34 shows the wavelength converter at Node 2 employed to convert data from wavelength λ_2 to λ_1. The new lightpath between Node 1 and Node 3 is set up by using the wavelength λ_2 on the link from Node 1 to Node 2, and then by using the wavelength λ_1 to reach Node 3 from Node 2. So the same lightpath in a wavelength convertible network is converted into a different wavelength along each of the links in its path. Thus, wavelength conversion enhances the efficiency in the network by resolving the wavelength conflicts of the lightpaths reducing blocking of connections. The impact of wavelength conversion on WDM wide area network (WAN) design is further discussed later on.

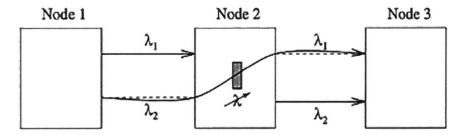

FIGURE 3.34 Wavelength conversion in the optical network.

In wavelength convertible network, the requirements of wavelength converter are given below:

1. fast set up time of output wavelength.
2. conversion to both shorter and longer wavelengths.
3. moderate input power levels.
4. possibility for same input and output wavelengths (i.e., no conversion).
5. insensitivity to input signal polarization.
6. low-chirp output signal with high extinction ratio and large signal-to-noise ratio, and simple implementation.

Wavelength conversion techniques are classified into two types: optoelectronic wavelength conversion, in which the optical signal must first be converted into an electronic signal, and all-optical wavelength conversion, in which the signal remains in the optical domain. All-optical conversion techniques may be subdivided into techniques which employ coherent effects and techniques which use cross modulation.

3.9.1 OPTO ELECTRONIC WAVELENGTH CONVERSION

Figure 3.35 represents an optoelectronic wavelength converter in which the optical signal of the wavelength to be converted is first changed into the signal in electronic domain using photodetectors [64,69]. The electronic bit stream is stored in the buffer (labeled FIFO for the First-In-First-Out queue mechanism). The electronic signal is then used to drive the input of a tunable laser (labeled T) for tuning to the desired wavelength of the output. This method is used for bit rates up to 10 Gbps. However, this method is complex and requires a lot more power than the other methods which

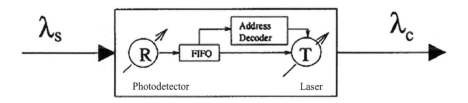

FIGURE 3.35 An optoelectronic wavelength converter.

are discussed below. All information in the form of phase, frequency and analog amplitude of the optical signal may be lost during the conversion process.

3.9.2 Wavelength Conversion Using Coherent Effects

Wavelength conversion methods using coherent effects are based on wave-mixing properties. Wave mixing arises from a nonlinear optical response of a medium generated from mixing of more than one wave. The generated wave intensity is proportional to the product of the interacting wave intensities, which depends on both phase and amplitude information. It allows simultaneous conversion of a set of multiple input wavelengths to another set of multiple output wavelengths and accommodates signal with high bit rates. Figure 3.36 shows a nonlinear wave mixing in which the value $n = 3$ refers to Four- Wave Mixing (FWM) and $n = 2$ refers to Difference Frequency Generation (DFG).

These techniques of wave mixing are described below:

Four-Wave Mixing (FWM: FWM is a third-order nonlinearity in silica fibers, which contributes three optical waves offer quenches f_i, f_j, and f_k(k# i,j) to interacting multichannel WDM system to produce a fourth wave of frequency:

FWM is also obtained in other passive waveguides such as semiconductor waveguides and in an active medium such as SOA. FWM is used [65] for wavelength conversion in optical networks owing to its ultra-fast response and high transparency to bitrate and modulation format. In this technique a 3-dB conversion range over 40nm (1535–1575 nm) is obtained with a flat conversion efficiency of −16dB and a polarization sensitivity of less than 0.3 dB.

Difference Frequency Generation (DFG): DFG is a consequence of a second-order nonlinear interaction of a medium with two optical waves: a pump wave and a signal wave [64]. DFG is free from satellite signals that appear in FWM-based techniques. This technique offers a full range of transparency without adding excess noise to the signal. It is also bidirectional and fast, but it suffers from low efficiency and high polarization sensitivity. The main difficulties in implementing this technique lie in the phase matching of interacting waves and in fabricating a low-loss waveguide for high convers ion efficiency.

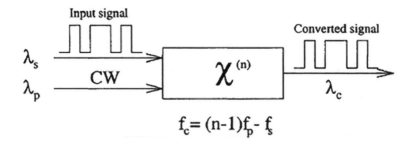

FIGURE 3.36 A wavelength converter based on nonlinear wave mixing effects.

3.9.3 Wavelength Conversion Using Cross Modulation

Figure 3.37 shows a cross-gain modulation (CGM)-based wavelength conversion, which is based on active semiconductor optical devices such as SOAs and lasers [70]. These techniques are known as optical-gating wavelength conversion. The principle of an SOA in CGM is based on intensity-modulated input signal modulating the gain in the SOA. A continuous-wave (CW) signal at the desired output wavelength is modulated by the gain variation so that it carries the same information as the original input signal. The CW signal is either incident on SOA in the same direction as the input signal (co-directional) or incident on SOA in the opposite direction as the input signal (counter-directional). The CGM scheme contributes a wavelength-converted signal inverted with respect to the input signal. The CGM scheme contributes penalty-free conversion at 10 Gbps. Its disadvantages are extinction ratio degradation for the converted signal due to inversion of the converted bit stream.

The wavelength conversion is also made by SOA in another way where the operation of a wavelength converter using SOA in cross-phase modulation (CPM) mode based on the refractive index of the SOA dependent on the carrier density in its active region. An incoming signal producing the carrier density modulates the refractive index, contributing phase modulation of a CW signal coupled into the converter. The SOA is used into an MZI so that an intensity-modulated signal format is obtained at the output of the converter. Figure 3.38 illustrates an asymmetric MZI wavelength converter based on SOA in CPM mode. With the CPM scheme, the converted output

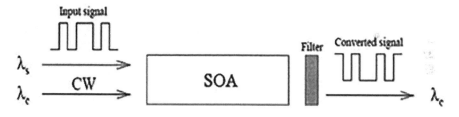

FIGURE 3.37 A wavelength converter using co-propagation based on SOA.

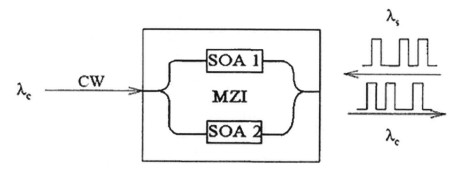

FIGURE 3.38 An MZI-based SOA wavelength conversion.

signal is either inverted or non-inverted, unlike in the CGM scheme where the output is always inverted. The CPM scheme is also more power efficient compared with the CGM scheme. A signal up-conversion utilizing a CPM effect in an all-optical SOA-MZI wavelength converter is demonstrated in which this scheme not only shows high conversion efficiency, polarization immunity and no increase in phase noise, but also linear signal up-conversion with a low optical power requirement.

3.9.3.1 Semiconductor Laser Based Wavelength Conversion

Figure 3.39 shows a wavelength converter based on single-mode semiconductor lasers in which lasing-mode intensity is modulated by input signal light through lasing-mode gain saturation. The obtained output (converted) signal is inverted compared with the input signal. This gain suppression mechanism has been employed in a Distributed Bragg Reflector (DBR) laser to convert signals at 10 Gbps. In the method using saturable absorption in lasers, the input signal saturates the absorption of carrier transitions near the bandgap and allows the probe beam to transmit.

3.9.3.2 All-Optical Wavelength Conversion Based on CPM in Optical Fiber

Wavelength conversion based on CPM effect in fibers and subsequent optical filtering has been focused due to having an all-optical wavelength conversion. The principle of the new technology is based on the generation of a frequency comb through CPM in an optical fiber with subsequent optical filtering of the desired tone. When incoming data propagate through the fiber along with a sinusoidally intensity-modulated high-power optical pump signal, the data acquire a sinusoidal phase modulation from the pump through the CPM, generating multiple wavelengths spaced at the modulation frequency on both sides of the incoming signal wavelength, i.e., generating sidebands around the incoming signal wavelength. By filtering out a portion of the sidebands, converted signal of the desired wavelength at the output is obtained. Compared with other techniques, this approach has several important advantages: First stability, operability in a wide signal wavelength range and short switching

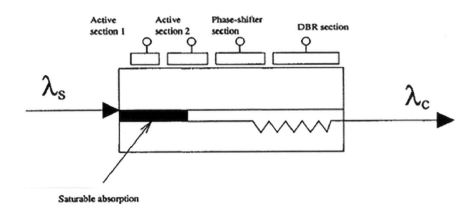

FIGURE 3.39   Semiconductor laser-based wavelength conversion.

window with relatively broad signal pulses. It contributes additional functions such as waveform reshaping and phase reconstruction.

3.10   HIGH-SPEED SILICON PHOTONICS TRANSCEIVER

High-speed optical transceivers are required to deal with the high demand for high data rate due to the skyrocketed increase of users and services day by day. These are generally placed the path forward to overcome these limitations, but no traditional optical technology can provide a low-cost solution. There is an opportunity for high-speed silicon photonics, where optical and electrical circuits are monolithically formed with conventional silicon-based technology.

3.10.1   SILICON PHOTONICS TRANSCEIVER ARCHITECTURE

Figure 3.40 shows an architecture of silicon-based transceivers consisting of three transmitters and three receivers with WDM multiplexer and demultiplexer. There are channels having control thermally to the MZI modulator and multiplexer/ demultiplexer. This is a TPM that denotes the thermal phase modulators to control these elements. It has a communication bus, digital control circuitry and a built-in self-test block for wafer of the transceiver functionality at full data rate by means of an in-house electrooptical probe station. The light sources are only external to the chip architecture having fiber coupled to the chip through grating couplers denoted as "TXn CW-in" (as shown in Figure 3.40).

The elements in the architecture are

- The high-speed MZI modulator operates with the principle of wave propagation and is a distributed-driver design based on carrier depletion in a silicon reverse-biased diode permitting high-speed, low-power operation. The MZIs consists of monitoring photo detector and low-speed phase modulators for closed-loop control at quadrature.
- Low-loss waveguides transmit signals around the chip.
- High-efficiency grating couplers put light in and out of the CMOS circuit at close to normal incidence.
- The WDM multiplexer and demultiplexer are used for three channel designs based on an interleaver architecture. The designed channel spacing is 200 GHz, with adjacent channel crosstalk of < -17 dB. The WDM elements are made tunable to remain locked automatically to the input laser wavelengths by a closed-loop control system.
- Ge-PIN waveguide diodes integrated to the system are set for high-speed signal detection as well as providing signals for closed-loop control systems. The extremely low capacitance of the diodes is used in receiver designs with high sensitivity.
- The architecture has all the high-speed analog circuits usually found in a transceiver (limiting amp, laser drivers, modulator driver, etc.) and large digital blocks to control the system and to communicate to the outside world through a digital bus.
- High-speed electrical interfaces are used to have the standard.

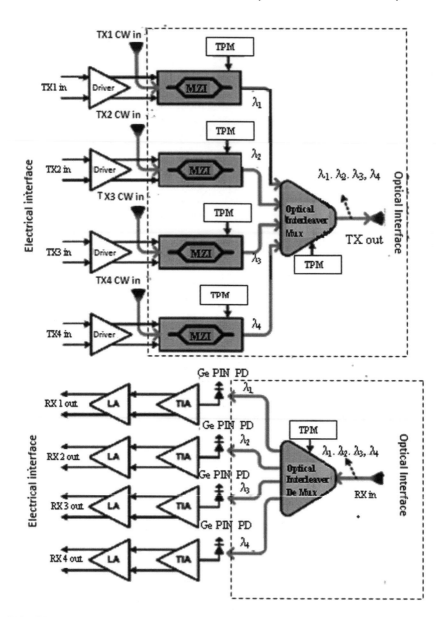

FIGURE 3.40   An architecture of silicon-based transceivers.

3.10.2 Performance

The transmitter and the receiver part of the system are important to include predict link budget margin by using a full bidirectional link with pairs of packaged transceiver devices. The tests performed demonstrated BER< 10^{-12} at 10 Gbps per channel for a $2^{31}-1$ pseudo-random bit sequence over extended testing of multiple days without any temperature control, only closed-loop control systems on chip.

SUMMARY

Optical network nodes consist of WDM/WDDM, optical switch, ADM/demultiplex-ers, traffic grooming devices, etc. In this chapter, we have discussed that all these devices used optical network nodes. Since these devices are made by using inte-grated optics for improved reliability, immunity to vibration and electromagnetic interference, low loss transmission, small size, light weight, large bandwidth (multi-plexing capability), low power consumption and batch fabrication economy, we have discussed basic components such as DC, TMI coupler, MMI coupler, MZ devices, and AWG used in the above integrated optical devices. We have also discussed opti-cal amplifier used for amplification signal in wide-area optical backbone. Since opti-cal amplifier makes non-uniform amplifications over all wavelength channels used in WDM optical network, we have also mentioned tunable gain equalizer for uni-form optical power of wavelength channels. We have discussed the different types of wavelength converter used in an optical network.

EXERCISES

3.1. Suppose we want to design a system with 16 channels, each channel with a rate of 1 Gbps. How much bandwidth is required for the system?

3.2. Suppose we have a fiber medium with a bandwidth of about 20 nm. The center wavelength is 0.82 µm. How many 10 GHz channels can be accom-modated by the fiber? Calculate the maximum number of channels for a center wavelength of 1.5 µm.

3.3. Consider an optical communication system in which the transmitter tuning range is from 1450 to 1600 nm, and the receiver tuning range is from 1500 to 1650 nm. How many 1 Gbps channels can be supported in the system?

3.4. Consider a WDM passive-star-based network for N nodes. Let the tuning range of the transmitters be 1550–1560 nm, and let the tuning range of the receivers be 1555–1570 nm. Assume that the desired bit rate per channel is 1 Gbps. Also assume that a channel spacing of at least 10 times the chan-nel bit rate is needed to minimize crosstalk on a WDM system. Find the maximum number of resolvable channels for this system.

3.5. Find the converted wavelength of a wavelength converter based on FWM in which the input wavelength and pumping wavelengths are 1.52 and 1.53 µm.

3.6. Four-stage point symmetric cascaded delay line filter with uniform distri-bution acts as an EDFA gain equalizer. Find the total length of the filter with delay line path difference $\Delta L = 10$ µm and height of 50 µm if the input and output waveguide length, coupling length of each coupler and central wavelength are 200, 20 and 1.55 µm, respectively.

3.7. Find population inversion of an EDFA gain equalizer when the popula-tion of Er^{++} ions in metastable and lower energy levels are 3×10^{10} and 1.6×10^{10}.

3.8. A 5-stage Y symmetric cascaded delay line filter with uniform distribu-tion acts as ADM. Find the total length of the filter with delay line path difference $\Delta L = 5$ µm and height of 30 µm if the input and output wave-guide length, coupling length of each coupler and the dropping wavelength are 500, 12 and 1.55 µm, respectively.

3.9. An optical amplifier delivers an output power in response to input power P_{in} described by

$$P_{out} = A \left[1 - \exp\left(-bP_{in}\right) \right]$$

where A and b are constants.

  i.   What is the saturation power of the amplifier?
  ii. Find the power gain of the amplifier for small input power.

3.10. How many channels can be accommodated/groomed in STS-192 SONET if each channel requires a data speed of 1.25 Gbps.

3.11. Find STS SONET if an optical link requires to have four channels of 625 Mbps and two channels of 1250 Mbps channels.

3.12. Find the minimum coupling length required for an MMI coupler to get cross and bar states if the propagation constants for fundamental mode, first-order mode and second-order mode are 6.542 (µm)$^{-1}$, 6.531 (µm)$^{-1}$ and 6.231 (µm)$^{-1}$ respectively.

3.13. Find the minimum coupling length required for a TMI coupler to get cross and bar states if the propagation constants for the fundamental and first-order modes are 6.572 (µm)$^{-1}$ and 6.541 (µm)$^{-1}$, respectively.

3.14. Find the length of wavelength multiplexer-based TMI coupler to multi-plex wavelength channels $\lambda_1 = 1.53$ µm and $\lambda_2 = 1.55$ µm, respectively. The propagation constants $\beta_{00}^{\lambda_1}$, $\beta_{01}^{\lambda_1}$, and $\beta_{01}^{\lambda_2}$ are $\sim 6.0556, 5.9807, 5.974$ and 5.9023 (µm)$^{-1}$, respectively and the input and output waveguide lengths are 200 and 210 µm, respectively.

3.15. Find the length of TOMZ switch having a heater length of 3 mm, and a coupler with beat length of 100 µm and transition length of 250 µm.

3.16. If each channel has a bit rate of B Gbps and 2B GHz bandwidth requiring for encoding and modulation efficiency of 2 Hz/bps and a channel spac-ing requiring six times channel rate, prove that the maximum number of resolvable channels of the network is given by

$$W = \frac{\Delta f + 6B}{8B}$$

3.17. If WDM has a central wavelength λ and tuning range of receiver/transmitter is $\Delta\lambda$, find the frequency range.

REFERENCES

1. S. Sohma, "Low switching power silica based super high delta thermo optic switch with heat insulating grooves," *IEE Electronics Letters*, vol. 38, no. 3, pp. 127–128, 2002.

2. R. Kasahara, et al., "New Structures of silica-based planar light wave circuits for low power thermooptic switch and its application to 8×8 optical matrix switch," *Journal of Lightwave Technology*, vol. 20, no. 6, pp. 993–1000, June 2002.

3. H. S. Hinton, "Photonic switching fabrics," *IEEE Communications Magazine*, vol. 28, pp. 71–89, April 1990.

4. P. P. Sahu, "Photonic switching using KDP based mechanically controlled directional coupler and its fiber optic networks," *Proceedings of ICOT-2004*, 2004, pp. 568–575.

5. A. Neyer, "Integrated optical multichannel wavelength multiplexer for monomode system," *IEE Electronics Letters*, vol. 20, no. 18, pp. 744–746, 1984.

6. P. P. Sahu, "Compact optical multiplexer using silicon nano-waveguide," *IEEE Journal of Selected Topics in Quantum Electronics*, vol. 15, no. 5, pp. 1537–1541, 2009.

7. B. J. Offrein, G. L. Bona, F. Horst, W. M. Salemink, R. Beyeler, and R. Germann, "Wavelength tunable optical add after drop filter with flat pass band for WDM networks," *IEEE Photonics Technology Letters*, vol. 11, no. 2, pp. 239–241, 1999.

8. P. P. Sahu and A. K. Das, "Reduction of crosstalk and loss of compact distributed cascaded Mach-Zehnder filter using lateral offset," *proc. of Comnam*, Kolkata, pp. 43–48, 2000.

9. P. P. Sahu, "Tunable optical add/drop multiplexers using cascaded Mach Zehnder coupler," *Fiber and Integrated Optics* (Taylor and Francis), vol. 27, no. 1, p. 24, 2008.

10. P. P. Sahu, "Polarization insensitive thermally tunable Add/Drop multiplexer using cascaded Mach Zehnder coupler," *Applied Physics: Lasers and Optics* (Springer), vol. B92, pp. 247–252, 2008.

11. H. Nishihara, M. Haruna, and T. Suhara, *Optical Integrated Circuits*, McGraw-Hill Book Company, New York, 1985.

12. M. Rajaranjan, B. M. A. Rajaranjan, and K. T. V. Grattan, "A rigorous comparison of the performance of directional couplers with multimode interference devices," *IEEE Journal of Lightwave Technology*, vol. 17, no. 2, pp. 243–248, 1999.

13. M. K. Chin, C. W. Lee, S. Y. Lee, and S. Darmawan, "High index contrast waveguides and devices," *Applied Optics*, vol. 44, no. 15, pp. 3077–3086, 2005.

14. P. P. Sahu and A. K. Das, "Compact integrated optical devices using high index contrast waveguides," *IEEE Wireless and Optical Communication*, IEEExplore Digital No-01666673, pp. 1–5, 2006.

15. S. Darmawan, S.-Y. Lee, C.-W. Lee, and M.-K. Chin, "A rigorous comparative analysis of directional couplers and multimode interferometers based on ridge waveguides," *IEEE Journal of Selected Topics of Quantum Electronics*, vol. 11, no. 2, pp. 466–475, 2005.

16. B. Deka and P. P. Sahu, "Transformation relationship between directional coupler, two Mode interference coupler and multimode interference coupler," *Journal of Optics*, vol. 38, no. 2, pp. 75–87, 2009.

17. M. R. Paiam and R. I. MacDonald, "Polarization insensitive 980/1550 nm wavelength de multiplexer using MMI couplers," *IEE Electronics Letters*, vol. 33, no. 14, pp. 1219–1220, 1997.

18. Y. Ma, S. Park, L. Wang and S. T. Ho, "Ultracompact multimode interference 3-dB coupler with strong lateral confinement by deep dry etching," *IEEE Photonics Letter*, vol. 12, no. 5, pp. 492–494, 2000.

19. M. Yagi, S. Nagai, H. Inayoshi, and K. Utaka, "Versatile multimodes interference photonic switches with partial index modulation regions," *IEEE Electronics Letters*, vol. 36, no. 6, pp. 533–534, 2000.

20. L. H. Spiekman, Y. S. Oei, E. G. Metaal, F. H. Groen, I. Moerman, and M. K. Smit, "Extremely small multimode interference coupler and utrashort bends on InP by deep etching," *IEEE Photonics Technology Letter*, vol. 6, no. 9, pp. 1008–1010, 1994.

21. T.-Y. Tsai, Z.-C. Lee, and M.-H. Cha, "A novel ultrashort two mode interference wavelength division multiplexer for 1.5 µm operation," *IEEE Journal of Quantum Electronics*, vol. 41, no. 5, pp. 741–746, 2005.

22. C. F. Janz, M. R. Paiam, B. P. Keyworth, and J. N. Broughton, "Bent waveguide couplers for demultiplexing of arbitrary broadly separated wavelengths using two mode interference," *IEEE Photonics Technology Letters*, vol. 7, no. 9, pp. 1037–1039, 1995.

23. P. P. Sahu, "Compact optical multiplexer using silicon nano-waveguide," *IEEE Journal of Selected Topics in Quantum Electronics*, vol. 15, no. 5, pp. 1537–1541, 2009.

24. D. Mercuse, "Directional coupler made of nonidentical asymmetric slabs, part-I: Synchronous coupler," *IEEE Journal of Light Wave Technology*, vol. 5, no. 1, pp. 113–118, 1987.

25. M. J. F. Digonnet and H. J. Shaw, "Analysis of tunable single mode optical fiber coupler," *IEEE Journal of Quantum Electronics*, vol-QE. 18, no. 4, pp. 746–754, 1982.

26. R. Utrich and G. Ankele, "Self imaging in homogeneous planar optical waveguides," *Applied Physics Letters*, vol.27, no. 6, pp. 337–339, 1975.

27. D. C. Chang and E. F. Kuester, "A hybrid method of paraxial beam propagation in multimode optical waveguides," *IEEE, Transactions on Microwave Theory and Techniques*, vol-MTT. 29, no. 9, pp. 923–933, 1981.

28. L. B. Soldano and E. C. M. Pennings, "Optical multimode interference devices on self imaging: Principle and applications," *IEEE Journal of Lightwave Technology*, vol. 13, no. 4, pp. 615–627, 1995.

29. P. P. Sahu, "Compact multimode interference coupler with tapered waveguide geometry," *Optics Communications*, vol. 277, no. 2, pp. 295–301, 2007.

30. P. P. Sahu, "A tapered structure for compact multimode interference coupler," *IEEE Photonics Technology Letters*, vol. 20, no. 8, pp. 638–640, 2008.

31. A. Neyer, "Integrated optical multichannel wavelength multiplexer for monomode system," *IEE Electronics Letters*, vol. 20, no. 18, pp. 744–746, 1984.

32. K. Okamoto, K. Moriwaki, and S. Suzuki, "Fabrication of 64 × 64 arrayed- waveguide grating multiplexer on silicon," *Electronic Letters*, vol. 31, pp. 184–186, February 1995.

33. K. Okamoto and A. Sugita, "Flat spectral response arrayed waveguide grating multiplexer with parabolic waveguide horns," *Electronic Letters*, vol. 32, pp. 1661–1662, August 1996.

34. R. Kasahara et al. "Low power consumption Silica based 2 × 2 thermooptic switch using trenched silicon substrate," *IEEE Photonics Technology Letters*, vol. 11, no. 9, pp. 1132–1134, September 1999.

35. Q. Lai, W. Hunziker, and H. Melchior, "Low power compact 2 × 2 thermooptic silica on silicon waveguide switch with fast response time," *IEEE, Photonics Technology Letters*, vol. 10, no. 5, pp. 681–683, 1998.

36. P. P. Sahu and A. K. Das, "Polarization-insensitive thermo-optic Mach Zehnder device based on silicon oxinitride waveguide with fast response time," *Fiber and Integrated Optics* (Taylor & Francis), vol. 29, no. 1, pp. 10–20, 2010.

37. W. J. Tomlinson, "Technologies for dynamic gain and channel power equalization," *Proceedings, OFC'03*, Atlanta, GA, p. 244, March 2003.

38. W. Stalling, *Data and Computer Communication*, PHI, 2003.

39. P. P. Sahu, "Thermooptic two mode interference photonic switch," *Fiber and Integrated Optics* (Taylor and Francis), vol. 29, pp. 284–293, 2010.

40. H. P. Chan, C. K. Chow, and A. K. Das, "A wide angle X-junction polymeric thermooptic digital switch with low crosstalk," *IEEE Photonics Technology Letter*, vol. 15, no. 9, pp. 1210–1212, 2003.

41. M. Yagi, S. Nagai, H. Inayoshi, and K. Utaka, "Versatile multimodes interference photonic switches with partial index modulation regions," *IEE Electronics Letters*, vol. 36, no. 6, pp. 533–534, 2000.

42. Z. Wang, Z. Fan, J. Xia, S. Chen, and J. Yu, "Rearrangable nonblocking thermoop-
    tic 4 × 4 switching matrix in silicon on insulator," *IEE Proceedings Optoelectronics*,
    vol. 152, no. 3, pp. 160–162, 2005.

43. A. Tzanakaki, I. Zacharopoulos, and I. Tomkos, "Broadband building blocks (optical
    networks)," *IEEE Circuits and Devices Magazine*, vol. 20, no. 2, pp. 32–37, March/
    April 2004.

44. A. K. Ghatak and K. Thyagarajan, *Optical Electronics*, Cambridge University Press,
    1993.

45. R. C. Alferness, "Titanium-diffused lithium niobate waveguide devices," in *Guided-
    Wave Optoelectronics* (T. Tamir, ed.), Springer-Verlag, New York, 1988.

46. L. Y. Lin and E. L. Goldstein, "Free-space micromachined optical switches with Sub-
    millisecond switching time for largescale optical crossconnects," *IEEE Photonics
    Technology Letters*, vol. 10, pp. 525–527, 1998.

47. C. Clos, "A study of non-blocking switching networks," *The Bell System Technical
    Journal*, vol. 32, no. 2, pp. 406–424, 1953.

48. L. Y. Lin and E. L. Goldstein, "Opportunities and challenges for MEMS in lightwave
    communications," *IEEE Journal on Selected Topics in Quantum Electronics*, vol. 8,
    pp. 163–172. 2002.

49. H. Zhu and B. Mukherjee, "Online connection provisioning in metro optical WDM
    networks using reconfigurable OADMs (ROADMs)," *IEEE/OSA Journal of Lightwave
    Technology*, vol. 23, no. 10, pp. 2893–2901, December 2005.

50. X. Zhang and C. Qiao, "An effective and comprehensive approach for traffic groom-
    ing and wavelength assignment in SONET/WDM rings," *IEEE/ACM Transactions on
    Networking*, vol. 8, no. 5, pp. 608–617, October 2000.

51. A. C. Baishya, S. K. Srivastav, and P. P. Sahu, "Cascaded Mach Zehnder coupler for
    dynamic EDFA gain equalization applications," *Journal of Optics* (Springer), vol. 39,
    no. 1, pp. 42–47, 2010.

52. G. Keiser, *Optical Fiber Communications*, McGraw-Hill Inc., New York, 1999.

53. S. Stanley, "Next-gen SONET silicon," Lightreading Report, June 2002.

54. S. Stanley, "Making SONET ethernet-friendly," Lightreading Report, March 2003.

55. F. B. Shepherd and A. Vetta, "Lighting fibers in a dark network," *IEEE Journal on
    Selected Areas in Communications*, vol. 22, no. 9, pp. 1583–1588, November 2004.

56. B. Mukherjee, *Optical WDM Networks*, Springer-Verlag, 2006.

57. V. V. Lysak, H. Kawaguchi, I. A. Sukhoivanov, T. Katayama, and A. V. Shulika,
    "Ultrafast gain dynamics in asymmetrical multiple quantum-well semiconductor opti-
    cal amplifiers," *IEEE Journal of Quantum Electronics*, vol. 41, no. 6, pp. 797–807, June
    2005.

58. M. Tachibana, R. I. Laming, P. R. Morkel, and D. N. Payne, "Erbium-doped fiber
    amplifier with flattened gain spectrum," *IEEE Photonics Technology Letters*, vol. 3,
    pp. 118–120, February 1991.

59. X. Zhou, C. Lu, P. Shum, and T. H. Cheng, "A simplified rnodel and optimal design
    of a multiwavelength backward-pumped fiber Raman amplifier," *IEEE Photonics
    Technology Letters*, vol. 13, no. 9, pp. 945–947, September 2001.

60. N. I. Mohammed, "Raman amplifiers for telecommunications," *IEEE Journal of
    Selected Topics in Quantum Electronics*, vol. 8, no. 3, pp. 548–559, May/June 2002.

61. M. Raisi, S. Ahderom, K. Alarneh, and K. Eshraghian, "Dynamic Micro-photonic
    WDM equalizer," *Proceedings, 2nd IEEE International Workshop on Electronic
    Design, Test and Applications*, Perth, Australia, pp. 59–62, January 2004.

62. P. P. Sahu, "Polarization independent thermally tunable EDFA gain equalizer using
    cascade Mach-Zehnder coupler," *Applied Optics*, vol. 47, no. 5, pp. 718–724, 2008.

63. P. P. Sahu, "Thermally tunable EDFA gain equalizer using point symmetric cascaded
    Mach-Zehnder Filter," *Optics Communications*, vol. 281, no. 4, pp. 573–579, 2008.

64. S. J. B. Yoo, "Wavelength conversion technologies for WDM network applications," *IEEE/OSA Journal of Lightwave Technology*, vol. 14, pp. 955–966, June 1996.

65. F. Forghieri, R. W. Tkach, A. R. Chraplyvy, and D. Marcuse, "Reduction of four-wave mixing crosstalk in WDM systems using unequally spaced channels," *IEEE Photonics Technology Letters*, vol. 6, no. 6, pp. 754–756, 1994.

66. B. Rarnamurthy and B. Mukherjee, "Wavelength conversion in optical networks: Progress and challenges," *IEEE Journal on Selected Areas in Communications*, vol. 16, no. 7, pp. 1040–1050, September 1998.

67. R. Sabella and E. Iannone, "Wavelength conversion in optical transport networks," *Journal of Fiber and Integrated Optics*, vol. 15, no. 3, pp. 167–191, 1996.

68. S. Subramaniam, M. Azizoglu, and A. K. Somani, "All-optical networks with sparse wavelength conversion," *IEEE/ACM Transactions on Networking*, vol. 4, pp. 544–557, August 1996.

69. J. M. Wiesenfeld, "Wavelength conversion techniques, *"Proceedings, Optical Fiber Communication* (OFC '96), San Jose, CA, vol. Tutorial TuP 1, pp. 71–72, 1996.

70. M. D. Vaughn and D. J. Blumenthal, "All-optical updating of subcarrier encoded packet header with simultaneous wavelength conversion of baseband payload in semiconductor optical amplifiers," *IEEE Photonics Technology Letters*, vol. 9, no. 6, pp. 827–829, June 1997.

4 Processing of Integrated Waveguide Devices for Optical Network Using Different Technologies

Since the origin of integrated optics' concept in 1970s, extensive studies have been performed on a variety of optical materials [1–64] such as silicon-based materials [1–10,16–18,21–23,27–36,39–43,49,52,58–63], lithium niobite (LiNbO$_3$) [1,11,19,20,37,38,50], polymers [44–48,51,53–57] and III–V semiconductors [1,12,14,15] for processing of integrated optical waveguide devices. These materials considered in integrated optics so far are identified as either low index contrast (e.g., GeO$_2$-SiO$_2$ [27–36], LiNbO$_3$ [19,20]) or high index contrast material (e.g., GaAsInP [14,15], silicon oxynitride (SiON)) [2–10].

In this chapter, we have mentioned fabrication process and characteristics for integrated optical waveguide devices by using SiO$_2$/SiON, SiO$_2$/GeO$_2$-SiO$_2$ and silicon on insulator (SOI) material. However, the same using other materials such as LiNbO$_3$, InP/GaAsInP and polymer is also described for comparison with the same material. It is seen that SiO$_2$/SiON material has many advantages such as wide index contrast of their waveguides, stability of the device, low losses, polarization insensitiveness, compatibility to the based conventional IC processing technology, etc.

4.1 FABRICATION AND CHARACTERISTICS OF SILICA (SiO$_2$)/SILICON OXYNITRIDE (SiON)-BASED DEVICES

In the last few years, growing attention has taken to silicon oxynitride (SiO$_x$N$_y$ or SiON in short) as a potential material for integrated optics [3,4] due to its excellent optical properties such as low absorption losses in the visible and near infrared region, good chemical inertness and low permeability. In addition, the index of refraction of SiON layers can be varied continuously over a wide range between 1.45 (SiO$_2$) and 2.0 (Si$_3$N$_4$), which also has an attractive property that allows fabrication of waveguides with desired characteristics of fiber match and compactness [3–5]. Moreover, the growth of SiON layers on the silicon substrate with refractive index control is compatible to silicon integrated circuit processing technology.

Amorphous silicon oxynitride layers are deposited using a variety of techniques. The most used growth techniques are low-pressure chemical vapor deposition (LPCVD) and plasma-enhanced chemical vapor deposition (PECVD). Nowadays, PECVD is normally used to deposit silicon- and germanium-based material due to the following reasons. First, in the LPCVD processes, the temperature of the operation is 700°C–1000°C [41], which limits its applicability in the structures where the temperature-sensitive components are involved, and in the case of PECVD method, the temperature of operation is lower (150°C–400°C) [59]. Second, the deposited films of PECVD method have good uniform refractive index and thickness over a large area in comparison with that of the LPCVD method.

4.1.1 Deposition of Thin Film SiON Layer by Using LPCVD

Figure 4.1a shows the schematic diagram of an LPCVD reactor consisting of a horizontal tube with heaters (called as hot wall reactors). The substrate is placed on holder in chamber and heated by a heater in a controlled manner. The substrate temperature is kept in between 700°C and 1000°C [41]. In this case, not only the wafers/substrates but also the reaction chamber walls get coated during processing. So, it requires frequent cleaning to avoid contamination during deposition. It has a large diffusion coefficient at low pressures (~100 Pa), leading to a growth limited by the rate of surface reactions rather than by the rate of mass transfer to the substrate. The surface reaction rate is very sensitive to temperatures, but the temperature is relatively easy to control. This method gives advantage of front and backside deposition. They find wide applications due to their economy, throughput and uniformity. Two main disadvantages are low deposition rate and relatively high operating temperatures.

De Ridder et al. [41] have demonstrated LPCVD deposition of SiON layer from SiH_2Cl_2, O_2 and NH_3 gases. The processed conditions are pressure = 100 mTorr and temperature = 900°C. The production of SiON film by LPCVD method with these specific parameters is mostly performed for the fabrication of waveguides in a higher refractive index range ($n \geq 1.7$). The deposition rate is lower than that of other methods such as PECVD method. The optical loss in the visible range is below 0.2 dB/cm. At wavelength 1550 nm the loss is higher due to hydrogen content of the film. The optical loss is reduced to < 0.2 dB after post-deposition annealing, at temperature 1150°C, which removes the N–H bonds. The basic reaction is given below:

$$SiH_2Cl_2 + O_2 + NH_3 \rightarrow SiO_xN_yH_z(\text{solid}) + H_2O\,(\text{gas}) + HCl\,(\text{gas})$$

$$\rightarrow SiO_xN_y(\text{solid}) + H_2O\,(\text{gas}) + HCl\,(\text{gas})$$

$$\left(\text{after annealing at temperature}\,1150°C\right)$$

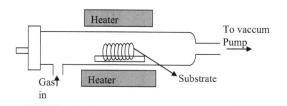

FIGURE 4.1   (a) An LPCVD machine.

(*Continued*)

where $x = \dfrac{X}{X+Y}$ = concentration of oxygen, $y = \dfrac{X}{X+Y}$ = concentration of nitrogen, X = flow rate of O_2 and Y = flow rate of NH_3. The table shows the decrease of refractive index with increase of flow rate of O_2. As flow rate of O_2 increases, x increases but y decreases. It means that in SiON compound, nitrogen content decreases and oxygen content increases. So, the refractive index decreases due to increase of flow rate of O_2 (as x increases).

4.1.2 DEPOSITION OF SiO₂/SiON LAYER BY USING PECVD

As stated earlier, PECVD system is used for the deposition of SiO₂/SiON layer due to lower temperature of operation and good uniformity of refractive index and thickness. The PECVD system used in processing of SiON layer is a parallel plate-type reactor, as shown in Figure 4.1b. The glow discharge (plasma) generated by radio frequency occurs between two electrodes having a gap of 2 cm. The plates have diameter 24 cm d, and RF power is required to the upper electrode establishing the plasma while the samples are kept on the bottom grounded electrode which is temperature stabilized in the range from 200°C to 350°C. The system can be worked at a pressure range of 0.01–10 Torr and the applied RF power (13.56 MHz) can have values up to 300 W.

The films in the device structure are grown by making use of an intermediate state, glow discharge which contains mainly ionized gas. The ionized gas has positive ions, negative ions and electrons. The electron energy takes values between 1 and 20 eV and density vary from 10^9 to $10^{12}/cm^3$. The first step in the process of deposition is generation of reactant species via impact reactions with plasma electrons which are created by glow discharge. As an example, electron impact reaction for SiH₄ is given below:

$$e^- + SiH_4 = SiH_2 + H_2 + e^{-1}$$

$$= SiH_3 + H_2 + e^{-1}$$

$$= SiH + 2H_2 + e^{-1}$$

This process is observed during the deposition as a glow emitted from the plasmas.

The process of deposition of solid films on a substrate using gaseous precursors involves several complicated steps. There are two extreme cases of film deposition

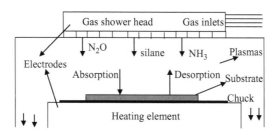

FIGURE 4.1 (CONTINUED) (b) A PECVD parallel plate plasma reactor with a capacitive coupling of RF power.

conditions. The first case is observed at the conditions of low pressure and high RF power. In this case, the number of reactive species arriving at the surface is less than the reaction rate. This is called as mass transfer limited reaction. In the second case, the number of reactive species arriving at the surface is more than the reaction rate. This is called as reaction limited process, which is observed at the low RF power and low temperature. For efficient growth of the desired film, the number of reactive species arriving at the surface should be equal to the reaction rate. For that, a careful optimization of the reaction chamber has to be conducted.

4.1.2.1   Silicon Dioxide (SiO₂)

The chemical reaction to deposit the silica using SiH_4 and N_2O is written as

$$SiH_4\,(gas) + 4N_2O = SiO_2\,(solid) + 2H_2O(gas) + 4N_2\,(gas) \qquad (4.1a)$$

The reaction of SiH_4 with N_2O is based on the oxidation of silane by molecular oxygen produced by dissociation of N_2O. The external RF power source accelerates the electrons in the reaction chamber as shown in Figure 4.1b. The primary initial electron-impact reactions between electron and reactant gases form ions and radical reactive species, i.e., molecular oxygen. This molecular oxygen reacted with silane gas gives SiO_2. Thus actually oxygen from N_2O produces SiO_2 for which refractive index becomes lower than that of SiON, which is formed by N_2O and NH_3.

After determination of the minimum sample size, the next concern is to grow silicon oxide films and analyze their thickness and refractive index properties. For this purpose, various films were grown at 250°C and 350°C with the above process parameters. The flow rate of N_2O was varied between 25 and 300 sccm. The index of the films is changed between 1.56 and 1.46. The stoichiometry of the grown layers has a deviation from SiO_2 due to having certain amount of hydrogen and nitrogen present in the film, indicating an index larger than 1.45. The layers are deposited for 30-minute process the value of the refractive index reaches saturation after a flow rate of N_2O of about 100 sccm.

In addition, 100 sccm flow value is the point where the deposition rate saturates too, after a steady increase. The observed behavior can be explained by the amount of nitrogen integrated into the film. After this point, silicon oxide formation begins to dominate over silicon nitride formation, since it is now available in excess amount and has a greater affinity for reacting with silane gas. The above characterizations were performed in 250°C and 350°C in order to compare the film qualities. The growth rate decreases as the process temperature is increased. The films grown at higher temperatures are denser and have less micro-voids, which were also verified by etch rates of these layers. Another note is about the relation of the N_2O flow to the film density. As it is increased, the reaction rate (thus the growth rate) increases, which results again in a lower density of layers.

4.1.2.2   Silicon Nitride

Silicon nitride, other than oxide, is a material, properties of which would be of interest as we expect it to be the other extreme for silicon oxynitride layers to be grown. The stochiometric ratio of silicon nitride is known to be 3/4 (Si_3N_4); however, again PECVD silicon nitride films deviate from this because of the incorporation of hydrogen into the layers. The basic reaction mechanism of formation of silicon nitride (Si_3N_4) is given by

$$SiH_4\,(gas) + 4NH_3 = Si_3N_4\,(solid) + 7H_2\,(gas) \qquad (4.1b)$$

The NH_3 reacted with silane and produces Si_3N_4 which has a higher refractive index than SiO_2.

4.1.2.3 SiON Layer

The optical and physical properties of SiON are expected to be between that of SiO_2 and Si_3N_4. Therefore, monitoring of the compositional characteristics of the mentioned layers should result in the better control of SiON film properties. The reduction of the film refractive index with increasing oxygen was observed because of oxygen's greater chemical reactivity compared with nitrogen.

From chemical reactions (4.1a) and (4.1b), both N_2O and NH_3 are involved in controlling refractive index. For the deposition of SiON film, both N_2O and NH_3 are needed. The chemical reaction of the deposition of SiON is written as

$$SiH_4 + N_2O + NH_3 \rightarrow SiO_xN_yH_z\,(solid) + H_2O(gas) + N_2\,(gas)$$

$$\rightarrow SiO_xN_y\,(solid) + H_2O(gas) + HCl$$

$$\big(gas\big)\big(after\,annealing\big) \qquad (4.1c)$$

Donald E. Bossi et al. [5] reported the wavelength dispersion curve of refractive index of SiO_2, SiON and Si_3N_4 for a wavelength range of 0.4–1.15 µm in Figure 4.2a.

In the figure, the uppermost curve indicates the wavelength dispersion curve of Si_3N_4 refractive index, whereas the lowermost curve represents the wavelength dispersion curve of SiO_2 refractive index. The figure also shows the refractive index of SiON as an increasing function of mole fraction of Si_3N_4, whereas refractive index

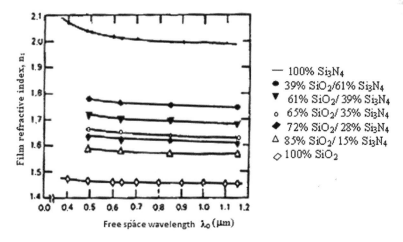

FIGURE 4.2 (a) Variation of film refractive index with free-space wavelength [5].

(Continued)

of SiON (at $\lambda_0 = 0.846\,\mu m$) decreases with mole fraction of the SiO_2 (X_{SiO_2}) related to mole fraction of Si_3N_4 ($X_{Si_3N_4} = 1 - X_{SiO_2}$). The refractive index of SiON shows a linear dependence of mole fraction of SiO_2 and Si_3N_4. The refractive index of SiON increases with an increase of nitrogen content and decreases with an oxygen content of SiON.

The SiON films are obtained by using silane, ammonia and nitrous oxide as reactant gases. The different process parameters reported in [59] for the deposition of SiON layer are shown in Table 4.1. The flow rate of silane (2% SiH_4/N_2: diluted in nitrogen because of being highly unstable at room temperature and tendency to burn in case of exposure to air) is fixed at 180 sccm, and the flow rate of N_2O and NH_3 is varied to get the desired refractive index of SiON. The layer is grown at 350°C with an application of RF power of 10 W. The applied RF frequency is 13.56 MHz.

Figure 4.2b shows the refractive index versus N_2O flow rate at NH_3 flow rates of 15 and 30 sccm for the process parameters shown in Table 4.2a [59]. The highest refractive index can be up to ~2.0 (Si_3N_4). The decrease in the film refractive index with increasing oxygen content was observed in all cases (as seen in Figure 4.2b), due to oxygen's greater chemical reactivity compared with nitrogen [59], whereas the flow rate of ammonia increases, and the film index is increased similarly due to their higher nitrogen content.

The film thickness values of the samples varied roughly between 4300 and 3000 Å. The increase in N_2O flow rate shows an increase of film growth rate too. The deposition rate is decreasing with an increase of ammonia flow rate as shown in Figure 4.2c. These properties are due to the oxygen's greater affinity for reacting with silane gas. As the reactive oxygen concentration increases, it starts to control the chemical reactions over nitrogen. For the decrease of the growth rate with increase in nitrogen concentration in the film, the nitrogen-related bonding has been increased so that the growth rate (GR) of silicon nitride >GR of silicon oxynitride >GR of silicon oxide.

In Table 4.2a [63], the refractive index and thickness of the SiON layer are mentioned for a processing time of 30 minutes with different flow rates of N_2O at a constant NH_3 flow rate of 15 sccm and SiO_2 flow rate of 180 sccm. The film deposition

---

TABLE 4.1

Process Parameters of SiON Film Used by Previous Author [63]

Parameters	Values
Si-substrate temperature	350°C
RF power @ 13.56MHz	10 W
Pressure	1000 mTorr
N_2O flow rate	20–450 sccm
SiH_4 flow rate	180 sccm
NH_3 flow rate	15–30 sccm

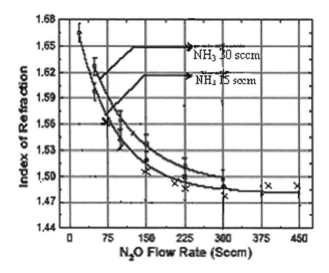

FIGURE 4.2 (CONTINUED) (b) Variation of index contrast of SiON layer with N_2O flow rate at NH_3 flow rate of 15 and 30 sccm [63].

(*Continued*)

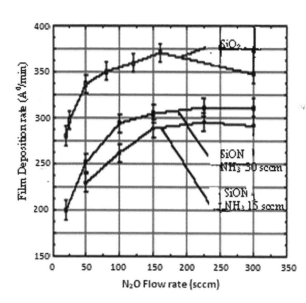

FIGURE 4.2 (CONTINUED) (c) Film deposition rate of SiON versus N_2O flow rate [63].

rate increases with N_2O flow rate and becomes saturated at flow rate ~150 sccm. But the refractive index of SiON layer decreases with an increase of N_2O flow rate.

The PECVD-deposited SiON layer has small amounts of O–H bonds, N–H bonds and Si–H bonds which causes optical absorption at 1.38, 1.48 and 1.51 μm, respectively, as reported by previous authors [59]. A widely used infrared spectroscopic

TABLE 4.2a

Refractive Index and Film Deposition Rate as a Function of Flow Rate of Reactant Gases, Processing Time of 30 minutes, Substrate Temperature ~350°C, RF Power @ 13.56 MHz ~10 W, Pressure ~1000 mTorr

Flow Rate of SiH₄(sccm)	Flow Rate of NH₃(sccm)	Flow Rate of N₂O (sccm)	Refractive Index of SiON Obtained	Film Deposition Rate of SiON Layer (Å/min)
180	15	20	~1.67	~200
		75	~1.5675	~271
		150	~1.515	~305
		225	~1.4875	~315
		300	~1.485	~316
		375	~1.48	-
		450	~1.4775	-

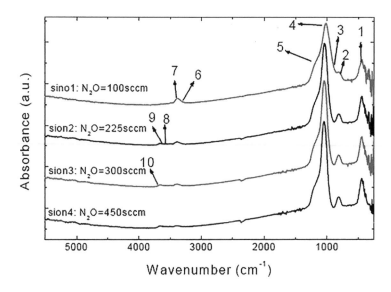

FIGURE 4.3 (a) IR absorbance of silicon oxynitride films grown with 15 sccm NH_3 and N_2O flow rates of 100, 225, 300 and 450 sccm, respectively [59].

(Continued)

technique, Fourier transform infrared (FTIR) spectroscopy is employed for finding the absorption and concentration of these bonds. Figure 4.3a shows an IR absorption of SiON layers of thickness ~4.5 μm with 15 sccm NH_3 flow rate and N_2O flow rates of 100, 225, 300 and 450 sccm, respectively.

The corresponding IR vibrations observed in SION samples [59] are shown in Table 4.2b.

It is reported by previous authors [59] that for SiON layer grown with a constant NH_3 flow rate of 15 sccm and N_2O flow rate varying from 100 to 450 sccm, the

TABLE 4.2b
Vibration Spectra

Vibration Type	Peak Frequency (cm⁻¹)			
	SiON1	SiON2	SiON3	SiON4
(1) Si–O rock	449	445	446	443
(2) Si–O bend	815	817	816	817
(3) Si–N stretch	923	983	-	-
(4) Si–O sym. stretch	1018	1042	1040	1044
(5) Si–O asym. stretch	1154	1144	1130	1167
(6) N–H...N stretch	3341	3345	3351	3358
(7) N–H stretch	3389	3396	3399	3403
(8) H–O–H stretch	3493	3499	3499	3499
(9) SiO–H stretch	3571	3578	3589	3589

concentration of N–H bonds ranged between 1.2×10^{22} and 0.37×10^{22} cm^{-3}. So, it is required to decrease or eliminate these bonds from the film structure. It is reported that these bonds can be eliminated by annealing treatment in an annealing furnace as shown in Figure 4.3b.

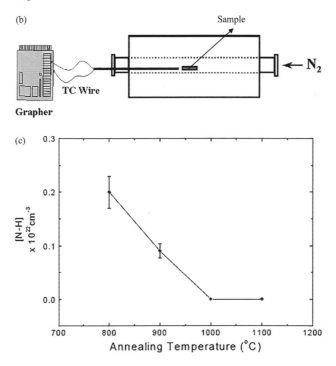

FIGURE 4.3 (CONTINUED) (b) Annealing furnace setup and (c) N–H bond concentration with annealing temperature [59].

(Continued)

Figure 4.3c shows N–H bond concentration with annealing temperature for SiON layer of 4.7 μm and index ~1.48 [59]. In the figure, the concentration of N–H bonds decreases from 0.2×10^{22} cm^{-3} at temperature 800°C to 0.09×10^{22} cm^{-3} at 900°C, and it goes below the detection limit after 1000°C. These bond concentration is measured by FTIR spectroscopy. The ellipsometer confirms that, after annealing at 900°C, there is no measurable change of refractive index, but the thickness of the films decreased by 2%. In our fabrication of devices, we have used the same annealing furnace setup and maintained the same temperature of 1000°C for the furnace.

The LPCVD method is preferred for the deposition of SiON layer of refractive index >1.7, whereas an SiON film of refractive index varying from 1.45 to 1.7 is preferred to develop by PECVD method. Moreover, in the PECVD processes, the temperature of the operation is lower than that in LPCVD processes as stated earlier. The operation at low temperatures in PECVD method is possible with the electrical energy to the environment, resulting in glow discharges.

4.1.3 Tuning of Refractive Index Using Thermooptic Effect

The refractive index can be changed with application of heat on the waveguide. This effect is called as thermooptic effect. SiO$_2$/SiON waveguides show a thermooptic effect. If the heating power P is applied on the waveguide via a thin film heater made on the waveguide, the refractive index of the waveguide increases due to the rise of temperature. The thermooptic phase change $\Delta\phi(P)$ due to the application of heating power P is expressed as [6]

$$\Delta\phi(P) = \left(2\pi/\lambda\right)\frac{dn}{dT}\Delta T_c(P)L_H \qquad (4.1d)$$

L_H = length of heater and $\dfrac{dn}{dT}$ = thermooptic temperature coefficient and $\Delta T_c(P)$ is the temperature rise due to application of heating power P. The thermooptic coefficient for SiO$_2$/SiON is $\sim 1 \times 10^{-5}$/°C. Figure 4.3d shows the variation of ΔT_c with heater length (L_H) for a thermooptic phase change of π obtained by using the equation (4.1d).

As L_H increases, ΔT_c decreases and is almost saturated at $L_H = 5$ mm. The black rectangle shows the experimental results demonstrated by other authors [18,60] using silica on silicon technology. For $\Delta T_c(P) = 15$°C and $\Delta\phi(P) = \pi$ at $\lambda = 1.55$ μm, L_H is calculated as ~5 mm by using the following equation:

$$L_H = \frac{\lambda}{2\Delta T_c(P)\dfrac{dn}{dT}} \qquad (4.1e)$$

4.1.4 Devices Fabricated and Demonstrated by Using SiO$_2$/SiON Material

SiO$_2$/SiON waveguides were fabricated and demonstrated by different authors [3–5], and the index contrast can be varied over a wide range up to 0.53. Using

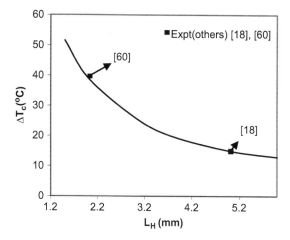

FIGURE 4.3 (CONTINUED) (d) Variation of ΔTc with LH for a phase shift of π for silica-based thermooptic phase shifter at wavelength $\lambda = 1.55\,\mu m$.

these waveguides, many authors have demonstrated different optical devices such as waveguide couplers, thermooptical space switch, wavelength tunable optical add/drop filter [16], adaptive gain equalizer [6], polarization-independent thermooptical phase shifter and polarization-insensitive MMI coupler.

4.1.5 PROPERTIES OF $SiO_2/SiON$

The $SiON/SiO_2$ waveguide material has the following properties which makes it suitable for waveguide-type integrated devices.

1. It shows thermooptic property.
2. Its stability is high.
3. It is a polarization-insensitive material.
4. Its index contrast range is wide.
5. Material cost is moderate.

The main disadvantage of SiON technology is that the propagation losses of the waveguides vary between 1.5 and 10 dB/cm ($\lambda = 1.55\,\mu m$) and are reduced to 0.2–0.36 dB/cm for thermally treated films as mentioned in Section 2.2.3 [5,6]. Still it is higher than that of GeO_2-SiO_2/SiO_2 technology.

4.2 FABRICATION AND CHARACTERISTICS OF SiO_2/GeO_2-SiO_2 WAVEGUIDE MATERIAL

One of the extensively used structures based on silicon technology is SiO_2/GeO_2-SiO_2 waveguides on silicon [7,8]. The advantages are silica has a stable well-controlled refractive index. Since optical fibers are made of silica, the use of index matching oil

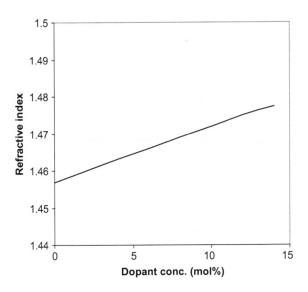

FIGURE 4.4   Variation of refractive index with dopant concentration of GeO_2 in SiO_2 [64].

between waveguides and optical fibers results in a non-reflecting interface that has a large core size. Because of these, it has allowed fiber coupling loss (~0.05 dB/facet) and low propagation losses (~0.1 dB/cm). Since both optical fiber and waveguides are also matched with the same thermal expansion coefficient because of same material, those can be fused for a reduction of fiber coupling loss.

The index variation is obtained by doping SiO_2 with GeO_2. K. J. Bales et al. [9] have reported that the refractive index of SiO_2 increases with an increase of doping concentration of GeO_2 in Figure 4.4. From the figure, it is seen that the index difference between the core and cladding layers varies for a maximum up to ~0.02 if the cladding is SiO_2 and the core is of SiO_2/GeO_2-SiO_2 material.

There are several methods of fabrication for this structure. Among these methods, PECVD and flame hydrolysis methods are normally used for fabrication of better-quality films of GeO_2-SiO_2.

4.2.1   Deposition of SiO_2/GeO_2-SiO_2 Layer Using PECVD

In the previous section, PECVD process for the deposition of SiON has been already described in earlier section. The PECVD process is used for the deposition of GeO_2-doped SiO_2 on Si-substrate. SiH_4 (silane), GeH_4 (Gelane) and N_2O are the main reactants for the deposition of GeO_2-doped SiO_2 layers. Doping is used to control the refractive index. The substrate is kept on the bottom electrode in which the temperature is kept in the range from 200°C to 350°C. The deposition of the GeO_2-doped SiO_2 is fabricated by using GeH_4, SiH_4 and N_2O gas mixture at a pressure range of 300–900 mTorr and with RF power (380 kHz) varying from 200 to 1000 W (applied to the surface). The reaction of SiH_4 with N_2O is based on the oxidation of SiH_4 and

TABLE 4.3

Process Parameters of (SiO$_2$-GeO$_2$) Layer [10]

Parameters	Values
Substrate temperature	200°C–350°C
RF power @ 380 kHz	200–1000 W
Pressure	300–900 mTorr
Deposition rate	1000–6000 Å/min
N$_2$O flow rate	1000–2000 sccm
SiH$_4$ flow rate	10–40 sccm
GeH$_4$ flow rate	1–3 sccm
SiH$_4$: N$_2$O ratio	1:15–1:100
$\Delta n\%$ (index contrast)	0.3%–0.7%

GeH$_4$ by molecular oxygen produced by dissociation of N$_2$O. The external RF power accelerates the electrons in the reaction chamber as shown in Figure 4.1. The molecular oxygen reacted with SiH$_4$ gas and GeH$_4$ gives SiO$_2$ doped with GeO$_2$, which is deposited on the substrate. The reaction for deposition of the silica is given below:

$$SiH_4\,(gas) + 4N_2O = SiO_2\,(solid) + 2H_2O\,(gas) + 4N_2\,(gas) \qquad (4.1f)$$

The basic reaction in doping impurity GeO$_2$ in SiO$_2$ is given below:

$$GeH_4\,(gas) + 4N_2O = GeO_2\,(solid) + 2H_2O\,(gas) + 4N_2\,(gas) \qquad (4.1g)$$

The doping impurity GeO$_2$ increases the refractive index of the guiding layer. The process parameters for the deposition of GeO$_2$–SiO$_2$ layer are given in Table 4.3. Like SiON layer, the PECVD deposited GeO$_2$–SiO$_2$ layer also has a certain amount of O–H bonds, N–H bonds and Si–H bonds that also causes optical absorption at 1.38, 1.48 and 1.51 μm, respectively [7,8]. The post-deposited annealing at 1000°C removes these bonds of the GeO$_2$–SiO$_2$ layer.

4.2.2 Deposition of SiO$_2$/GeO$_2$-SiO$_2$ Material Using Flame Hydrolysis

The origin of this process comes from the optical fiber manufacturing, and this process can produce a thick layer (~100 μm) of doped silica at high deposition rates. In addition, the deposition and consolidation process are intrinsically planer form, hence providing excellent cladding uniformity over closely spaced cores. The chamber of FHD system [7,8] is shown in Figure 4.5a, where the mixture of gas is burnt in O$_2$/H$_2$ torch to produce fine particles, which stick onto a substrate fixed on a rotating table. SiCl$_4$ and GeCl$_4$ are used to produce SiO$_2$ doped with GeO$_2$, respectively. A small amount of Cl$_2$ and BCl$_3$ is added in the chamber for lowering the temperature of synthesized glass particles. After deposition, the heating to a temperature

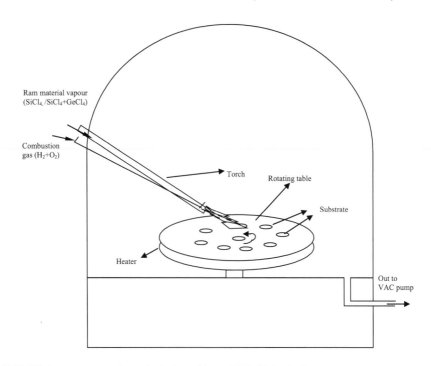

FIGURE 4.5   (a) Flame hydrolysis deposition of SiO_2/SiO_2-GeO_2.

(Continued)

of around $1100°C$–$1300°C$ [8] consolidates the material on substrate. The chemical reaction in FHD for the deposition of SiO_2 and $SiO_2 + GeO_2$ is given below:

SiO_2 deposition: $SiCl_4 + O_2 + 2H_2 = SiO_2 + 4HCl$

$SiO_2 + GeO_2$ deposition: $SiCl_4 + GeCl_4 + 2O_2 + 4H_2 = SiO_2 + GeO_2 + 8HCl$

The deposition rate of FHD method (~1 μm/min) is much faster than that of PECVD (1 μm/hour) because of low-pressure cracking system [8]. The PECVD method is not preferred for thick layer deposition due to more time requirement than that of FHD method. FHD deposition is cheaper in comparison to PECVD. The temperature of deposition in the case of FHD method is higher than that of PECVD.

4.2.3 TUNING OF REFRACTIVE INDEX USING THERMOOPTIC EFFECT

Like $SiO_2/SiON$ material, the refractive index can be changed with application of heat on SiO_2/GeO_2-SiO_2 waveguides. If the heating power P is applied on waveguide via a thin film heater made on the waveguide, the refractive index increases with an increase of temperature. The phase change $\Delta\phi(P)$ due to application of heating power P is expressed as in equation (4.1d). The thermooptic temperature coefficient for SiO_2/GeO_2 is ~$1 \times 10^{-5}/°C$ [2].

4.2.4 Devices Fabricated and Demonstrated by Previous Authors Using SiO₂/GeO₂-SiO₂ Material

Using SiO_2/SiO_2-GeO_2 waveguides, many authors reported different optical devices such as thermooptic Mach Zehnder (MZ) matrix switches, reconfigurable gain equalizer, add–drop filters [30,31], array waveguide grating multiplexer [32] and low-loss integrated optic dynamic chromatic dispersion compensators [33].

R. Kasahara demonstrated a low-power thermooptic MZ switch using SiO_2-GeO_2/SiO_2 material with an index contrast of 0.75% for an operating wavelength of 1550 nm [18] as shown in Figure 4.5b and c. T. R. Schlipf et al. reported an optical delay line circuit using the same index contrast for a reconfigurable EDFA gain equalizer [29] as shown in Figure 4.5d. H. H. Yaffe et al. reported resonant couplers acting as add–drop multiplexer with an index contrast of 0.35% for adding/dropping wavelengths of 1310 and 1550 nm [30] as shown in Figure 4.5e. From these studies by other authors it is seen that SiO_2/GeO_2-SiO_2 material has a lower index contrast than $SiO_2/SiON$ material.

4.2.5 Properties of SiO₂/GeO₂-SiO₂

The SiO_2/GeO_2-SiO_2 waveguide material has the following properties which makes it suitable for waveguide-type integrated devices.

1. It shows thermooptic property, and the thermooptic coefficient is $\sim 10^{-5}/^0K$.
2. Its stability is high.
3. It is a polarization-insensitive material.
4. Material cost is moderate.

The low index contrast limits the minimum bending radius to ~5 mm.

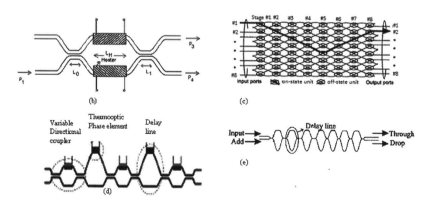

FIGURE 4.5 (CONTINUED) (b) Thermooptic MZ switch, (c) thermooptic matrix switch based on MZ switch [18], (d) thermooptic EDFA gain [6] equalizer based on delay line structure and (e) add/drop multiplexer based on delay line structure [16].

4.3 FABRICATION AND CHARACTERISTICS OF SOI WAVEGUIDE MATERIAL

SOI is another technology in which silicon is used as a waveguide [25]. This technology was initially developed for isolation applications in electrical circuits. SOI's implementation in integrated optics is possible since silicon is transparent in the near-IR region ($\lambda > 1.2\,\mu$m). So, silicon can be used as a core of the SOI waveguide and silica is used as the cladding.

4.3.1 FABRICATION OF SOI WAFER

There are several methods [9] for preparing SOI wafer. Most important methods are

1. Bond and Etch Back SOI (BESOI) method
2. Separated by Implanted Oxygen (SIMOX) method.

4.3.1.1 BESOI Processing

Here the silicon fusion bonded wafer process begins with an oxide layer (typically about 1 μm) grown on a standard silicon wafer. It is then bonded to another wafer, with the oxide in between the two silicon wafers [2]. For the bonding, no mechanical pressure or other forces are applied for avoiding strain inside the waveguide. Then it is annealed at 1100°C for 2 hours in ambient nitrogen, making a strong bond, at atomic level, between the two wafers. One of the wafers is then lapped to a desired thickness using mechanical polishing and lapping. The complete processing steps are shown in Figure 4.6.

The BESOI also has removal of a certain part of the device wafer by wet chemical etching. This method is also capable of making high-quality silicon films with thickness variations of about 7% at thickness as low as 100 nm. In this case a highly selective etch stop layer is formed in the seed (device wafer) prior to bonding by conventional lithography and dry etching. The etchback procedure results in the removal of almost all the seed wafers, except for the thin layer ahead of the peak of the etch mark.

4.3.1.2 SIMOX Method

Figure 4.7a shows the different processing steps used in the SIMOX process [36]. The substrate is bulk silicon. Implantation of oxygen into this substrate is then carried out by implant energy used to control the depth of penetration of oxygen ions in Si substrate, and also thickness of the oxygen-free layer of silicon above the implanted region (active Si layer). After the implantation of wafer with Si, thermal treatments are needed to anneal out implantation damage in the active Si layer as well as to enforce permanent Si–O bonding in the implanted region. Figure 4.7b shows the fabrication steps used for the formation of SOI waveguides.

The characteristics of wafer developed by BESOI method are not unique in comparison to that by using SIMOX [25].

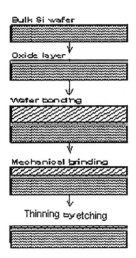

FIGURE 4.6   Steps for the BESOI process.

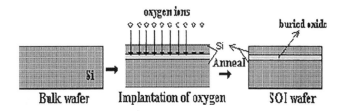

FIGURE 4.7   (a) Fabrication of SOI wafer by SIMOX method.

(Continued)

4.3.2   DEVICE FABRICATED AND DEMONSTRATED BY PREVIOUS AUTHORS USING SOI MATERIAL

Using SOI technology, many authors have reported integrated optic devices such as MMI coupler, thermooptic MZ switch and MMI-based thermooptic matrix switch. R. L. Espanola et al. have demonstrated a low-power thermooptic MZ switch with faster response time and an insertion loss of 32 dB using SOI technology for an operating wavelength of 1550 nm [34], as shown in Figure 4.7c. R. Jalali et al. have reported asymmetric MZ coupler and star coupler with an insertion loss of 9.5 dB for a wavelength filter using SOI technology. Z. Wang et al have reported a thermooptic 4 × 4 switching matrix with an insertion loss of 10.1 dB for an operating wavelength of 1550 nm [35]. W. E. Zhen et al. have reported a 2 × 2 MMI coupler of coupling length 3618 μm with large tolerance for a power splitter [36]. C. K. Tang et al. [52] have reported SOI optoelectronic devices based on rib structure in an operating wavelength region of 1300–1550 nm. Although higher index contrast waveguide device can be fabricated using SOI technology, insertion loss of the device is more than that of the devices using other waveguide material technology.

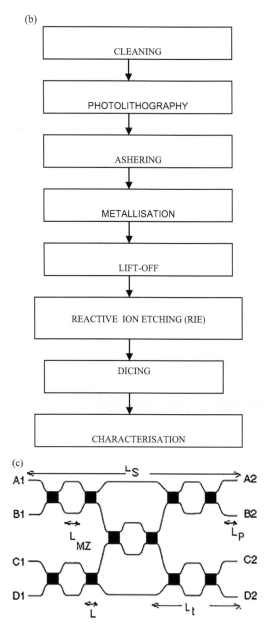

FIGURE 4.7 (CONTINUED)   (b) Steps for fabrication of SOI waveguide and (c) add/drop multiplexer based on delay line structure [35].

4.3.3 PROPERTIES OF SOI

Reported optical losses of SOI waveguides vary between 0.1 and 1 dB/cm (depending on the geometrical structure) and the insertion loss values are of the order of

1–2 dB/facet, which is a major drawback of this technology [25]. The properties of SOI waveguide are given below:

1. It shows thermooptic property.
2. Its stability is high.
3. Its index contrast is high.
4. Material cost is moderate.

4.4 FABRICATION AND CHARACTERISTICS OF Ti:LiNbO$_3$ WAVEGUIDE MATERIAL

LiNbO$_3$ technology became an attractive material for waveguide-type integrated device due to its electrooptic and acoustooptic properties as well as its high transparency in the near-infrared region. The refractive index differences between substrate and diffused waveguides are in the range of 10^{-3}–10^{-2} [1]. The waveguide core size operating a single mode at $\lambda = 1.55\,\mu m$ is around 6–8 μm, which is similar to the characteristics of single-mode optical fibers. This is the reason for very low coupling losses of LiNbO$_3$-guidedwave devices (~0.5 dB/facet). The fabricated waveguides with low propagation loss (≤ 1 dB/cm) were reported [1]. Lithium niobite is a brittle material, transparent in the visible region.

4.4.1 PROCESSING OF LiNbO$_3$-BASED WAVEGUIDE

There are several fabrication methods for LiNbO$_3$ waveguides, mentioned below:

a. Sputtering method: single crystal LiNbO$_3$ film is made on a glass substrate [1,2].
b. Epitaxial growth: single crystal LiNbO$_3$ film is epitaxially grown on LiTaO$_3$ crystal with the same crystal structure.
c. Out-diffusion method: By heating LiNbO$_3$ in a vacuum at nearly 1000°C, Li$_2$O is out diffusion from the crystal surface, providing a higher index layer near the surface [1].
d. Ion exchange method: The ion exchange is made by immersing LiNbO$_3$ in molten salts such as AgNO$_3$ and molten benzoic acid [1,2].
e. Thermal in-diffusion method: A metal film is deposited on LiNbO$_3$ followed by heating the crystal under the flow of Ar and N$_2$ or O$_2$ at about 1000°C. As a result, the metal is diffused into LiNbO$_3$ crystal [1].
f. Proton exchange method: It is a annealed step by step process.

Out of these processing techniques, thermal in-diffusion and proton exchange methods are mainly used for the fabrication of LiNbO$_3$ waveguides.

4.4.1.1 Thermal in Ti-Diffusion Method

Commercially available LiNbO$_3$ is typically used for a 2-inch diameter substrate of 0.5–2 mm thickness. The optical axes are specified with respect to the polished surface according to the design of the functional waveguide device. Additionally, the

z-plane of the crystal is chemically etched with HNO_3 in a much shorter period. This will help to find the positive and negative direction of optical axes, which determine the sign of the index change with respect to the polarity of an applied voltage. The purchased $LiNbO_3$ wafer is cut and polished. The surface of $LiNbO_3$ is cleaned in the following manner.

4.4.1.1.1   Cleaning of LiNbO₃ Substrates

The wafer cleaning requires the removal of chemical impurities from the surface without damaging the substrate surface. Dry-physical, wet-chemical and vapor-phase methods can be used to achieve these objectives. It is usually done by rinsing in hot organic solvents such as trichlorethylene, acetone, methanol and ultrasonic agitation [1].

4.4.1.1.2   Thin Film Ti Coating Using Physical Vapor Deposition

The box coater electron beam unit is used for coating titanium over wafer. It consists of two units –vacuum chamber and E-beam unit [1]. The wafer is cleaned and placed on the circular holder. It is loaded in a vacuum chamber in which vacuum is created using turbo pump and diffusion pump. After 2 hours, vacuum will reach to 2×10^{-6} mbar. Then, the coating process is started.

E-beam gun supply consisting of 5 HT and LT transformer is put on. The power I/p to the gun is the product of beam voltage and emission current which is 6–8 kV and 0.1 A max., respectively. When the current increases and the material reaches its melting point, then the material will melt, and vacuum will lower down. So again, it is waited for vacuum to improve. Then the shutter is opened, and the thickness monitor starts counting the rate of deposition and thickness. The rate at about 1–2 Å/s and coating up to 1000 Å for titanium deposition is maintained and similarly for gold evaporation can be carried out. After the required thickness is achieved, beam current should be fixed to zero. Then the temperature is reduced below 50°C. Then the coated wafer is removed for the next process.

4.4.1.1.3   Waveguide Pattern Transfer

After cleaning the coated wafer using acetone and methanol, the photolithography is used for transfer of pattern using a mask having the core of the device. The negative photoresist (methyl ethyl ketone/methylisobutyl ketone) is coated on a substrate with spinner at 4000 rpm for 40 seconds. An exposure time with UV light is ~1.5 seconds, after pre-baking of sample at 90°C for 20–25 minutes. The photoresist is developed for 60 seconds, and then the sample is kept for post-baking in an oven at 115°C for 35 minutes in order to make the hardening of the exposed portion of photoresist.

4.4.1.1.4   Titanium Etching

Titanium strips are formed on $LiNbO_3$ wafer through chemical etching of titanium to create the core of the waveguide. The composition of the wet etchant of titanium is 0 mL Ethylene diamine Tetra Acetic Acid (EDTA) + 5 mL H_2O_2 + 10–15 drops of NH_3 (30°C–40°C). Wet etching removes titanium where there is no photoresist. So, titanium strips are formed. Further the photoresist is removed using ashing/plasma etching. Figure 4.8 shows the different steps used for making titanium strips on $LiNbO_3$ crystal.

4.4.1.1.5 Thermal in Ti Diffusion of $LiNbO_3$

Thermal Ti in diffusion is a standard technique for fabricating $LiNbO_3$ waveguides. The titanium strips formed on $LiNbO_3$ was thermally diffused in a horizontal furnace. This furnace (Carbonite, STF) is programmable and has got very good temperature stability. The central tube is made of quartz with an internal diameter of 4 inches. Ti in diffusion was carried out at 1000°C for 10 hours. When Ti-coated wafer is heated to 1000°C, the undesired Li_2O out diffusion takes place during Ti in diffusion. The refractive index n_e increases in both Li_2O out diffusion and Ti in diffusion. As a result, there is a significant increase of scattering loss because of non-uniform change of refractive index. Li_2O out diffusion can be eliminated by diffusing Ti in moistened/

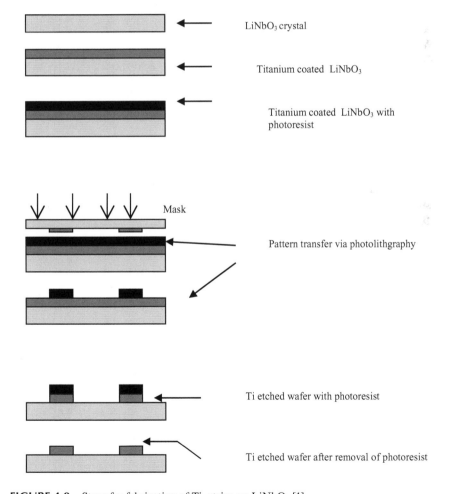

FIGURE 4.8 Steps for fabrication of Ti–strips on $LiNbO_3$ [1].

wet Ar and O_2 gases. This was kept so as to minimize the out diffusion of lithium, which can lead to a planar waveguide on the substrate. $LiNbO_3$ crystal is kept in a small quartz tube of 5 cm long and 15 mm diameter, which itself contained a 1-m-long quartz tube. The furnace used for diffusion has temperature stability within ±2°C [1]. The diffused channels could be seen as the strip regions are slightly above the surrounding substrate. The gas is bubbled through a water bath at 60°C. After passing through bubbler the humidity of the gas is nearly 80% depending on the gas flow rate. In addition to prevent audience of water drops to the quartz tube, winding a tape heater is made around a portion of the quartz tube outside the furnace. The gas flow rate is adjusted so that silicon oil is placed at the gas outlet bubbles slightly. If the flow rate is high, the temperature becomes non-uniform near the center of the quartz tube. The temperature of the furnace should be increased to 1000°C within 30 minutes. The Ti in diffusion is made at 1000°C in flowing wet Ar for 4–8 hours depending on the thickness of diffusion. The Ti in diffusion is continued at the same temperature in flowing wet O_2 gas for 1–2 hours to compensate for the lack of O_2 in $LiNbO_3$. After diffusion, the $LiNbO_3$ is kept at room temperature for 15–20 minutes for cooling.

4.4.1.1.6   Index Contrast of Ti-Diffused $LiNbO_3$

The index contrast Δn_e and diffusion depth depend on coated Ti film thickness. Figure 4.9 shows the variation of Δn with Ti film thickness [1]. If the temperature and diffusion time are constant, the index contrast Δn is linearly proportional to the Ti-film thickness. So, the Δn_e can be easily controlled by Ti-film thickness. The Ti- in diffusion provides a larger index change of extraordinary wave than that of an ordinary wave.

On the other hand, there is a small variation of diffusion depth with Ti-film thickness as shown in Figure 4.10 [1]. In the case of extraordinary wave, diffusion depth is more than that of ordinary wave. Since, in Ti in-diffusion waveguide, no light streak is observed, the propagation loss is small (~0.5 dB/cm).

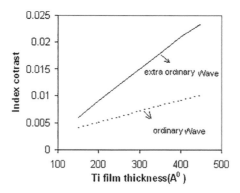

FIGURE 4.9   Variation of index contrast of ordinary (dotted line) and extraordinary waves (solid line) with Ti-film thickness [1].

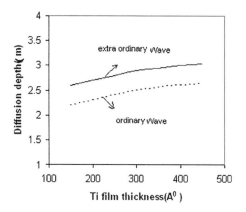

FIGURE 4.10 Variation of diffusion depth of ordinary (dotted line) and extraordinary waves (solid line) with Ti-film thickness [1].

4.4.1.2 Proton Exchange Method

The fabrication of annealed proton exchange (APE) waveguides is a step-by-step process spanning over a considerable amount of time [1]. The fabrication process can be made in three main stages. Figure 4.11a shows the flow chart. First, standard photolithography is used for the transfer of waveguide pattern on LiNbO$_3$ substrate. Then proton exchange method is followed for this part of the process, benzoic acid was used as the proton source. Benzoic acid is non-toxic. But above 200°C it gives fumes and penchant smells. If the proton exchange is performed, the fumes will escape and solidify at low temperature. So, proton exchange cannot be performed in an ordinary furnace. A jig is designed to perform the proton exchange in a closed chamber. The samples to be fabricated were suspended above the acid melt inside the closed chamber during warm up and cool down using a glass holder. The temperature of the melt was monitored with the help of a thermocouple inserted directly in the melt. Once the desired temperature was reached, the glass holder was moved down so that the samples were dipped in the acid melt. After a specified time of exchange, the glass holder is moved up and the whole setup is allowed to cool naturally. Then the metal

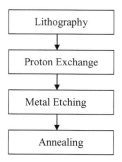

FIGURE 4.11 (a) Fabrication flow chart of proton exchange.

(*Continued*)

is removed by using metal etchant. Finally, it is put inside a temperature-controlled furnace at a temperature of 350°C for annealing.

The LiNbO₃ waveguide developed by proton exchange method is less polarization sensitive than that by thermal Ti in-diffusion method.

4.4.2 Tuning of Refractive Index Using Electrooptic Effect

The refractive index is changed with the application of electric field using electrooptic effect. LiNbO₃ shows an electrooptic effect. Here refractive index changes with the application of electric field. If the electric field V/w is applied along the z axis (c axis of LiNbO₃) via a gold electrode formed on the waveguide, the index contrast of extraordinary wave along the z axis is changed as [10]

$$\Delta n_e = n_e^3 r_{33} V/d \tag{4.2}$$

d = width of the waveguide and r_{33} = electrooptic coefficient = 30.8 pm/V for LiNbO₃ and V = voltage applied on the waveguide. Normally, the switching speed of digital optical switches based on electrooptic effect using Ti:LiNbO₃ material is ~2.5 MHz [37,50].

4.4.3 Devices Fabricated and Demonstrated by Previous Authors Using LiNbO₃ Material

Using LiNbO₃ technology, many authors have reported different integrated optical devices such as two-mode interference (TMI) couplers, directional couplers and digital optical switches. A. Never et al. [19] have reported a TMI coupler with coupling length ~6 mm, $\Delta n \sim 0.6\%$ and w_{mmi} ~2.7 μm using Ti:LiNbO₃ technology for wavelength multiplexer/demultiplexer device with wavelength range 0.57 to 0.59 μm as shown in Figure 4.11b. M. Papuchon et al. [20] have implemented a TMI coupler of coupling length ~5 mm, $\Delta n \sim 1\%$ and w_{mmi}~2.7 μm using Ti: LiNbO₃ technology for an optical bifurcation device with operating wavelength ~0.5145 μm as shown in Figure 4.11c. R. Krahenbuhl et al. reported Y branch digital optical switches using an index contrast of 0.6% with switching voltage ~9V, switching speed of 2.5 MHz and fiber-to-fiber loss of 4 dB [37]. H. S. Hinton reported directional couplers using Ti: LiNbO₃ technology for the application of photonic switching [38]. H. A. Haus et al. have demonstrated an optical directional coupler (taper) with an index contrast of 1% for a 2 × 2 electrooptic switch [50]. From the studies of other authors, it is seen that LiNbO₃ material forms a lower index contrast waveguide.

4.4.4 Properties of LiNbO₃

The LiNbO₃-based waveguide material has the following properties which makes it suitable for waveguide-type integrated devices.

1. It shows electrooptic and acousto-optic properties.
2. Its stability is high.

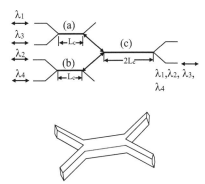

FIGURE 4.11 (CONTINUED) (b) Four channel cascaded multiplexer/demultiplexer and (c) TMI coupler.

3. It is a polarization-sensitive material.
4. Its index contrast range is moderate.
5. Material cost is high.

The major disadvantages of $LiNbO_3$ technology are as follows:

1. There are difficulties in the integration of lithium niobite with other active optical devices that prevent large-scale integration.
2. Lithium niobite waveguide devices have large sizes due to the low achievable index contrast.

4.5 FABRICATION AND CHARACTERISTICS OF InP/GaAsInP WAVEGUIDE MATERIALS

Work on III–V semiconductors was motivated mainly by the possibility of construction of monolithic OICs. Material systems such as GaAs/AlGaAs and InGaAsP/InP were used for the construction of basic devices [12,13]. The optical waveguides, besides connecting other devices, were also incorporated in many functional waveguide devices. For fabrication of waveguides, the refractive index of the guiding region is greater than the index of its surroundings. The III–V (or II–VI) ternary and quaternary compounds are optically active, and their energy bandgap is also changed over a wide range by altering the relative concentration of elements. Its unique property is that the lattice constants of GaAs and InP are almost identical (5.65 and 5.82 Å, respectively), which is useful for the fabrication of devices.

4.5.1 PROCESSING OF InP/InGaAsP WAVEGUIDE

The InP/InGaAsP waveguide device is grown using the molecular beam epitaxy (MBE) growth system. Before growing the InP/InGaAsP waveguide, it is required to prepare the substrate and clean the same. The substrate used for the InP/InGaAsP waveguide is InP. The substrate is cut and polished to the desired size. The substrate cleaning has the following steps:

1. The polished surface is degreased to remove residual waxes.
2. HCl is used to remove other surface contaminants.
3. It is rinsed in deionized water and dried in O_2 environment.

4.5.1.1 Deposition of GaAsInP and InP Layers Using MBE Growth System

MBE involves the reaction of one or more thermal beams of atoms or molecules with crystalline surface under ultra-high vacuum condition (10^{-8} Pa). It has precise control in both chemical compositions and doping profiles. Single crystal multilayer structures of dimension in the order of atomic layers can be created by MBE. The fabrication of semiconductor heterostructures having thin layers from a fraction of micron down to a nanometer is also possible. The MBE growth rates are quite low, and for GaAs, a value of 1 μm/h is typical. The system has film deposition control, cleanliness and in situ chemical characterization. In Figure 4.12a, the schematic diagram of a conventional MBE system chamber [13] consisting of an arrangement of effusion ovens (source) and substrate is shown.

The separate effusion ovens of pyrolytic boron nitride are placed as source of deposited material for each constituent of the layer and dopants. The temperature of each oven is adjusted to give the desired evaporation rate. The main parts of the

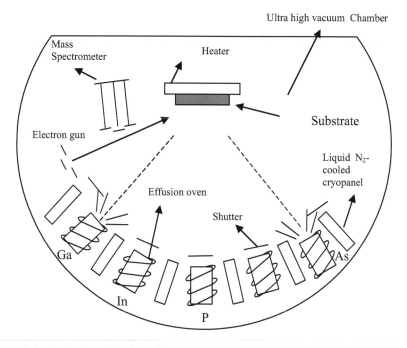

FIGURE 4.12   (a) MBE chamber with arrangement of sources and substrate.

(*Continued*)

MBE system are vacuum creating system, effusion cells, flux monitor, residual gas analyzer, substrate manipulator and analysis chamber.

4.5.1.1.1 Vacuum Creating System

For MBE process, ultra-high vacuum is created because system must be free from any type of contaminants and unwanted gas molecules to achieve high purity and precision. Since the operating ranges of different pumps are different from each other, no single pump creates the degree of vacuum which is needed. Therefore, different pumps are used in various combinations as given in Table 4.4.

Effusion cells: Eight effusion cells constructed with tantalum and PBN material are provided – out of these two cells have 14 cc capacity, three cells have 40 cc capacity and one cell has 125 cc capacity. The maximum outgassing temperature is 1600°C and maximum operating temperature is 1400°C.

Flux monitor: A nude ionization gauge is provided to measure the flux of various atomic beams. The flux monitor is mounted on a bellow which enables to push the gauge right in front of the wafer during flux measurement.

Residual gas analyzer: A quadrupole mass spectrometer with 1–200 a.m.u. is provided for residual gas analysis. The minimum detectable partial pressure is 2×10^{-13} Torr. The spectrometer is controlled by a micro-computer that can control four such spectrometers.

Residual gas analysis (quadrupole mass spectrometer): QMS analyses residual gases by separating each of them depending on their mass to charge ratio and measuring their relative abundance.

4.5.1.1.2 Analysis Chamber

This chamber is used for the analysis of the surface of grown layers or substrates. This chamber consists of the following: Ion gun is provided to etch an area of $1\,cm^2$ to remove surface contaminates such as carbon also to acquire the depth profile of the material. This system allows the elemental analysis of the grown crystal and observation of contamination level on the surface of the substrate.

Substrate manipulator: It has X, Y, Z mount with rotary motion. The manipulator can handle a wafer up to a 3″ diameter. The substrate holder is made for the production of uniform layers so that the substrate manipulator is rotated by a stepper motor at the slow rate of 3–5 rpm. The general construction of such an assembly is of a refractory metal block (usually Mo), which is heated either resistively or by

TABLE 4.4
Different Vacuum Pumps Used in MBE System

Types of Vacuum	Range	Pumps
Rough vacuum	760 to 1×10^{-3} Torr	Rotatory vane oil sealed mechanical pumps, dry vacuum pumps
High vacuum	1×10^{-3} to 1×10^{-8} Torr	Oil-diffusion pumps, mechanical cryo-pumps, turbo molecular pumps
Ultra-high vacuum	$<1 \times 10^{-8}$ Torr	Ion pumps, titanium sublimation pumps

radiation. The substrate is usually held on the Mo block as shown in Figure 4.12a. Mo holder holds the wafer with the help of few Ta pins. The heating of GaAs substrate mounted on holder must be done directly by thermal radiation from the heater. The GaAs wafer is suspended by a few thin Ta or Mo wires onto a Mo gasket-like holder, on which the back surface of the wafer can receive direct radiation from the heater. Mo holder can be transferred and mounted onto the heater assembly just as the conventional Mo holder. Since the backside of the GaAs wafer mounted on Mo block receives the radiation from the heater, the top portion of the back surface of GaAs can decompose particularly at high temperatures.

This can be reduced if the back surface of the GaAs facing the heater is surrounded by enclosure housing, since the excess pressure of as generated by the dissociation of GaAs is maintained inside the housing and prevents further decomposition of Ga. A thermocouple or a pyrometer usually measures the substrate temperature. To achieve high uniformity of film thickness and alloy composition, in most MBE systems the substrate holder is equipped with a rotation mechanism. Effusion cells have an operating temperature range up to 1400°C, which makes possible short outgassing sequence at 1600°C. Although these temperatures are more than adequate for III–V materials, in practice, more cells are limited to operating temperatures ≤1200°C, which is only just within the range of that required for Ga, Al, In and P (dopant source) evaporation. This is because of the outgassing problem caused principally by the large temperature difference between the heater and the actual temperature of the charge in the cell. A means of rapidly opening and closing line of sight between the beam sources and substrate must be available for each beam source. These shutters must then meet rigorous mechanical design criteria in order to satisfy the performance requirements necessary to fully exploit the precise layer thickness defined in MBE.

The presence of the charge on the crucible makes the closing of an externally controlled shutter over the source oven. The thermal equilibrium is required prior to the growth with the shutter closed. The shutter-induced thermal transient affects directly the amounts of material flux from the charge. In the superlattice structures, some layers need less than one minute for the deposition.

The thermal transients are removed by increasing the separation between the mouth of the crucible and the shutter or varying the angle of shutter with the plane of the crucible opening. The composition of the deposited layer in MBE depends on flux ratios, substrate temperature and chemical species, since all these factors affect the surface incorporation kinematics on a substrate. The fluxes are controlled via the source temperature to grow to the desired composition. In the case of III–V alloy growth, the group III species have higher sticking coefficient at a substrate temperature below 600°C.

The stoichiometric control of $III_x III_{1-x} VV$ alloys result in control of bandgap, refractive index, etc. For example, $In_xGa_{1-x}AsP$ of bandgap from 1.28 to 1.43 eV is grown by choosing x from zero to one [13]. In the case of InP/GaAsInP waveguide, InP layer is acted as a substrate and GaAsInP is acted as a guiding layer. The InP substrate is placed on the holder. The temperature of the substrate is kept at ~580°C. In Figure 2.11a, Ga, In, As and P are placed in a respective oven for the growth of GaAsInP. The refractive index of InP ~3.17 [12] and that of GaAsInP is ~3.32 [12].

4.5.1.2 InP/GaAsInP Waveguide Fabrication

For the fabrication of waveguide, the waveguide pattern has to be transferred on the GaAsInP guiding layer using negative electron beam resist. After completion of the patterning process, reactive ion etching (RIE) is used to make the core of the waveguide [14]. The etchant used for GaAsInP is CH_4 (20%) $+H_2$ (80%). The sample is etched to a depth of 0.5 µm (if total etch depth is ~1 µm). The oxygen plasma is used to clean the chamber walls of the resist residue. The sample is again reinserted to etch the remaining portion of the sample. At this point, subsequent to the RIE process, the sample is given a short wet etch in saturated Br in water (SBW) H_2O: H_3PO_4 (2:15:1) in order to remove residual and etch-damaged material from side walls.

4.5.2 TUNING OF REFRACTIVE INDEX OF InP/GaAsInP WAVEGUIDE

In order to tune the refractive index, the gold contacts are made on the top of the device, where tuning of refractive index is desired. The current is injected through the gold contact. This current injection leads to carrier-related plasma effect which modifies the negative refractive index change. The good metal contacts are advantageous in order to avoid heating; otherwise, it leads to a temperature-related refractive index change with an opposite sign. The refractive index change is a function of applied current on the gold contacts. The refractive index change is approximately written as [15]

$$\Delta n = bI \tag{4.3}$$

where I = applied current and $b = -1.8 \times 10^{-4}$/mA. The constant b depends on the device implementation. The plasma effect leads the free carrier absorption, which leads to increased propagation losses with increasing refractive index change. This is formulated as follows [15]:

$$\alpha = aI \tag{4.4}$$

where $a = 7$ (mA)$^{-1}$ (cm)$^{-1}$ = loss constant.

4.5.3 DEVICES FABRICATED AND DEMONSTRATED BY PREVIOUS AUTHORS USING InP/GaAsInP MATERIAL

The InP/GaAsInP materials are used in integrated optical devices such as multimode interference (MMI) couplers, TMI coupler, MZ switches, and power splitters. Leghold et al. [15] have reported an MMI coupler of Δn ~13% and w_{mmi} of 11.3 µm (w_{mmi} = width of MMI coupler) using InP/GaAsInP waveguide for tunable power splitter, respectively. Levy et al. [39] have demonstrated a 2 × 2 MMI coupler with Δn ~13% and w_{mmi} of 9.3 µm for 3-dB power splitting as shown in Figure 4.12b. M. Yagi et al. [22] have reported a 3 × 3 MMI coupler of beat length L_{π} ~432 µm using Δn ~ 16.7% for versatile switching with partial index modulation. Yong Ma et al. [14] have demonstrated an MMI coupler with coupling length ~20.5 µm and

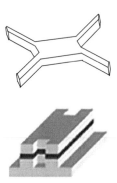

FIGURE 4.12 (CONTINUED)   (b) 2×2 MMI coupler and (c) 2×2 MMI coupler [23].

$\Delta n \sim 13\%$ using InP/GaAsInP waveguide for a 3-dB coupler. Darmawan et al. [23] have reported an MMI coupler of lower L_π of 35 μm using $\Delta n \sim 16.7\%$ for mode filtering as shown in Figure 4.12c. From the studies of other authors, it is seen that InP/GaAsInP waveguide device has a higher index contrast.

4.5.4 PROPERTIES OF InP/GaAsInP

The InP/GaAsInP-based waveguide material has the following properties which makes it suitable for waveguide-type integrated devices.

1. Its stability is high.
2. It is a polarization-insensitive material.
3. Its index contrast range is wide.

III–V semiconductor technology has some drawbacks at the standpoint of OIC applications. Since the fabrication process requires sophisticated technology, the cost of production is quite high. The cost of the material is high, and the insertion loss is greater than other materials.

4.6 FABRICATION AND CHARACTERISTICS OF POLYMERIC WAVEGUIDE MATERIAL

The polymer material group arises as a hybrid technology for integrated optics. Polymer films are found to be superior to other organic films because of their easy processing [2]. They are produced by simply dissolving the polymer powder in a solvent and then spin coating it on any type of substrate before being dried in an oven to get rid of the residual solvent. Optical waveguides need the fabrication of multilayers with cladding and core polymers. The control of the refractive index, thickness of the film and its resulting mode geometry structure is made with adjustment of the chemical deposition and viscosity of polymers. Furthermore, the control of the index of refraction of amorphous polymer films may be realized by selective photo polymerization in which the ultraviolet light is irradiated through a photomask on a plastic polymer film. The index changes due to photo polymerization ranges between 0.001 and 0.04, while the refractive indices of polymer films are roughly between 1.3 and 1.8 (at $\lambda = 1.32\,\mu m$) [1]. The flexibility in fabrication of films of good quality and their compatibility with almost all substrate materials such as Si, GaAs, InP and glass [2] are advantageous.

4.6.1 Fabrication of Polymeric Waveguides

Single-mode waveguides with polymeric materials is difficult due to controlling of the indexes with a precise, reliable and predictable manner. The techniques are conventional mask-based photolithography approaches [54,55], injection molding [56], ion implantation [57], laser-based micro-fabrication technique [53], etc. for fabrication of single-mode polymeric waveguide devices.

Figure 4.13 shows a laser writing setup for fabrication of polymeric waveguide with precise controlling refractive index of core. It is built around a Helium–Cadmium laser (Kimmon IK3351R-G) emitting a 41 mW TEM00 beam at a wavelength of 325:0 nm. The beam power is controlled with a holographic optical attenuator and can be monitored with a power meter that receives a fraction of the beam through a weak beam splitter (glass slide). The optical axis is folded several times with mirrors, to bring the focused beam out of the plane and vertically onto the sample which is mounted on an x-y-translation stage. An electronic mechanical shutter can be used to eclipse the beam. The beam was focused with a 160 mm lens. The optical power at the focus was about 1 mW. Development was performed by rinsing in acetone. The waveguides are written in photopolymer materials. Normal materials used in this technique is PMMA (Polymethyl methacrylate)/diacrylate-based polymer by using technique.

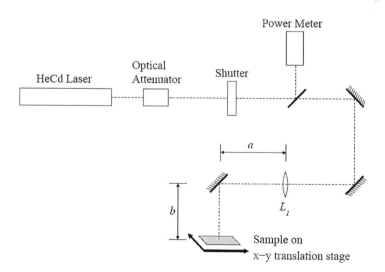

FIGURE 4.13 Laser-written setup for a polymeric waveguide.

The advantages are rapid and inexpensive prototyping, direct writing features within restricted regions of a sample without affecting the surrounding area, etc.

4.6.2 Tuning of Refractive Index Using Thermooptic Effect

Like silica-based material, polymeric waveguides show a thermooptic effect. If the heating power P is applied on the waveguide via a thin film heater formed on the waveguide, the refractive index of the waveguide increases due to the rise of temperature. The thermooptic phase change $\Delta\phi(P)$ due to the application of heating power P is expressed as [6]

$$\Delta\phi(P) = \left(2\pi/\lambda\right)\frac{dn}{dT}\Delta T_c(P)L_H$$

L_H = length of heater, $\dfrac{dn}{dT}$ = thermooptic temperature coefficient and $\Delta T_c(P)$ is the temperature rise due to application of heating power P. The thermooptic coefficient for polymeric material is ~1×10^{-4}/°C which is higher than that of silica-based material. Due to higher value of $\dfrac{dn}{dT}$ for polymer in comparison to silica, the length of heater is obtained by using equation (2.1e) to achieve $\Delta T_c(P) = 15$°C for thermooptic phase change of π as follows:

$$L_H = \frac{\lambda}{2\Delta T_c(P)\dfrac{dn}{dT}} = \frac{1.55 \times 10^{-3}}{2 \times 15 \times 1 \times 10^{-4}} = 0.5 \text{ mm}$$

4.6.3 Devices Fabricated and Demonstrated by Previous Authors Using Polymer Technology

Using polymer material, many integrated optical devices such as thermooptic MZ switches, optical add/drop filter and waveguide grating multiplexer are reported by different authors. M. B. J. Diemer et al. have reported a polymeric thermooptic optical waveguide switch with a switching time of 10 ms using polyurethane core and PMMA cladding with index contrast 7% [44]. R. Mooseburger et al. have demonstrated a digital optical switch with an insertion loss of 4 dB using polymer cyclotene 3022™ rib waveguides for an operating wavelength 1.55 μm [45]. H. P. Chan et al. have demonstrated a wide-angle X junction polymeric thermooptic digital switch of BCB (Benzo cyclo-butene) core index 1.54, an upper cladding index of 1.51 and a lower cladding of 1.44 with a low crosstalk of 24 dB for an operating wavelength of 1550 nm [46] as shown in Figure 4.12. C. Kostrzewa et al. have reported a tunable add/drop filter with polymer cyclotene 3022™ waveguide for multiwavelength networks with a center wavelength of 1550 nm [47]. N. Keil et al. have reported a thermal all-polymer arrayed waveguide grating multiplexer with a wavelength shift <0.05 nm in the 25°C–65°C temperature range [48]. Y. Hida et al. have reported an acrylic polymer waveguide thermooptic switch using an index contrast of 0.3% with a low electric power consumption of 4.8 mW and a low insertion loss of 0.6 dB at an operating wavelength of 1300 nm [51]. L. Eldada et al. [53] have reported Laser-fabricated, low-loss, single-mode raised rib waveguiding devices using polymers with an index contrast of 1.15% for power splitting at wavelength 1550 nm. The studies of these authors show that polymeric waveguide device has low power consumption, moderate range of index contrast and low insertion loss.

4.6.4 Properties of Polymeric Material

The polymeric waveguide material has the following properties which makes it suitable for waveguide-type integrated devices.

1. It shows thermooptic properties.
2. It is a polarization-insensitive material.
3. Its index contrast range is ~0.21.
4. Its stability is low.

4.7 COMPARATIVE STUDY OF INTEGRATED WAVEGUIDE MATERIALS

After mentioning integrated optical waveguide device processing using waveguide materials mentioned in Table 4.5, we have compared those in terms of advantages and disadvantages. It is seen from the table that a high index contrast is obtained by using $SiO_2/SiON$, InP/GaAsInP, SOI and polymeric waveguide materials, whereas a low refractive index is obtained by using Ti:$LiNbO_3$ and SiO_2/SiO_2-GeO_2 materials. Silicon-based materials show polarization-insensitive property in comparison to Ti:$LiNbO_3$ material because of crystal structure. Further polymeric and silicon-based materials provide thermooptic properties. The polymeric materials provide higher thermooptic coefficient and also have easy processing of devices, whereas silicon-based materials are highly stable and compatible to conventional IC processing technology.

The SOI shows a higher index contrast of fixed value ~2. On the other hand, a wide variation of the index contrast (maximum up to 0.53) is obtained by varying the nitrogen and oxygen content in SiON material. In the case of SOI waveguide device, the reported propagation loss of SOI waveguides is 0.1 dB/cm [59] and the fiber to chip coupling loss is of the order of 2–5 dB/facet [59], whereas in $SiO_2/$ SiON, the propagation losses are less than that of SOI materials and fiber to chip coupling loss (order of 1 dB per facet) is lower than that of SOI material [59]. The $SiO_2/SiON$ material also shows more chemical inertness property than other waveguide materials.

Except having a lower index contrast in comparison to SOI material, $SiO_2/SiON$ material has many advantages such as availability of wide index contrast, compatibility with conventional silicon-based IC processing, polarization insensitiveness, low losses, stability and chemical inertness.

TABLE 4.5

Optical Properties of Some Materials for Waveguide-Type Integrated Devices

Materials with Range of Refractive Index	Available Δn (max)	Taken by Different Authors	Properties				
			Thermooptic Coeff. $\left(\alpha = \dfrac{dn}{dT}\right)$	Electro-Optic Coeff. (r_{33})	Polarization Sensitivity/ Birefringence	Stability	Material Cost/ Processing Cost
SiO$_2$/SiON index range ~ (1.45–1.98)	~0.53	0.033[16] 0.103[17]	10^{-5}/°C [12]	-	Polarization-insensitive [2]/10^{-6}	High [2]	Moderate/high
GeO$_2$-SiO$_2$/SiO$_2$ index range ~ (1.45–1.47)	~0.02	0.0075[18] 0.0025[21]	10^{-5}/°C [2]	-	Polarization insensitive [2]/10^{-5}	High [2]	Moderate/high
SOI(3.4767)	~2.026	-	1.84×10^{-4}/°C [2]	-	Polarization insensitive/10^{-4}	High	Moderate/high
Ti: LiNbO$_3$ index range ~ 2.15–2.21	~0.06	0.006[19] 0.01[20]	-	30.8 pm/V [13]	Polarization sensitive [12]/10^{-2}	High [12]	High/high
InP/GaAsInP index range ~ (3.13–3.5)	~0.33	0.13[22] 0.167[23] 0.15[14,15]	-	-	Polarization insensitive [23]/2.5×10^{-4}	Stable [23]	High/high
Polymer index range ~ (1.44–1.65)	~0.21	0.03–0.1 [2]	10^{-4}/°C [1]	10–200 pm/V	Polarization insensitive/10^{-2} to 10^{-6}	Low–moderate [1]	Low/low

SUMMARY

In this chapter, we have described fabrication steps/process and characteristics for integrated optical waveguide devices by using $SiO_2/SiON$, SiO_2/GeO_2-SiO_2 and SOI material. However, the same using other materials such as $LiNbO_3$, InP/GaAsInP and polymer are also described for the comparison with the same material. It is seen that $SiO_2/SiON$ material has many advantages such as wide index contrast of their waveguides, stability of the device, low losses, polarization insensitiveness and compatibility to the based conventional IC processing technology.

EXERCISES

4.1. Find out heater length LH to obtain and $\Delta\phi(P) = \pi$ for an MZ switch having SiON waveguide core for a temperature difference of the arms $\Delta_c(P) = 15°C$ at $\lambda = 1.55\,\mu m$.

4.2. Find out heater length L_H to obtain and $\Delta\phi(P) = \pi/2$ for an MZ switch having SiO_2-GeO_2 waveguide core for temperature difference of the arms $\Delta T_c(P) = 7.5°C$ at $\lambda = 1.55\,\mu m$.

4.3. Find the flow rate of reactant gases of PECVD for getting both the core and cladding layers of $SiO_2/SiON$ waveguide having index contrast 5%.

4.4. Find the flow rate of reactant gases of PECVD for getting both the core and cladding layers of $SiO_2/SiON$ waveguide having index contrast 2%.

4.5. Find the flow rate of reactant gases of PECVD for getting SiO_2 layer.

4.6. Find out heater length L_H to obtain and $\Delta\phi(P) = \pi$ for an MZ switch having an acrylic polymeric waveguide core for a temperature difference of the arms $\Delta T_c(P) = 15°C$ at $\lambda = 1.55\,\mu m$.

4.7. Find out the voltage to obtain the index contrast of an extraordinary wave along the z axis of value 0.1% d = width of Ti-$LiNbO_3$ waveguide = $5\,\mu m$ and r_{33} = electrooptic coefficient = 30.8 pm/V for $LiNbO_3$ and V = voltage applied on the waveguide and $n_c = 1.42$.

REFERENCES

1. H. Nishihara, M. Haruna, and T. Suhara, *Optical Intagrated Circuits*, McGraw-Hill, New York, 1989.
2. M. Quillec, *Materials for Optoelectronics*, Kluwer Academic, Boston, MA, 1996.
3. M. J. Rand and R. D. Standley, "Silicon oxynitride films on fused silica for optical waveguides," *Applied Optics*, vol. 11, p. 2482, 1972.
4. R. German, H. W. M. Salemnik, R. Beyeler, G. L. Bona, F. Horst, and B. J. Offrein, "Silicon oxynitride layers for optical waveguide applications," IBM-Zurich Research Report, RZ 3101 (# 93147), 1999.
5. D. E. Bossi, et al., "Optical properties of silicon oxinitride dielectric waveguides," *Applied Optics*, vol. 26, no. 4, pp. 609–611, 1987.
6. B. J. Offrein, F. Horst, G. L. Bona, R. Germann, H. W. M. Salemink, and R. Beyeler, "Adaptive gain equalizer in high-index-contrast SiON technology," *IEEE Photonics Technology Letters*, vol. 12, no. 5, pp. 504–506, 2000.
7. B. R. Singh, "Silica based planar lightwave circuits: Status, perspective and future," *IETE Journal of Research*, vol. 45, no. 5–6, pp. 345–353, 1999.

8. Y. P. Li and C. H. Henry, "Silica based optical integrated circuits," *IEE Proceedings: Optoelectronics*, vol. 143, no. 5, pp. 263–280, 1996.

9. K. J. Bales, et al., "A review of glass fibers for optical communications," *Physics and Chemistry of Glass*, vol. 21, no. 1, pp. 5–21, 1980.

10. J. Canning, et al., "Birefringence control in/plasma enhanced chemical vapour deposition planar waveguide by ultra violet irradiation," *Applied Optics*, vol. 39, pp. 4296–4299, 2000.

11. A. K. Ghatak and K. Thyagrajan, *Optical Electronics*, Cambridge University Press, Cambridge, 1994.

12. H. P. Zappe, *Introduction to Semiconducor Integrated Optics*, Artech House, Boston, MA, 1995.

13. B. G. Streetman, *Solid State Electronic Devices*, Prentice Hall International Inc., Englewood Cliffs, NJ, 1990.

14. Y. Ma, et al., "Ultracompact multimode interference 3-dB coupler with strong lateral confinement by deep dry etching," *IEEE Photonics Letter*, vol. 12, no. 5, pp. 492–494, 2000.

15. Leothold, et al., "Multimode Interference couplers with tunable power splitter ratios," *IEEE, JLT*, vol. 19, no. 5, pp. 700–707, 2001.

16. B. J. Offrein, et al., "Wavelength tunable optical add after drop filter with flat pass band for WDM networks," *IEEE Photonics Technology Letters*, vol. 11, no. 2, pp. 239–241, 1999.

17. C. F. Janz, et al., "Bent waveguide couplers for demultiplexing of arbitrary broadly separated wavelengths using two mode interference," *IEEE Photonics Technology Letters*, vol. 7, no. 9, pp. 1037–1039, 1995.

18. R. Kashahara, et al., "New structures of silica based planar light wave circuits for low power thermooptic switch and its application to 8 × 8 optical matrix switch," *Journal of Light Wave Technology*, vol. 20, no. 6, pp. 993–1000, 2002.

19. A. Neyer, "Integrated optical multichannel wavelength multiplexer for monomodesyctem," *IEE Electronics Letter*, vol. 20, no. 18, pp. 744–746, 1984.

20. M. Papuchon, et al., "Electrically active optical bifurcation: BOA," *Applied Physics Letters*, vol. 31, no. 4, pp. 266–267, 1977.

21. N. Takato, et al., "Silica based integrated optic Mach Zehnder multi/demultiplexer family with channel spacing of 0.01–250 nm," *IEEE Selected Areas Communications*, vol. 8, no. 6, pp. 1120–1127, 1990.

22. M. Yagi, et al., "Versatilemultimodes interference photonic switches with partial index modulation regions," *IEE Electronics Letter*, vol. 36, no. 6, pp. 533–534, 2000.

23. S. Darmawan, S.-Y. Lee, C.-W. Lee, and M.-K. Chin, "A rigorous comparative analysis of directional couplers and multimode interferometers based on ridge waveguides," *IEEE Journal of Selected Topics of Quantum Electronics*, vol. 11, no. 2, pp. 466–475, 2005.

24. Y. Hida, et al., "Polymer waveguide thermooptic swich with low electric power consumption at 1.3 μm," *IEEE Photonics Technology Letters*, vol. 5, no. 7, pp. 782–784, 1993.

25. B. Jalali, S. Yegnanarayenan, T. Yoshimoto, I. Rendina, and F. coppinger, "Advances in silicon-on-insulator optoelectronics," *IEEE Journal of Selected Topics in Quantum Electronics*, vol. 4, p. 938, 1998.

26. E. Fluck, F. Horst, B. J. Offrein, R. Germann, H. W. M. Salemink, and G. L. Bona, "Compact versatile thermooptical space switch based on beam steering by a waveguide array," *IEEE Photonics Technology Letters*, vol. 11, no. 11, pp. 1399–1401, 1999.

27. B. J. Offrein, et al., "Polarization-independent thermooptic phase shifters in silicon-oxinitride waveguides," *IEEE Photonics Letters*, vol. 16, no. 6, pp. 1483–1485, 2004.

28. M. R. Paiam, et al., "Polarization insensitive 980/1550 nm wavelength de multiplexer using MMI couplers," *IEE Electronics Letter*, vol. 33, no. 14, pp. 1219–1220, 1997.

29. T. R. Schlipf, et al., "Design and analysis of a control system for an optical delay line circuit used as reconfigurable gain equalizer," *IEEE Journal of Lightwave Technology*, vol. 21, no. 9, pp. 1944–1952, 2003.

30. H. H. Yaffe, C. H. Henry, M. R. Serbin, and L. G. Cohen, "Resonant couplers acting as add-drop filters made with silica on silicon waveguide technology," *IEEE Journal of Lightwave Technology*, vol. 12, no. 6, pp. 1010–1014, 1994.

31. A. Schumacher, et al., "Fully reconfigurable 20 channel optical add-drop multiplexer with integrated variable optical attenuators and power montitors," *IEEE OFC*, vol. 1, pp. 171–172, 2003.

32. N. Ooba, et al., "A thermal silica based arrayed waveguide grating multiplexer using bimetal plate temperature compensator," *IEE Electronics Letters*, vol. 36, no. 21, pp. 1800–1801, 2000.

33. S. Suzuki, "Low loss integrated optic dynamic chromatic dispersion compensators using lattice: Form planar lightwavecircuits," *IEEE OFC*, vol. 1, pp. 176–177, 2003.

34. R. L. Espinola, et al., "Fast and low power thermooptic switch on thin silicon on insulator," *IEEE Photonics Letters*, vol. 15, no. 10, pp. 1366–1368, 2003.

35. Z. Wang, et al., "Rearrangable nonblocking thermooptic 4×4 switching matrix in silicon on insulator," *IEE Proceedings: Optoelectronics*, vol. 152, no. 3, pp. 160–162, 2005.

36. W. E. Chen, et al., "Silicon on insulatorbased 2×2 multi mode interference couplner with large tolerance," *Chinese Physics Letter*, vol. 18, no. 2, pp. 245–247, 2001.

37. R. Krahenbuhl, et al., "Performances and modeling of advanced Ti:LiNbO$_3$ digital optical switches," *IEEE Journal of Lightwave Technology*, vol. 20, no. 1, pp. 92–99, 2002.

38. H. S. Hinton, "Photonic switching using directional couplers," *IEEE Communication Magazine*, vol. 25, no. 5, pp. 16–25, 1987.

39. D. S. Levy, et al., "Fabrication of ultra compact 3 dB2 × 2 MMI power splitter," *IEEE Photonics Letters*, vol. 11, no. 8, pp. 1009–1011, 1999.

40. T. Y. Tsai, et al., "A novel ultra compact two mode interference wavelength division multiplexerfor 1.5 μm operation," *IEEE Journal of Quantum Electronics*, vol. 41, no. 5, pp. 741–746, 2005.

41. R. M. De Ridder, et al., "Siliconoxinitride planar waveguiding structures for application in optical communication," *IEEE Journal of Selected Topics in Quantum Electronics*, vol. 4, no. 6, pp. 930–936, 1998.

42. Y. Hida, et al., "Wavelength division multiplexer with wide passband and stop band for 1.3 μm/1.55 μm using silica based planar waveguide circuit," *IEE Electronics Letters*, vol. 31, no. 16, pp. 1377–1378, 1995.

43. N. Takato, et al., "Silica-based integrated optic Mach-Zehnder multi/demultiplexer family with channel spacing of 0.01–250 nm," *IEEE Journal on Selected Areas in Communications*, vol. 8, no. 6, pp. 1120–1127, 1990.

44. M. B. J. Diemeer, et al., "Polymeric optical waveguide switch using the thermooptic effect," *IEEE Journal of Lightwave Technology*, vol. 7, no. 3, pp. 449–453, 1989.

45. R. Moosburger, et al., "Digital optical switch based on an oversized polymer rib waveguides," *IEE Electronics Letters*, vol. 32, no. 6, pp. 544–545, 1996.

46. H. P. Chan, et al., "A wide angle X junction polymeric thermooptic digital switch with low cross talk," *IEEE Photonic Technology Letters*, vol. 15, no. 9, pp. 1210–1212, 2003.

47. C. Kostrzewa, et al., "Tunable polymer optical add/drop filter for multwavelengthnetworks," *IEEE Phtotonic Technology Letters*, vol. 9, no. 11, pp. 1486–1488, 1997.

48. N. Keil, et al., "Athermal al polymer arrayed waveguide grating multiplexer," *IEE Electronics Letters*, vol. 37, no. 9, pp. 579–580, 2001.

49. M. Kawachi, "Recent progress in silica based planar lightwave circuits on silicon," *IEE Proceedings: Optoelectronics*, vol. 143, no. 5, pp. 257–261, 1996.
50. H. A. Haus and N. A. Whitaker Jr. "Elimination of cross talk in optical directional couplers," *Applied Physics Letters*, vol. 46, no. 1, pp. 1–2, 1985.
51. Y. Hida, et al., "Polymer waveguide thermooptic switch with low electric power consumption at 1.3μm," *IEEE Phtotonic Technology Letters*, vol. 5, no. 7, pp. 782–784, 1993.
52. C. K. Tang, et al., "Development of a librsary of low loss silicon on insulator optoelctronicdevices," *IEE Proceedings: Optoelectronics*, vol. 143, no. 5, pp. 312–314, 1996.
53. L. Eldadaet, et al., "Laser fabricated low loss single mode raised rib waveguiding devices in polymers," *IEEE Journal of Lightwave Technology*, vol. 4, no. 7, pp. 1704–1712, 1996.
54. C. F. Kane and R. R. Krchnavek, "Benzocyclobutene optical waveguides," *IEEE Photonics Technology Letters*, vol. 7, pp. 535–537, 1995.
55. A. J. Beuhler, D. A. Wargowski, K. D. Singer and T. Kowalczyk, "Fabrication of low loss polyamide optical waveguides using thin film multichip module process technology," *Elec. Comput. Technol. Proc.*, pp. 618–620, 1994.
56. A. Neyer, et al., "Fabrication of low loss polymer waveguides using injection molding technology," *IEE Electronics Letters*, vol. 29, pp. 399–401, 1993.
57. D. M. Ruck, et al., "Optical waveguides in polymeric material by ion implantation," *Surface Coat Technology*, vol. 51, pp. 318–323, 1996.
58. K. Worhoff, et al., "Siliconoxinitride in integrated optics," *IEEE Proceedings of ECOC*, vol. 2, pp. 370–371, 1998.
59. F. Ay, "Silicon oxinitride layers for applications in optical waveguides," MS thesis, Bilkent University, 2000.
60. Q. Lai, W. Hunziker, and H. Melchior, "Low power compact 2×2 thermooptic silica on silicon waveguide switch with fast response," *IEEE Photonics TechnologyLetters*, vol. 10, no. 5, pp. 681–683, 1998.
61. B. J. Offrein, et al., "Polarization independent thermooptic phase shifter in silicon oxinitridewaveguides," *IEEE Photonics Technology Letters*, vol. 16, no. 6, pp. 1483–1485, 2004.
62. Y. Inoue, K. Katoh, and M. Kawachi, "Polarization sensivity of a silica waveguide thermooptic phase shifter for planar lightwavecircuits," *IEEE Photonics Technology Letters*, vol. 4, no. 1, pp. 36–38, 1992.
63. P. P. Sahu, "Studies on themooptic integrated optic devices using high index contrast $SiO_2/SiON$ waveguides for fiber optic networks," PhD thesis, Jadavpur University, 2007.
64. G. Keiser, *Optical Fiber Communications*, Mc-Graw Hill Inc, New York, 1999

5 Data Link Control for Optical Network

Since optical network transmits data [1], it is required to discuss link control for transmission of data. This chapter deals with the transmission control of data. For successful data communication/data exchange between two users, much more is required to control over sending and recovering data successfully from signal transmitted through a communication link. For effective data communication the data link control protocol must be designed by including the following objectives [2,3].

- **Frame synchronization:** Data are transmitted as blocks called as frames. The beginning and end of each frame are represented with a bit pattern for frame synchronization.
- **Flow control:** This control is required not to transmit frames at a rate faster than the rate of absorption at the receiving end.
- **Error control:** If the error is detected in the frame, steps must be taken either for correction or for resending the frames successfully after rejection of uncorrected frames.
- **Addressing:** The addressing of frames is needed for the frames to reach the destination where the source address of the frame must be known.
- **Link management:** The starting, maintenance and termination of sustained data exchange require the amount of coordination and cooperation among stations. This arrangement is basically called network management.

5.1 FRAME SYNCHRONIZATION

Two approaches are commonly used for achieving the desired synchronization – asynchronous and synchronous transmission.

5.1.1 Asynchronous Transmission

In case of asynchronous transmission [4,5] each character is synchronized as shown in Figure 5.1. When no character is transmitted, the line between the transmitter and receiver is basically in an idle state. The beginning of a character is represented by start bit with a value of binary 0. It is followed by a character of 5 bit of 8 bits that actually make the character (in case of ASCII coder character has 8 bits). The character is followed by the parity bit evaluated from character bits, where even or odd parity represented by convention is used. This bit is used for the detection of error in character. The final bit is used as a stop bit which is binary 1.

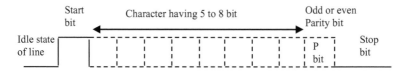

FIGURE 5.1　Asynchronous frame format.

Example 5.1

Find link utilization and overload for sending 100 characters using ASCII coder,

$$\text{Number of Data bits} = 100 \times 8 = 800 \text{ bits}$$

Each character synchronization includes start bit, parity bit and stop bit. Total extra bits are 3 bits

$$\text{Number of extra bits required for sending 100 characters} = 100 \times 3 = 300 \text{ bits}$$

$$\text{Link utilizations} = \frac{\text{Number of transmitted data bits}}{\text{Number of transmitted data bits} + \text{Extra bits}} = \frac{800}{800 + 300} = 72.7\%$$

$$\text{Overload} = \frac{\text{Extra bits}}{\text{Number of transmitted data bits} + \text{Extrabits}} = \frac{300}{800 + 300} = 27.3\%$$

5.1.2　Synchronous Transmission

In asynchronous transmission, link utilization is 72.7% needs to be increased. Instead of having one character for synchronization [4], one can have a group of characters in a frame for synchronization. It can increase link utilization/throughput and at the same time also decrease overload for a data communication. This approach of synchronization is called as synchronous transmission. Figure 5.2 shows the synchronous frame format in which a group of characters is used for synchronous transmission. The frame is started with a preamble bit pattern called Flag field 01111110, and the same flag field is used as a post-amble bit pattern. After preamble bit, there is a control field having 16 bits consisting of an address field and a frame sequence number. It is followed by a data field having variable number of bits as per the number of characters present in the frame. After data field there is a control field having 16 bits consisting of an error control bit that helps to detect the error in the frame. As discussed earlier, the final field is the post amble bit which is a flag bit pattern 01111110.

Bit length	8	16	Variable (> 16)	16	8
	8 bit Flag	Control fields	Data field	Control fields	8 bit Flag

FIGURE 5.2　A HDLC frame.

Example 5.2

Find link utilization and overload for sending 100 characters using ASCII coder for synchronous transmission,

$$\text{Number of Data bits} = 100 \times 8 = 800 \text{ bits}$$

Each character synchronization includes preamble bit, control fields and post-amble bit. Total extra bits are 48 bits

$$\text{Number of extra bits required for sending 100 characters} = 48 \text{ bits}$$

$$\text{Link utilizations} = \frac{\text{Number of transmitted data bits}}{\text{Number of transmitted data bits} + \text{Extra bits}} = \frac{800}{800 + 48} = 94.33\%$$

$$\text{Overload} = \frac{\text{Extra bits}}{\text{Number of transmitted data bits} + \text{Extra bits}} = \frac{48}{800 + 48} = 5.66\%$$

It is seen that the link utilization of synchronous transmission is more than that of asynchronous transmission. Most of the data communication protocol follows a synchronous transmission approach.

5.2   FLOW CONTROL

Data flow control guarantees that a transmitting data flow rate does not overcome the receiving data flow rate. When data reaches the destination, the receiver should do some amount of processing before passing the data to the higher-level layer. First we mention the mechanisms of flow control in the absence of errors. There are two flow control techniques used in data communications – stop and wait flow control and sliding window flow control [2,4],

5.2.1   STOP AND WAIT FLOW CONTROL

The simple flow control is the stop and wait flow control system in which the source transmits a small block of data, and after reception, the destination has its willing-ness to get another frame by sending back an acknowledgment of this block received by the destination. The source has to wait until it receives acknowledgment before sending another block. These small blocks are called as frames sent instead of large blocks of data [4],

- Limitation in buffer size of the destination
- **Sharing:** It permits one station to occupy the medium for an extended period for avoiding long delays for other sending stations.

Figure 5.3 illustrates stop and wait flow control showing frames transmitted by source and receiver by destination with timing. In the figure, "a" represents the prop-agation time and "1" indicates transmission time of a frame. There are two types

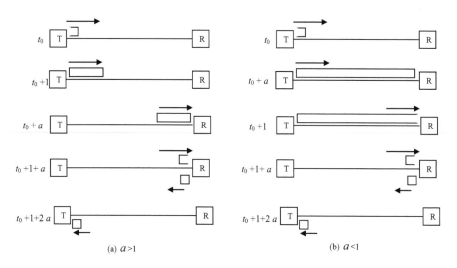

(a) $a > 1$ (b) $a < 1$

FIGURE 5.3 Timing diagram of stop and wait flow control (transmission time = 1 and propagation time = a).

of situations – $a > 1$ and $a < 1$. Figure 5.3a shows stop and wait flow control for the situation $a > 1$ in which the total time required to send a frame and receive the same at the receiver is $2a + 1$, which includes the time required for receiving the acknowledgment of the frame by the source. So the link utilization is written as

$$\frac{\text{transmission time}}{\text{total time required to complete the cycle of transmission}} = \frac{1}{2a+1}$$

Figure 5.3b represents stop and wait flow control for the situation $a > 1$, in which the total time required to send a frame and receive the same at the receiver is also $2a + 1$, which includes the time required for receiving the acknowledgment of the frame by the source. So the link utilization is also derived as the same as that for the situation $a < 1$.

The link utilization of stop and wait flow control is low as no transmission takes place during the transmission of acknowledgment of receipt of a frame.

5.2.2 Sliding Window Flow Control

The drawback of stop and wait flow control shows that only one frame can be transmitted at a time during which no transmission of other frames takes place and at the same time during the transmission of acknowledgment of received frame also no frame is in transit. This gives a poor efficiency of the system. The efficiency is enhanced by allowing multiple frames to be transmitted at a time. This type of data flow is possible by using a full duplex link [4].

To explain this flow control, we consider two stations – source A and destination B as shown in Figure 5.4. The stations are connected by a full duplex link. In this case the destination station B must have a buffer space of n frames which are allowed

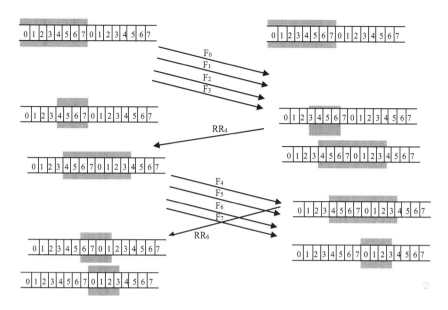

FIGURE 5.4   Sliding window flow control.

to send by source station A, not waiting for any acknowledgment. The acknowledgment of frames is tracked so that all the frames have been acknowledged and each is labelled with a sequence number so that B acknowledges a frame with acknowledgment that includes the frame sequence number. This acknowledgment shows to source A that B is ready to receive the next frame. We consider that B could receive frames F_0, F_1, F_2 and F_3 but withhold acknowledgment until frame F_4 has arrived, by then returning to an acknowledgment RR_4 with sequence number 4, automatically B acknowledges F_0, F_1, F_2 and F_3 at one time. At the same time A maintains a list of sequence numbers that is allowed to send. Each of these lists is basically a window of frames. This process goes on and looks like the window is sliding. The operation is named as sliding window flow control.

Since the maximum length in the figure is eight frames, after completion of one window, it goes to the next window of sliding. If the destination station is not ready and wishes to cut off the receipt of frames then it must issue no acknowledgment with receive not ready (RNR) at the frame sequence. For example, RNR5 states that the receiver is not ready for a frame with sequence 5.

For sliding window flow control, the efficiency or link utilization is a function of window size N and the value of propagation time a. To find the link utilization, the transmission time of frame is considered to be 1 for convenience. There are two types of situation – (a) $N > 2a + 1$ and (b) $N < 2a + 1$. Figure 5.5 shows the point (station A) to point (station B) data flow using sliding window flow control. Station A begins to transmit a sequence of frames at time t_0. The leading edge of the first frame reaches station B at $t_0 + a$. The first frame is received fully by station at time $t_0 + a + 1$. After receiving the first frame, station B starts sending acknowledgment ACK1, which is received by station A at time $t_0 + 1 + 2a$ (neglecting the small

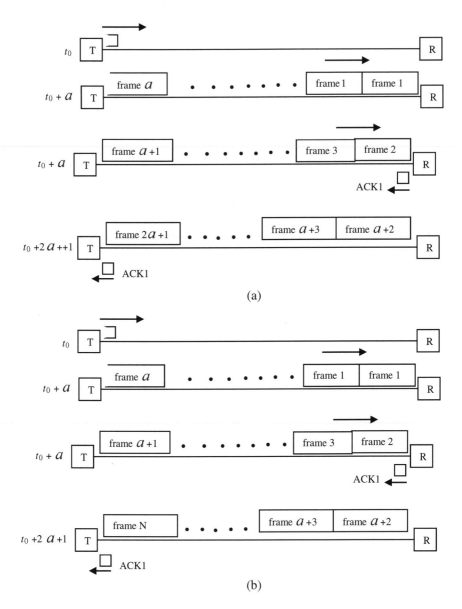

FIGURE 5.5   Sliding window flow control. (a) $N > 2a + 1$, (b) $N < 2a + 1$.

transmission time of acknowledgment). To estimate the performance for the above two cases, the following procedure follows as shown in Figure 5.5.

Case1: $N > 2a + 1$: The acknowledgment for frame 1 reaches A before A has already transmitted $2a + 1$ number of frames at time $t_0 + 1 + 2a$. So the link utilization is equal to $U = \dfrac{\text{No of transmitted frame within time } 2a + 1}{\text{Total time required for transmiting these frames}} = \dfrac{2a + 1}{2a + 1} = 1.0$

Case2: $N < 2a + 1$: The acknowledgment for frame 1 reaches A before A has already transmitted N number of frames at time $t_0 + 1 + 2a$. So the link utilization is equal to

$$U = \frac{\text{No of transmitted frame within time } 2a + 1}{\text{Total time required for transmiting these frames}} = \frac{N}{2a + 1} = 1.0,$$

Both cases of flow control work under ideal cases where there is no noise/distortion in the received signal during transmission of the signal. In fact, all the transmission media introduce the noise or distortion which provides the error in the signal which cannot be tolerated in data transmission. So it is required for detection of error in the frame and, after detection of error, steps should be taken so that data should be received successively without error. Along with flow control, one should have error control for successful transmission of data from source to destination.

5.3   ERROR DETECTION AND CONTROL

Before discussion of error control, one should know about error detection in the frame. There are two approaches adapted for error detection – approach where a particular error bit can be located, and once it is located it can be corrected easily. In another approach, error can be detected but it cannot locate which bit is an error.

5.3.1   Error Detection

As discussed earlier, the design of transmission system should include error detection because error results error/change in one or more bits in the transmitted frame [3,4]. Figure 5.6 shows the block diagram of an error detection approach used normally in data transmission system. For a given frame, additional bits required to get an error detecting code (E) are included as error control bits. The E is evaluated from data bits in frame and is f (Data). After receiving frame having data bits and error detecting bits, again the error detecting codes are evaluated by using the same technique that in the transmitted and is denoted as E′. After comparison between E and E′, error frame is detected in the following manner. The bits in E′ will be the same as that in E. If there is no error included in data during transmission, the bits in E′ will not be the same as that in E. After comparison between E and E′, an error frame is detected.

The simplest approach is used with parity check in case of asynchronous data transmission. The parity bit (even/odd parity) is estimated from ASCII bits of a character. In the similar way as shown in Figure 5.6, the error of asynchronous frame is detected. This parity check is not foolproof error detection.

The error detecting codes are estimated by different ways in case of synchronous transmission. Here, two approaches are discussed and normally used for data transmission – vertical and horizontal redundancy check (VRC and HRC) and cyclic redundancy check (CRC) which are described below.

5.3.1.1   Vertical and Horizontal Redundancy Check

The VRC and HRC are applied to a matrix of data bits for estimation of VRC and HRC codes [4] that are transmitted along with data as error detecting codes (E)

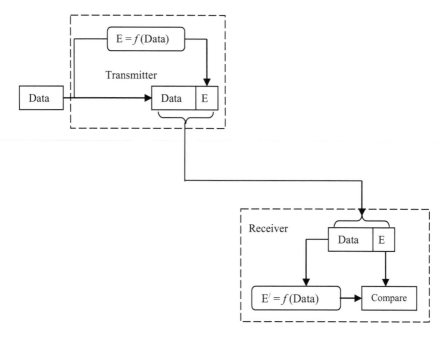

FIGURE 5.6   Error detection.

through transmission media to the receiver (as shown in Figure 5.6). The data bits $a_{11}, a_{12} \ldots a_{1n}, a_{21}, a_{22} \ldots a_{2n} \ldots a_{m1}, a_{m2} \ldots a_{mn}$, are represented in matrix form as given below:

The error detecting VRC codes are estimated as

$$VRC_1 = a_{11} \oplus a_{21} \oplus \cdots \oplus a_{m1}$$

$$VRC_2 = a_{12} \oplus a_{22} \oplus \cdots \oplus a_{m2}$$

$$\ldots\ldots\ldots\ldots\ldots\ldots\ldots\ldots\ldots\ldots$$

$$\ldots\ldots\ldots\ldots\ldots\ldots\ldots\ldots\ldots\ldots$$

$$VRC_n = a_{1n} \oplus a_{2n} \oplus \cdots \oplus a_{mn}$$

The error detecting HRC codes are estimated as

$$HRC_1 = a_{11} \oplus a_{21} \oplus \cdots \oplus a_{1n}$$

$$HRC_2 = a_{12} \oplus a_{22} \oplus \cdots \oplus a_{2n}$$

$$\ldots\ldots\ldots\ldots\ldots\ldots\ldots\ldots\ldots\ldots$$

$$\ldots\ldots\ldots\ldots\ldots\ldots\ldots\ldots\ldots\ldots$$

$$HRC_m = a_{m1} \oplus a_{m2} \oplus \cdots \oplus a_{mn}$$

The error detecting codes are written as

$$E = \text{VRC}_1, \text{VRC}_2 \ldots \text{VRC}_n, \text{HRC}_1, \text{HRC}_2 \ldots \text{HRC}_m$$

To make fixed error detecting codes bits, the bits of data matrix should be kept constant. This means that the number of data bits in the frame remains constant. In this technique there are two error bits in the same row and two errors in the same column (as for an example $a_{11}, a_{12}, a_{21}, a_{22}$ bits are the error bits), the error bit remains undetected because there is no change of VRC or HRC codes. So this technique is not foolproof. One of the advantages of this technique is that it can locate which bit is an error bit. It is required to find the most powerful technique that can detect the error frame with foolproof. In the next section, we discuss the foolproof error detection scheme.

5.3.1.2   Cyclic Redundancy Check

One of commonly used error detecting technique is CRC [4] in which n bit frame check sequence (FCS) is generated as error detecting codes from data having $k + n$ bits (where k = no. of data bits and n = no. of bits obtained from decimal number 2^n) by using pattern (P) of $n + 1$ bits, which is known as predetermined divisor. The pattern should have most and least significant bit of always bit 1. The receiver then divides the data with FCS bits by the same pattern P. If there is no remainder, there is no error in the frame [6].

For description of this technique, we define the following entity

$T = (k + n)$ bits having k bit message bits to be transmitted in frame $(k > n)$.
$M = k$ bit message which is first k bit of T.
P = pattern having $n + 1$ number of bits.
$F = n$ bit FCS which last n bits of T.

The T is written as

$$T = 2^n M + F$$

F is determined from $2^n M$ divided by pattern,

$$\frac{2^n M}{P} = Q + \frac{R}{P}$$

Q = quotient and R = remainder. The above division is binary, and the remainder is one bit less than that of P. The FCS is also one bit less than that of pattern. So we can use the remainder R as FCS. So we can write $T = 2^n M + R$. The T is divided by P as given below:

$$\frac{T}{P} = \frac{2^n M + R}{P} = \frac{2^n M}{P} + \frac{R}{P} = Q + \frac{R}{P} + \frac{R}{P}$$

Any binary number added with itself (modulo2) gives zero. So we can write

$$\frac{T}{P} = Q + \frac{R}{P} + \frac{R}{P} = Q$$

To explain CRC, we take the simple example given below:

M = 1010101101
P = 110101
FCS F = ??

The message M is multiplied by 2^5 giving 101010110100000, which is divided by P to find F

$$P \rightarrow 110101\,)101010110100000\,(1101101101 \leftarrow Q$$

```
              110101
            --------
              111111
              110101
            --------
               101010
               110101
             --------
                111111
                110101
              --------
                 101000
                 110101
               --------
                  111010
                  110101
                --------
            R→111100
                 110101
               --------
                    1001
```

So FCS is estimated as 11100 and T is obtained as 101010110101001.

$$P \rightarrow 110101\,)101010110101001\,(1101101101 \leftarrow Q$$

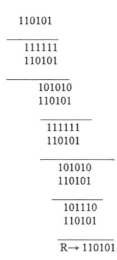

$$\frac{110101}{\begin{matrix}111111\\110101\end{matrix}}$$

$$\frac{101010}{110101}$$

$$\frac{\overline{111111}}{110101}$$

$$\frac{\overline{101010}}{110101}$$

$$\frac{\overline{101110}}{110101}$$

$$\overline{R \rightarrow 110101}$$

Since there is no error in T, T is perfectly divisible with P and there is no remainder.

5.3.1.2.1 Polynomial Form of CRC

Pattern $P = 110101$ is represented as
$P(X) = X^5 + X^4 + X^2 + 1$ and the message $M = 1010101101$

$$M(X) = X^9 + X^7 + X^5 + X^3 + X^2 + 1$$

The CRC process can be described as

$$\frac{X^5 M(X)}{P(X)} = Q(X) + R(X)/P(X)$$

The $T(X)$ can be written using CRC as

$$T(X) = X^5 M(X) + R(X)$$

Three versions of pattern $P(X)$ are normally used for the estimation of FCS as error controlling codes of CRC [IEEE 802 standards]. These are given below:

$$CRC - 16 = X^{16} + X^{15} + X^2 + 1$$

$$CRC - CCITT = X^{16} + X^{12} + X^5 + 1$$

$$CRC - 32 = X^{32} + X^{26} + X^{23} + X^{22} + X^{16} + X^{12} + X^{11} + X^{10} + X^8 + X^7 + X^5$$

$$+ X^4 + X^2 + X + 1$$

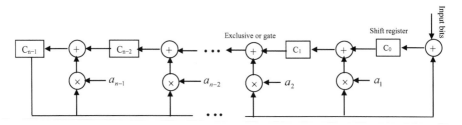

FIGURE 5.7 CRC architecture for pattern $P(X) = 1 + a_1 X + a_2 X^2 + \cdots + a_{n-1} X^{n-1} + X^n$.

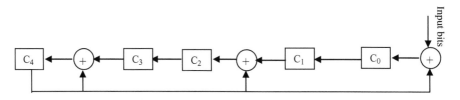

FIGURE 5.8 CRC architecture for pattern $P(X) = 1 + X^2 + X^4 + X^5$.

5.3.1.2.2 Digital Logic Circuit of CRC

The CRC process is designed by using dividing circuit having exclusive-or gates and shift register. The shift register has a string of 1-bit storage device in which each device has an output line that indicates the value currently stored and an input line. At each clock time/discrete time instant, the value stored in the device is changed by the value coming via its input line. The entire register is controlled by clock, simultaneously causing a 1-bit shift along the entire register.

Figure 5.7 shows a general architecture of n number of single-bit shift registers and exclusive-or gates. In the figure, the general pattern $P(X)$ is represented in terms of polynomial as $P(X) = \sum_{i=0}^{n} a_i X^i$, where $a_0 = a_n = 1$ and all other values of coefficients are either 0 or1 depending on bit pattern of $P(X)$. In the figure, there is no multiplication in feedback for input and final feedback as $a_0 = a_n = 1$. Initially all the shift registers are cleared. One by one message bits with $2^n(2^n M(X))$ come to the input circuit starting with the most significant bit of message bits, and all the shift registers are updated and stored [6].

Figure 5.8 shows a CRC circuit using pattern $P = 110101$, which can be written in polynomial as $P(X) = 1 + X^2 + X^4 + X^5$. The process begins with all shift registers C_0, C_1, C_2, C_3 and C_4 cleared (all zeros). The message bits are $M = 1010001101$. The $2^5 M = 101000110100000$ is entered starting with the most significant bit (one bit at a time). Table 5.1 shows the step-by-step operation in which by one bit comes and the shift registers C_0, C_1, C_2, C_3 and C_4 are updated and stored till last bit comes to the input of CRC architecture. Finally the bits stored in C_0, C_1, C_2, C_3 and C_4 are evaluated as FCS, which is obtained as 01110.

TABLE 5.1
FCS Estimation

FCS estimation

Steps	C_4	C_3	C_2	C_1	C_0	$C_4 \oplus C_3$	$C_4 \oplus C_1$	$C_4 \oplus Input$	Input	
0	0	0	0	0	0	0	0	1	1	
1	0	0	0	0	1	0	0	0	0	
2	0	0	0	1	0	0	1	1	1	Message
3	0	0	1	0	1	0	0	0	0	bits
4	0	1	0	1	0	1	1	1	1	
5	1	0	1	0	1	1	1	1	0	
6	1	1	1	1	1	0	0	0	1	
7	0	1	0	1	0	0	0	0	1	
8	1	0	1	1	1	1	0	1	0	
9	1	1	0	1	1	0	0	0	1	
10	0	0	0	1	0	0	1	1	0	
11	0	0	1	0	1	0	0	0	0	Five
12	0	1	0	1	0	1	1	0	0	Zeros
13	1	0	1	0	0	1	1	1	0	
14	1	1	1	0	1	0	1	1	0	
15	0	1	0	0	1	0	0	1		

5.3.2 ERROR CONTROL

Error control refers to steps taken by data transmission system as and when error is detected in the frames. If the error bits are located in the frames, then one can correct the errors in the frames. There are many coding techniques such as block check sequence (BCS) coding and convolution codes used for error detection and correction. But no technique gives a 100% foolproof error detection and correction. When the error frames are detected and the locations of error bits are found, then it is difficult to correct the frame, and for that, it is required to control data transmission so that at the destination corrected frames should be received and correct message should be recovered [4]. There are two ways in which error can arise

- The noise burst can damage the frame in which the error may be included.
- Errors can be included due to distortion in the signal during transmission.

As discussed, steps should be taken, if error frames are detected (but no location of error in the frame). There are three types of situations in the destination.

1. If the frame does not have any error, then positive acknowledgments are sent to the source by the destination station.
2. If the frames detected errors, the destination sends back a negative acknowl-edgment to the source, which means the error is detected in the frames. The source retransmits these frames.

3. The source transmits the frames, but the acknowledgment for these frames is not received in the predetermined time. In this situation, the frames are retransmitted till the source receives the acknowledgment.

In the second situation, the frames detected errors in the destination and an automatic repeat request (ARQ) are sent to the source for retransmission of frames. These mechanisms are called as ARQ [7]. There are three versions of ARQs adapted with flow control.

- Stop and Wait ARQ
- Go-back-N ARQ
- Selective-reject (SREJ) ARQ.

5.3.2.1   Stop and Wait ARQ

Stop and Wait ARQ system follows Stop and Wait flow control, which is already discussed earlier [4]. The source transmits a single frame and must then wait for an acknowledgment. During this time no frame is transmitted until the source receives the acknowledgment sent by the destination. This procedure is only stop wait flow control without error control. The error control can be made using ARQ technique.

Two types of error occurs – for the first sort of error, the frames that reached to destination are damaged and detected by using the error detecting technique. The second sort of error is a damaged acknowledgment, even if the frame is received at destination without error.

Figure 5.9 shows the diagram that explains stop and wait ARQ. In the figure, station A sends the frames and station B receives the same. If the frame received by station B is correct, then it sends acknowledgment (ACK). The ACK is damaged in transit and is not recognized by A which resends the same frame after timeout. The duplicate frame arrives to the station B; and station B has accepted the two same frames as if they are separate. To avoid this, the frames are labelled as 0 or 1 and positive ACK is a form of ACK0 and ACK1. If the frame received by station B is a detected error, then ACK0 is sent to station A as negative acknowledgment.

It is seen that maximum link utilization for stop and wait ARQ is obtained with no errors and is $U_{max} = \dfrac{1}{1+2a}$. But we have to determine the link utilization with the possibility that some frames are repeated because of errors. The link utilization U can be defined as $U = \dfrac{T}{T_{tot}}$, where T = time required for the transmitter to emit a single frame, T_{tot} = total time that is engaged for the transmission of single frame to the destination by source and transmission of acknowledgment of receipt of the frame to the source by destination $= T + 2T_p$, T_p = propagation time. If errors occur, the link utilization expression is modified as

$$U = \frac{T}{N_r T_{tot}} = \frac{1}{N_r(1+2a)}$$

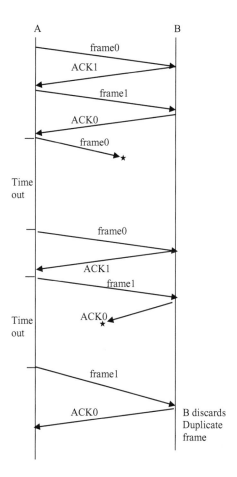

FIGURE 5.9   Stop and wait ARQ.

where N_r = the expected number of transmission of a frame and $a = T_p/T$. The N_r is derived by considering the probability P that a single frame has an error. The probability that single frame has no error is $(1-P)$. It is also considered that ACKs and NAKs are never in errors. The N_r is derived as

$$N_r = E[\text{transmissions}] = \sum_{i=1}^{\alpha}(iP^{i-1}(1-P)) = \frac{1}{1-P}$$

By substituting the value of N_r in the expression of U, we can write

$$U = \frac{1-P}{(1+2a)}$$

The advantage of stop and wait with ARQ is its simplicity, but its main disadvantage is poor link utilization because flow control is not efficient.

5.3.2.2 Go-Back-N ARQ

As discussed earlier, the problem described in stop and wait ARQ is that only one frame is transmitted at a time during which no transmission of other frames takes place and at the same time during transmission of acknowledgment of received frame also no frame is in transit. This gives serious inefficiencies of the system. The efficiency can be greatly improved by using a sliding window flow control with ARQ (error control). The simple form of error control combined with sliding window flow control concept is Go-back-N ARQ [4,7].

In Go-back-N ARQ, source station transmits a series of frames one by one with sequence number. When no errors occur, the destination acknowledges with receive ready (RR). If destination station finds an error in a frame, it sends a negative acknowledgment with REJ (reject) for that frame. The destination station rejects the frame and all future incoming frames until the frame in error is correctly received. Thus the source station, after receiving an REJ, must retransmit the frame in error plus all the succeeding frames that were transmitted interim. Figure 5.10 shows the timing diagram that describes the operation of Go-back-NARQ in which station A transmits the frames to station B. In the figure, frames are transmitted beginning from frame 0 sequentially. After receiving frame 1, RR2 acknowledgment is sent. In the figure, frame 5 is the detected error and then REJ5 is sent to the source A for

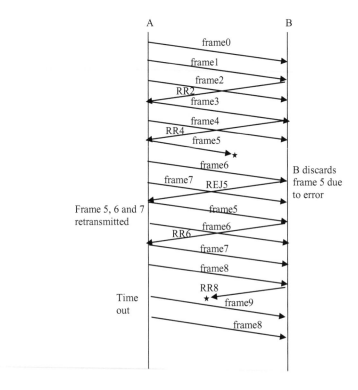

FIGURE 5.10   Go-back-N ARQ.

retransmitting the same. After receiving REJ 5, by that time, the frames 6 and 7 are already received by station B, and all these frames sent after frame 5 are rejected and go back to frame 5 and begin sending the frames from frame 5 onwards.

The link utilization of Go-Back-N ARQ is written as

$$U = \frac{1}{N_r(1+2a)}$$

By using the reason of Go-back-N ARQ, the expected number of transmissions of a frame is written as

$$N_r = E[\text{Number of transmitted frames to sucessively transmit one frame}]$$

$$= \sum_{i=1}^{\alpha} f(i)P^{i-1}(1-P)$$

where $f(i) = 1 + (i - 1)K = 1 - K = K_i$. The N_r is written as

$$N_r = \sum_{i=1}^{\alpha} f(i)P^{i-1}(1-P) = (1-K)\sum_{i=1}^{\alpha} P^{i-1}(1-P) + K\sum_{i=1}^{\alpha} iP^{i-1}(1-P)$$

$$= (1-K) + \frac{K}{1-P} = \frac{1-P+KP}{1-P}$$

By taking $K = 1 + 2a$ for $N > 1 + 2a$ and $K = N$ for $N < 1 + 2a$, the link utilization is written as

$$U = \begin{cases} \dfrac{1-P}{(1+2aP)} & N > 1 + 2a \\[4mm] \dfrac{N(1-P)}{(1+2a)(1-P+NP)} & N < 1 + 2a \end{cases}$$

5.3.2.3 SREJ ARQ

The efficiency can be greatly improved by using selection of error frames and rejection of those asking for retransmission of those from the source station with ARQ (error control). This form of the concept is SREJ ARQ [4,7].

In SREJ ARQ, like Go-back-N ARQ, source station sends a series of frames sequentially one by one with sequence number. When no error occurs, the destination acknowledges with RR. If destination station finds an error in a frame, it transmits a negative acknowledgment with SREJ for that frame. The destination station will reject that error frame and request the source station to retransmit the same by sending an SREJ message with select and reject the same frame. Figure 5.11 shows the diagram describing the operation of SREJ ARQ in which station A sends the frames to station B. In the figure, frames are sent starting from frame 0 sequentially. After receiving frame 1 an RR2 acknowledgment is sent. In the figure frame 5 is

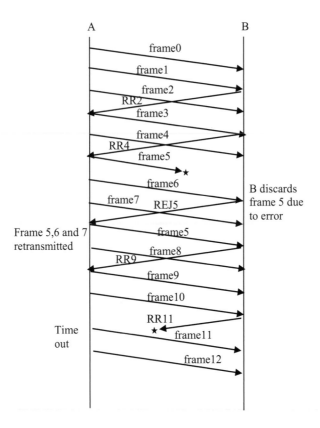

FIGURE 5.11   SREJ ARQ.

detected error and then SREJ5 is sent to the source A for retransmitting the same. When SREJ5 is received, by that time, the frames 6 and 7 are already received by station B. Even, these frames sent after frame 5 are not rejected and instead of go back to frame 5, only the frame 5 is sent by the source A. Again the source starts sending the frames from frame 8 onwards.

The link utilization is written as

$$U = \frac{1}{N_r(1+2a)}$$

By using reason of SREJ ARQ, the expected number of transmissions of a frame is written as

$$N_r = E[\text{Number of transmitted frames to sucessively transmit one frame}]$$

$$= \sum_{i=1}^{\alpha} iP^{i-1}(1-P)$$

The N_r is written as [8]

$$N_r = \sum_{i=1}^{\alpha} iP^{i-1}(1-P) = \frac{1}{1-P}$$

The link utilization is written as

$$U = \begin{cases} 1 = P & N > 1 + 2a \\ \dfrac{N(1-P)}{(1+2a)} & N < 1 + 2a \end{cases}$$

For transmission of only one frame, the operation of both Go-back-N ARQ and SREJ ARQ becomes same as that of stop and wait ARQ. If N is more, we can notice a significant improvement of SREJ ARQ over that of Go-back-N ARQ.

5.4 HIGH-LEVEL DATA LINK CONTROL (HDLC)

The most important data link control is high-level data link control (HDLC) [8,9] which is widely used. It is standardized by using standard ISO 33009/ISO 4335. It is followed in many other important data link control protocols that use almost the same format of frame, and same mechanisms are followed. This protocol may also be used in lower level of optical networks.

5.4.1 TYPES OF STATION

There are three types of stations – primary, secondary and combined stations.

The primary station manages the operation of link and frames transmitted by primary station are commands. The secondary station operates under the control of primary stations, and the frames transmitted by secondary stations are responses. For this, the primary stations maintain a separate logical link with each secondary station. The combined station maintains the features of both primary and secondary stations.

5.4.2 TYPES OF CONFIGURATIONS

Two types of configurations – unbalanced configuration having one primary station and one or more secondary stations supporting full duplex and half duplex transmissions and balanced configuration having two combined stations supporting both full duplex and half duplex transmissions.

5.4.3 TYPES OF DATA TRANSFER MODES

- **Normal response mode (NRM):** It works with unbalanced configurations in which the primary station starts data transfer to the secondary station and after receiving data, the secondary station transmits data as responses.

- **Asynchronous balanced mode (ABM):** It works with balanced configurations in which either of combined stations may initiate data transfer to a combined station without receiving permission from other combined stations.
- **Asynchronous response mode (ARM):** It works with unbalanced configurations in which the secondary stations may initiate data transfer to the primary station without permission, but the primary station takes responsibility.

5.4.4 HDLC Frame Format

The HDLC follows a synchronous transmission in which synchronization is made for a group of characters (called as frames) using flag fields [4]. Apart from the flag fields, it has an address field and control fields that precede data field having characters' bits. These flag fields, address fields and control fields are known as a header. Another control field having FCS and flag fields (end field) is known as a trailer. Figure 5.12 shows the structure of HDLC frame format. HDLC has three types of frames – information frame named as I-frame, supervisory frame named as S-frame and an unnumbered frame known as U-frame.

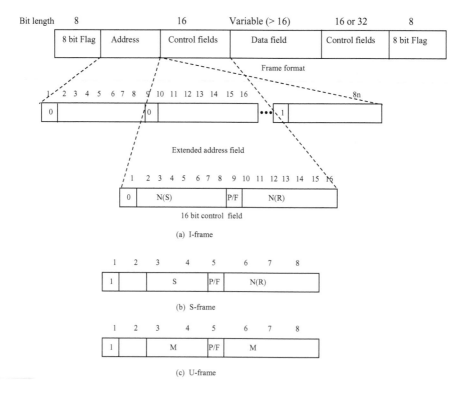

FIGURE 5.12 HDLC frame format (I – Information, S – Supervisory, U – Unnumbered) $N(S)$ – Send sequence number and $N(R)$ – Receive sequence number.

- **Flag field:** Flag fields are synchronization bits placed at both start and end of the frame with a unique pattern 01111110. On both user–network interfaces, the receivers are continuously hunting for a flag sequence for synchronization at the start and end of the frame. There may be a possibility to have pattern 01111110 inside the frame. This possibility destroys frame synchronization. To remove this problem, the procedure known as bit stuffing is used. In between start flag and end flag signal, the transmitter inserts 0 bit after five 1s inside the frame. In the receiver side, after start flag, it monitors the bit stream and removes 0 after every five 1s. Figure 5.13a shows the bit patterns before and after bit stuffing.
- **Address field:** The address field identifies the secondary station that is to receive the frame transmitted by the primary station. This field is not required for point-to-point data transmission but it is included for

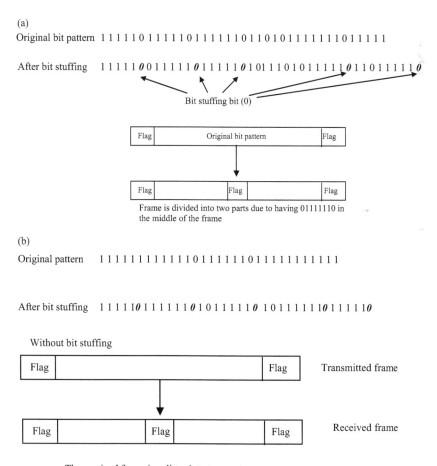

FIGURE 5.13   (a) Bit stuffing and the effect on frame without bit stuffing. (b) Bit stuffing of original and effect on data bit pattern without bit stuffing.

uniformity only. The address field is eight bit length, but for prior agreement it is extended with an address length which is a multiple of seven bits. The least significant bit of each octet is 1 or 0 depending on whether it is not the last octet of the address field. The remaining seven bits of each bits of each octet is part of the address. The single octet address of 11111111 is represented as all the stations address in both basic and extended formats. It is used to allow the primary station to broadcast a frame for reception by all secondary stations.

- **Control fields:** HDLC has three types of frames having different control fields. The I-frames having the data to be transmitted for the user should support flow control and error control (ARQ mechanism) and are piggy-backed. The S-frames provide an ARQ mechanism only but piggyback-ing is not used. The U-frames provide supplemental link control functions. First one or two bits represent the types of frame, and the remaining bit positions are organized into subfields as indicated in Figure 5.12. Their uses are explained below the operation of HDLC. The basic control field for S-frame and I-frame uses 3 bit sequence numbers. In case of an extended frame, the frame sequence is a 7 bit sequence number. U-frames always use an eight bit control field.

- **Information fields:** The information field must be included in I-frames and may be included in some of the U-frames. It is an integral number of octets. The length of the information field is variable up to a system-defined maximum.

- **Frame check sequence field:** The control field before end flag field is the FCS field which is determined by using CRC as an error controlling code. The length of FCS codes is governed by CRC pattern. The 16 bit and 32 bit FCS's are determined by using 16 bit CRC-CCITT (Consultative Committee for International Telephony and Telegraphy) and CRC-32 respectively.

5.4.5 Operation of HDLC

HDLC operations are mainly based on the exchange of I-frames, S-frames and U-frames. The various commands and responses used for these operations are mentioned in Table 5.2. The table also shows the different frames with their purposes and functions.

There are three phases of HDLC operation for data transfer – Initialization, data transfer and disconnect [4].

5.4.5.1 Initialization

Initialization can be made by using one of six set mode commands as mentioned in Table 5.2. These commands serve the following purposes to state:

- Other side initialization requested
- One of three modes (NRM, ABM and ARM) requested
- Whether 3 or 7 bit sequence numbers are used

TABLE 5.2
HDLC Command and Response

Name	Commands (C)/ Response (R)	Function and Description
Information (I)	C/R	Exchange user data
	Supervisory (S)	
RR	C/R	Positive acknowledgment ready to receive I-frame
RNR	C/R	Positive acknowledgment not ready to receive I-frame
Reject (REJ)	C/R	Negative acknowledgment Go back-N
SREJ	C/R	Negative acknowledgment SREJ
	Unnumbered (U)	
Set normal response/extended mode (SNRM/SNRME)	C	Set mode, extended = 7 bit sequence numbers
Set asynchronous response/extended mode (SARM/SARME)	C	Set mode, extended = 7 bit sequence numbers
Set asynchronous balance/extended mode (SABM/SABME)	C	Set mode, extended = 7 bit sequence numbers
Set initialization mode (SIM)	C	Initialize link control functions in address station
Disconnect (DISC)	C	Terminate logical link connection
UA	R	Acknowledge acceptance of the set mode commands
Disconnect mode (DM)	C	Terminate logical link connection
Request disconnect (RD)	R	Request for DISC command
Request initialization mode (RIM)	R	Request for SIM command
Unnumbered information (UI)	C/R	Used to exchange control information
Unnumbered poll (UP)	C	Used to solicit control information
Reset (RSET)	C	Used for recovery: resets $N(R)$ and $N(S)$
Exchange identification (XID)	C/R	Used to request and report status
Test (TEST)	C/R	Exchange identical information fields for testing
Frame reject	R	Reports receipt of unacceptable frame

If other side admits this request, then the HDLC module transmits an unnumbered acknowledgment (UA) back to the initiating side. If the request is rejected, then a disconnected mode (DM) frame is sent.

5.4.5.2 Data Transfer

When the initialization is processed and admitted, then a logical connection is set up. Both sides start sending user data I-frames starting with sequence number 0. The $N(S)$ and $N(R)$ of I-frames indicating sequence number make flow control and error control. An HDLC module sending a sequence of I-frames provides the number sequentially using either 3 or 7 bits. The $N(S)$ is the sequence number with which I-frame is transmitted, whereas $N(R)$ is the sequence number with which I-frames are received, indicating I-frame number is expected to be received. S-frames are used for flow and error control. The RR is used to acknowledge the last frame received, indicating the next station to be received. RNR acknowledges I-frame and request the sender to suspend I-frame transmission. The REJ with frame number is issued in Go-back-N ARQ if the frames detected error. The SREJ with number $N(R)$ is issued in SREJ ARQ if the frames detected error.

5.4.5.3 Disconnect

The disconnect is initiated either if there is a fault (link or node) or if data transfer is over. For the above operation, HDLC issues DISC command to request for disconnection. The other side must admit the disconnection by replying with UA.

5.4.6 Examples of HDLC Operations

There are fundamental operations of HDLC-lick setup and disconnect two-way data exchange, busy condition and one-way data transmission. Figure 5.14a shows the HDLC operation on link setup and disconnect. In the figure, the station A transmits set asynchronous balance mode (SABM) command to the other side and starts the timer. It issues SABM again after time out till UA response is sent by Station B. Once station A receives UA response, the link is set up. After data transfer is over, the station A issues DISC commands for the disconnection of link setupand after receiving the same station B issues UA response.

Figure 5.14b shows the HDLC operation on two-way data exchange. After establishment of link setup the station A transmits I-frames (I,1,1; I,2,1: so on from A to B) to the other side. The I-frames (I,1,3; so on from B to A) are sent by Station B. In addition of that, it also issues S-frames such as UA.

Figure 5.14c shows the HDLC operation on one-way data transfer with busy condition. After establishment of link setup the station B transmits the frames (I,1,0; I,2,0 etc.) from B to A. Since station A is not ready, station A transmits RNR 3 to B showing frame I,3,0 is not ready to receive. The RNR from A represents busy condition of frame. Again the station B issues RR, 0, P with p-bit set representing poll bit. Station A sends RNR to station B till it is ready to receive and at the same time station B issues RR with p-bit. Once A is ready, it sends RR, 3 and B and then transmits I,3,0.

Figure 5.14d shows the HDLC operation on one-way data exchange with error control using SREJ command. After establishment of link setup the station A transmits I-frames (I,1,1; I,2,1: so on from A to B) to the other side. The station B detects error in frame I,3,0 and issues SREJ 3 to station A. Before receiving SREJ 3, Station A already sends frame I,4,0. Once SREJ 3 is received by A, it again sends I,3,0. After transmission of frame I,3,0, it starts transmission of frame from I,5,0 and so on.

Figure 5.14e shows the HDLC operation on one-way data exchange with control due to link failure. After establishment of link setup the station A transmits I-frames (I,1,1; I,2,1: so on from A to B) to the other side. The frame I,3,0 does not reach due to link failure and ifI,3,0 is not received, station B sends RR3 representing

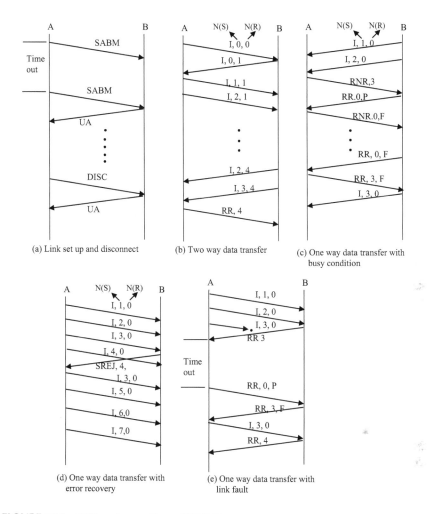

FIGURE 5.14   Different operations of HDLC.

RR for I,3,0 frame. After time out station A issues RR0. F representing ready to send the same frame, before receiving the same by B, station B sends again RR 3. Then station A sends frame I,3,0 again. After receiving frame I,3,0 station B sends RR4 and the process goes on till data transfer is over.

5.5   OTHER LINK CONTROL PROTOCOL

5.5.1   LAPB

Link access procedure balanced (LAPB) was standardized by ITU-Ts part of packet switching network link control. It is part of HDLC in which only ABM is followed for data transmission. The frame format of LAPB is same as that of HDLC shown in Figure 5.15a.

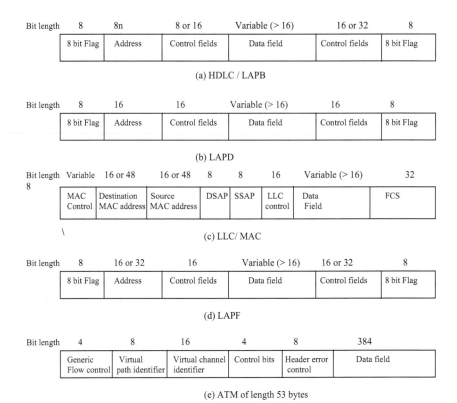

FIGURE 5.15   Different data link control frame formats.

5.5.2 LAPD

Link access procedure D channel (LAPD) was standardized by ITU-T as part of ISDN (Integrated services digital network) link control. LAPD provides data link control for D channel, which is a logical channel at the interface of ISDN [4]. It also supports only ABM for data transmission. The frame format of LAPD is shown in Figure 5.15b. It is the same as that of HDLC, and the difference is address field of length 16 bit which is $8n$ in HDLC.

5.5.3 LLC/MAC

Logical link control (LLC) is part of IEEE 802 family of standards normally for local area network (LAN) which does not follow HDLC. The main difference is medium access control (MAC) which is an essential part of LLC. Figure 5.15c shows an LLC frame format. The details of LLC are discussed in the next chapter.

5.5.4 LAPF

Link access procedure frame mode bearer services (LAPF) was made standardized by ITU-I as part of frame relay link control of X.25 packet switching architecture [4].

LAPD provides data link control for frame relay designed to provide a streamlined capability to use over high-speed packet-switched networks. It also uses only ABM for data transmission. The frame format of LAPD is shown in Figure 5.15d. It is almost the same as that of HDLC.

As with X.25, frame relay makes multiple connections over a single link. These multiple connections are data link connections in which each is indicated by a unique data link connection identifier (DLCI). Data transfer involves the following steps:

1. Set up a logical connection between two end points and allot a unique DLCI to the connection.
2. Swap information in data frames, and each frame has a DLCI field to identify the connection.
3. Discharge the logical connection.

The establishment and release of a logical connection are made by the exchange of messages over a logical connection dedicated to call control with DLCI = 0. A frame with DLCI = 0 has a call control message in the information field. At a minimum, four message types are used: SETUP, CONNECT, RELEASE and RELEASE COMPLETE.

One of the sides/stations start request for the establishment of logical connection using SETUP message. If the other side admits the request, it replies with a CONNECT message; otherwise it replies with a RELEASE COMPLETE message. The station sending the SETUP message may assign the DLCI by choosing an unused number and including this value in the SETUP message; otherwise, the DLCI value is assigned by the accepting station in the CONNECT message. Either station may request to clear a logical connection by sending a RELEASE message. The other station after receipt of this message replies with RELEASE COMPLETE message. Table 5.3 shows the call control messages used and defined in ITU-/I standard Q.931 for frame relay. These messages are used for common channel signaling between a user and an ISDN. The procedures and parameters for exchange are made in a frame relay for D-channel. The SETUP, CONNECT and CONNECT ACK messages are used for establishment of the same channel.

Once the access connection is set up, message exchange occurs between the end users for setting up frame mode connection of B-channel. For establishment of this channel, the SETUP, CONNECT and CONNECT ACK messages are used.

The values of DLCI indicate different frame relay connections. The DLCI 0 represents a frame relay connection between user and frame handler. The DLCI 8191 indicates the management procedures. The representations of different values of DLCI are mentioned in Q.931 and Q.933.

5.5.5 ATM

Like LAPF, asynchronous transfer mode (ATM) provides a streamlined capability to use it over high-speed packet-switched networks [4]. The frame of ATM is called ATM cell with a fixed bit length of 53 bytes. The ATM is based on cell relay. The details of ATMs discussed are as follows.

TABLE 5.3

Messages for a Frame Relay Connection

Message	Direction	Functions
	Access Connection Establishment Messages	
ALERTING	u→n	Represents that user alerting has begun
CALL PROCESSING	Both	Represents that access connection establishment is initiated
CONNECT	Both	Represents access acceptance by terminal equipment
		Represents that user has been in access connection
CONNECT ACKNOWLEDGMENT	Both	Reports progress of an access connection in the event of internetworking with a private network
PROGRESS	u→n	Represents access connection establishment
SETUP	Both	
	Access Connection Clearing Messages	
DISCONNECT	Both	Sent by user to request connection clearing; sent by network to indicate connection clearing
		Indicates intent to release the channel and call reference
RELEASE	Both	Represents release of the channel and call references
RELEASE COMPLETE	Both	
	Miscellaneous Messages	
STATUS	Both	Sent in response to a STATUS ENQUIRY or at any time to report an error
STATUS ENQUIRY	Both	Solicits STATUS message

5.5.5.1 ATM Protocol

ATM is based on cell relay concept, which is the same as packet switching using X.25. Like cell relay based on X.25, ATM involves the transfer of discrete chunks of data called as ATM cell with a fixed size of 53 bytes. It also allows multiple logical connections multiplexed over a physical interface. The information flow on each logical connection is organized with fixed-size ATM cells.

ATM has a streamlined protocol with minimum error and flow control capabilities. These capabilities decrease the overhead of the processing of ATM cells. It also decreases the overhead bits of the processing ATM cells [10]. Figure 5.16 shows an ATM protocol architecture consisting of physical layer, ATM layer and the ATM adaptation layer (AAL) along with the interface planes between the user and network. The functions of these layers are given below:

- **Physical layer:** The physical layer has specifications of transmission medium and signal encoding scheme. There are mainly two types of data rate specified in physical layers – 155.52 and 622.08 Mbps. Other data rates (more and less than these rates) are also possible.

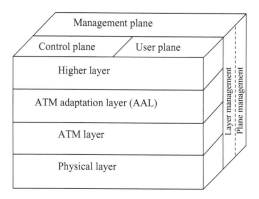

FIGURE 5.16   ATM protocol architecture.

- **ATM layer:** ATM layer is common for all the service and provides packet transfer capabilities. It does the transmission of data in fixed-size cells with the use of a logical connection, which is explained later in this section.
- **ATM adaptation layer:** The AAL uses information transfer protocols. The AAL maps higher-layer information into ATM cells to be transported over an ATM network via ATM layer and also collects information from ATM cells passed through ATM layers for delivery of the same to higher layers.
- **Higher layers:** This layer consists of three planes – user plane, control plane and management plane.
  - **User plane:** It provides for user information transfer along with associated controls which are flow control and error control
  - **Control plane:** It does call control and connection control functions
  - **Management plane:** It also does management functions related to the system as a whole, providing coordination between all the planes and layer management managing functions relating to the resources and parameters residing in its protocol entities.

5.5.5.2   ATM Logical Connections

Logical connections in ATM are mainly virtual channel (VC) connections (VCC) which mainly make switching in an ATM network [4,10,11]. The connection is made between two end users through the ATM network so that a variable rate full duplex flow of fixed size cells is transferred over the connections The VCCs are also required for the user–network exchange (control signaling) and network–network exchange (network management and routing).

The bundle of VCCs make a virtual path connection (VPC) having same end points. Thus all the cells transmitting over all the VCCs in a single VPC are switched together. The virtual path concept provides grouping of connections that share common paths through the network into a single unit, as shown in Figure 5.17. The virtual path concept was developed in response to efficient link utilization of high-speed networking in which the control cost of the network is becoming an increasingly higher proportion of the overall network cost. Network management

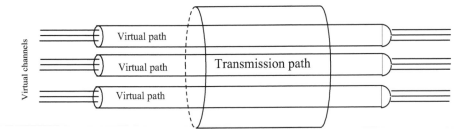

FIGURE 5.17   ATM logical connections.

actions are applied to a small number of connections instead of a large number of individual connections. The large numbers of virtual paths are combined to send those to transmission paths.

- **Virtual path/virtual connection characteristics**
  ITU-T recommends the following characteristics of VCC
  - **Quality of service:** A VCC is assigned to user with quality of service (QoS) specified by the parameters such as cell loss ratio (ratio of cells lost to cell transmitted and cell delay variation)
  - **Switched and semi-permanent virtual characteristics:** All connections switched needs call control signaling and dedicated in nature.
  - **Cell sequence integrity:** The sequence of the transmitted cell within VCC preserved traffic parameter negotiations and usage monitoring: Traffic parameters such as average rate, peak rate, burstiness and peak duration are negotiated between a user and the network, for each VCC is monitored by the network to ensure that negotiated parameters are not violated.
    The above four characteristics are also adapted by VPC.
  - **VC identifier (VCI) restriction within a VPC:** One or more VCI or the numbers may not be available to the user of VPC but maybe reserved for network use.
- **Control signaling**
  In ATM a mechanism is required to set up and relieve VPC and VCCs. The exchange of information involved in this connection and release process are referred to as control signaling, which takes place on separate connections from those that are being managed. There are four methods for the establishment and release process of VCC.
  1. Semi-permanent VCCs are used for station to station transfer. In this case there is no control signaling required for this method.
  2. If there is no pre-established call control signaling channel, then one must be set up. For no pre-established call, control signaling exchange must occur between the user and network. So there is a need for permanent channel preferably at a low data rate and can be used to set up VCCs that are used for call control. This type of channel is known as meta-signaling channel.

3. The meta-signaling channel is used to set up VCCs between the user and network for call control signaling. This user-to-network signaling VC is used to set up VCCs to carry user data.

4. The meta-signaling channel is used to set up user-to-user signaling VC. Such a channel is set up within a pre-established VPC. It is used for two end users.

   For VPC, three methods are adapted

1. Semi-permanent VPC is used on prior agreement. There is no control signaling required.

2. VPC establishment is controlled by the user. In this case the customer uses a signaling VCC to request the VPC from the network.

3. VPC establishment and release is network controlled. In this case the network sets up a VPC for its own convenience.

- **Call establishment**

  Figure 5.18 represents call establishment flowchart having VCs and virtual paths. The process of setting up VPC is combined with the process of setting up individual VCCs. The virtual path control process has estimation of

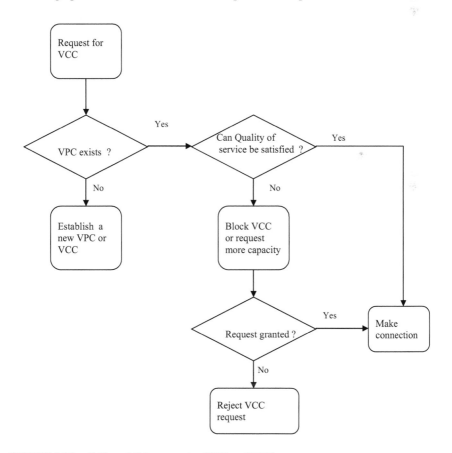

FIGURE 5.18   Call establishment using VPC and VCC.

routes, allocating capacity and storing connection state information. To set up a VC, there must be a VPC to the destination node with sufficient capacity to support VC with necessary QoSs. The following steps are adapted for all establishments.

1. For a VCC request, VPC is asked. If VPC exists, then ask for QoS. If VPC does not exist, then ask for a new VPC or VCC.
2. If QoS is satisfied, then make the connection; otherwise, go to the next step.
3. If QoS is not satisfied, then block VCC or request for more capacity. If request is granted, then make connection; otherwise go to the next step.
4. It request is not granted, then reject VCC request.

5.5.5.2.1 ATM Cells

ATM makes fixed-size ATM cells consisting of a 5 octet header and 48 bytes information as shown in Figure 5.16d. There are several advantages of the use of small fixed-size cells. The use of these small size cells reduces queue delay for high-priority cells [12].

Header format: The header format is an important part of ATM cell. Figure 5.19 shows the header field at the user–network interface, whereas the header field of ATM cell at the network–network interface is shown in Figure 5.18b. The only difference is that generic flow control is only present in cell header format of the user–network interface but not present at the network–network interface. Other fields such as virtual path identifier (VPI), VCI, payload type (PT), cell loss priority (CLP) and header error control (HEC) in header format remain the same for both ATM cell types.

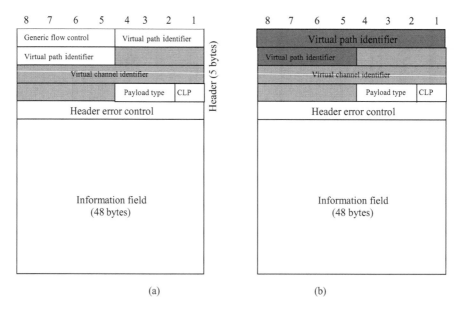

FIGURE 5.19   ATM cell format.

Generic flow control: The generic flow control does end-to-end functions used to control cell flow only at the local user–network interface. This field having 4 bits is used to assist the user in controlling the flow of traffic with different QoSs. The use of this field represents multiple priority levels to control flow of information in a service-dependent manner.

VPI: The VPI field has a routing field of the network. It is 8 bits for the user–network interface and 12 bits for the network–network interface. The additional 4 bits for network–network interface permit to have more virtual paths in the network.

VCI: The VCI field having 16 bits length is used for routing to and from the end user. Its functions are more used for a service access point. It identifies a particular VC link for a given VPC. The VCI value is assigned in ATM cells so that these cells can reach the point of destination. A group of VCIs are assigned with a common value of VPI [10].

PT: The PT field having three bits indicates the type of information in the information field. The first bit indicates user information, the second bit indicates whether the congestion is experienced or not and the third bit is known as ATM user to ATM user (AAU) indication bit. The value 1 of the first bit indicates management/maintenance information, whereas the value 0 represents the data cell. The value 0 of the second bit indicates an ATM cell not having overcrowding experienced and vice versa for value 1 of the second bit.

CLP: The CLP [10] provides the guidance to the network in the event of congestion. A value of 0 represents a cell of relatively higher priority, which is rejected only when there is no alternative. A value 1 indicates that the cell is subject to reject within the network and the cell is delivered only when there is no overcrowding.

HEC: Each ATM cell contains an 8 bit HEC field that is found from the remaining 32 bits of the header. The pattern polynomial used to form the HEC code is $X^8 + X^2 + X + 1$ representing bit pattern 100000111). Figure 5.20 shows the operation of HEC algorithm at the receiver for detection of error in the header. As each cell is received, the HEC calculation and comparison are performed as shown in the figure. When an error is detected, the receiver will correct the error if it is a single-bit error or it will detect that multibit error shave occurred. For multibit error cells are discarded. The receiver remains in the detection mode as long as error cells are received.

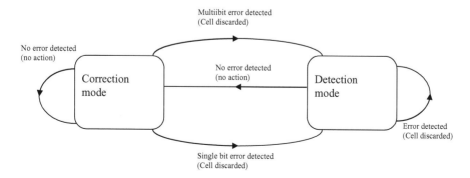

FIGURE 5.20 HEC algorithm operations at the receiver.

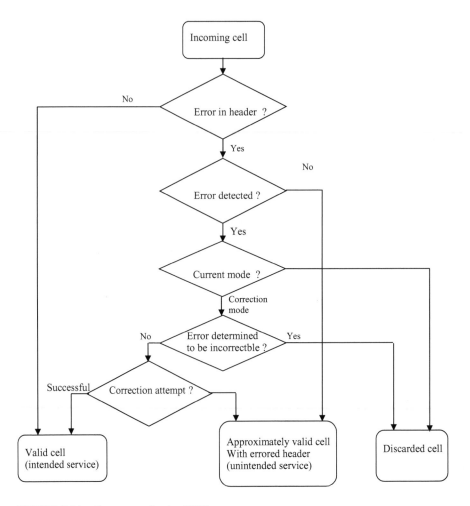

FIGURE 5.21   Error control using HEC.

Figure 5.21 shows the error control flowchart for HEC operation. If there is no error of the frame or error can be successfully corrected in frame, the cells are valid cells. If the detection of error in frame is made and the errors are not corrected successfully, the frames are discarded. If the frames are detected error and are unsuccessful in correction, then the frames are apparent valid cells having error in the header.

5.5.5.3 Transmission of ATM Cells

The ITU-T recommendations for broadband ISDN give the data rates and synchronization technologies for ATM cell propagation along the user–network interface [13]. The BISDN (Broadband integrated services digital network) requires specification that ATM cells are to be transferred at a rate of 155.52 or 622.08 Mbps. So two approaches namely cell-based and SDH (Synchronous digital hierarchy) -based physical layers are used and specified in L.413.

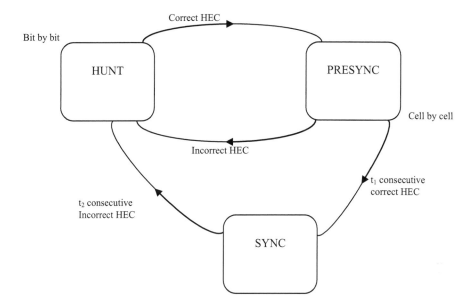

FIGURE 5.22   Call delineation algorithms.

In case of cell-based physical layer, no framing is required. The interface structure has a continuous stream of 53 octet cells. Since there is no external frame required in the cell-based approach, synchronization is made as per the HEC field. Figure 5.22 shows the following steps of synchronization. There are three states – HUNT, PRESYNC and SYNC states.

1. In the HUNT state, a cell delineation algorithm is carried out bit by bit to determine whether the matching between received and estimated HEC is made or not. Once a match is obtained, one header has been found and then it goes to the next state.
2. In the PRESVNC state, a cell structure is made. The cell delineation algorithm is carried out cell by cell until the match rule has confirmed it t_1 times.
3. In SYNC state, the HEC is taken for error detection and correction. Cell delineation is lost if the matching is identified as incorrect for t_2 times consecutively.

The values of t_1 and t_2 affect delay in establishment of synchronization. Greater values of these cause longer delays in synchronization. The advantage of cell-based transmission scheme can be used for both transmission and transfer mode functions on common structure.

5.5.5.3.1   SDH-Based Physical Layer

ATM cells can be carried over a line using SONET (Synchronous optical network)-based synchronous digital hierarchy. The framing in this technique is made by using STM-1 (STS-3, 155 Mbps). In this case payload capacity (2, 340 octets) is not an integral multiple of the cell length (53 octets), and the cell goes beyond a payload boundary. But this arrangement is more efficient and cost effective using 622 Mbps ATM stream.

SUMMARY

In this chapter, we have discussed data link control schemes along with error detection for data communication network, and these schemes can also be used for optical networks. Both flow control and error control schemes have been described. We have also mentioned HDLC with CRC, which can be used in optical LAN. We have also discussed ATM data link control, which can be linked with optical backbone via IP over WDM interface and SDH.

EXERCISES

5.1. A channel has a data rate of 2 kbps and a propagation delay of 10 ms. For what range of frame sizes does stop and wait give an efficiency of at least 50%.

5.2. Consider the use of 2000 bit frames on a 2 Mbps satellite channel of 540 ms delay. What is the maximum link utilization for
  a. Stop and wait flow control.
  b. Continuous flow control with a window size of 7.
  c. Continuous flow control with a window size of 127.
  d. Continuous flow control with a window size of 255.

5.3. Consider a half duplex point-to-point link using a stop and wait scheme,
  a. What is the effect on link utilization of increasing message size so that fewer messages will be required? Other factors remain constant.
  b. What is the effect on link utilization of increasing number frames for a constant message size?
  c. What is the effect on link utilization of increasing frame size?

5.4. Using CRC-CCITT polynomial, generate CRC code/FCS for message signal 100010010101010.
  a. With long division
  b. With shift register circuit

5.5. Find VRC and HRC bit for the following messages considering 5×4 matrix
  a. 10101101000100100010
  b. 11011101010101000101
      Find FCS for the above messages considering pattern $P = 1101001$ by using long division and shift register circuit.

5.6. What is necessary to have NAK0 and NAK1 for stop and wait ARQ.

5.7. Consider a frame consisting of two characters of four bit search. Assume that the probability of bit error is 10^{-3} and that it is dependent for each bit.
  a. What is the probability that the received frame contains at least one bit error?
  b. What is the probability if we add a parity bit to each character?

5.8. Assume that the primary HDLC station in NRM has sent six I-frames to a secondary station. The primary's $N(s)$ count was three (011 binary) prior to sending the six frames. If the poll bit is on the sixth frame, what will be the $N(R)$ count back from the secondary after the last frame? Assume an error-free operation.

5.9. Consider several physical links that connect two stations. We would like to use a multilink HDLC that makes efficient use of these links by sending frames on a FIFO basis on the next available link. What enhancements to HDLC are needed?

5.10. A channel has a data rate of R bps and a propagation delay of t seconds per kilometer. The distance between sender and receiver nodes is L kilometers. Nodes exchange fixed-size frames of B bits. Find a formula that gives the minimum sequence size of frame as a function of R, t, B and L (considering maximum utilization). Assume that ACK frames are negligible in size and the processing at the nodes is instantaneous.

5.11. What is the need for bit stuffing in HDLC frame. How can you make bit stuffing by giving an example?

5.12. Although ATM does not include any end-to-end error detection and control functions on the user data, it is provided that a HEC field detects and corrects header errors. Considering the value of this feature, the bit error rate is taken to be B. If the errors are uniformly distributed, then the probability of an error in the header is

$$\frac{h}{h+i} \times B.$$

and the probability of an error in the data field is

$$\frac{i}{h+i} \times B.$$

Where h is the number of bits in the header and i is the number of bits in the data field.

a. Suppose that the errors in the header are not detected and not corrected. In this case, a header error may result in misrouting of the cell to the wrong destination; therefore i bits will arrive at the incorrect destination and I bits will not arrive at the correct destination. What is the overall bit error rate $B1$? Find an expression for multiplication effect on the bit error rate $M1 = B1/B$.

b. Consider that the errors in header are detected but not corrected. In that case, i bits will not arrive at the correct destination. What is the overall bit error rate $B2$? Find an expression for multiplication effect on the bit error rate $M2 = B2/B$.

c. Consider that the errors in header are detected but corrected. In that case, i bits will not arrive at the correct destination. What is the overall bit error rate $B3$? Find an expression for multiplication effect on the bit error rate $M3 = B3/B$.

5.13. One key design decision for ATM was whether to use fixed or variable length cells. Let us consider this decision from the point of view of efficiency. Transmission efficiency as

$$N = \frac{\text{no of message bytes}}{\text{No of message bytes} + \text{no of overhead bytes}}$$

a. Consider the use of fixed length packets. In this case, the overhead consists of header bytes. Take the following terms

L = data field size of the cell in bytes

H = Header size of the cell in bytes

X = No. of message bytes

Derive an expression for N.

b. If the cells have variable length, then the overhead is determined by the header plus the flags to delimit the cells of an additional length field in the header. Let H_v = additional overhead bytes required to enable the use of variable length cells. Derive an expression for N in terms of X, H and H_v.

5.14. Compare sustainable cell rate and burst tolerance, as used in ATM networks with combined information rate and excess burst size as used in a frame relay network. Do the respective terms represent the same concept?

5.15. A proposed congestion control technique is known as arithmetic control. In this method, the total number of frames in transit is fixed by inserting a fixed number of permits into the network. These permits calculate at random through the frame relay network. Whenever a frame handler wants to relay a frame just given to it by an attached user. It must first capture and destroy a permit. When the frame is delivered to the destination user by the frame handler to which it attaches, that frame handler reissues the permit. List three potential problems with this technique.

5.16. How can you make frame relay connection using commands such as CONNECT, CONNECT ACKNOWLEDGMENT and SETUP.

5.17. Considering multiple physical links connected between two stations, multilink HDLC is used for sending frames with FIFO. What arrangements are needed for HDLC for the above?

5.18. Consider the primary HDLC station in NRM has sent six I-frames to a secondary. The primary $N(S)^*$ count has three 011 binary prior sending the six frames. If the poll bit is on the sixth frame what will be $N(R)$ count back from the secondary after the last frame. Assume an error-free operation.

REFERENCES

1. B. Mukherjee, *Optical WDM Networks*, Springer-Verlag, New York, 2006.
2. U. Black, *Physical Level Interfaces and Protocols*, IEEE Computer Society Press, Los Atlantis, CA, 1995.
3. U. Black, *Data Link Protocols*, Prentice Hall, Englewood Cliffs, NJ, 1993.
4. W. Stalling, *Data and Computer Communication*, Prentice Hall, Upper Saddle River, NJ, 2003.

5. J. Boudec, "The Asynchronous transfer mode: a tutorial," *Computer Networks and ISDN Systems*, vol. 24, pp. 279–309, 1992.
6. T. Ramabadran and S. Gaitonde, "A tutorial for CRC computations," *IEEE Micro*, vol. 8, pp. 62–75, 1988.
7. S. Lin, D. Costello, and M. Miller, "Automatic repeat request error control schemes," *IEEE Communication Magazine*, December 1984.
8. J. Walrand, *Communication Networks: A First Course*, McGraw Hill, New York, 1998.
9. W. Bux, K. Kummerle and H. Thurong, "Balanced HDLC procedures: a performance analysis," *IEEE Transactions on Communication*, vol. 28, pp. 1889–1898, 1980.
10. W. Goralski, *Introduction to ATM Network*, Mc-Graw Hill, New York, 1995.
11. M. Prycker, *Asynchronous Transfer Mode: Solutions for Broad Band ISDN*, Eilla Horwood, New York, 1993.
12. D. Bertsekas and R. Gallager, *Data Networks*, Prentice Hall, Englewood Cliffs, NJ, 1992.
13. History of ATM Technology: http://www.atmforum.com.

6 Data Communication Networks Having No Optical Transmission

The current scenario of up-to-date society is going to be changed tremendously due to the influence of "internet" on various aspects of necessity like communication, education, health and entertainment. We can say this initiative is just 20 years old. In this direction a computer network started with copper cables and wireless with different medium access controls (MACs), in LANs [1–7], Wireless LANs [8–10] and ATM [11–12]. Later on to increase the coverage area of the network, data speed and to reduce bit error rate, optical fiber cables are used in the network without all optical devices such as wavelength division multiplexer and optical switch. This chapter discusses all these networks which are based on different topology using copper cables, wireless media and optical fiber cables without all optical devices [13–20].

6.1 HISTORY AND BACKGROUND OF NETWORKING-DIFFERENT GENERATIONS

The Advanced Research Project Agency Network (ARPANET) is a network used for communication with an ARPANET host, attached to another ARPANET Interface Message Processor (IMP) [6]. The additional packet-switching network (other than ARPANET) is ALOHANET, which is a satellite network linking together different centers of university situated in Hawaiian islands. The ALOHANET is a packet-based radio network allowing multiple remote sites on the Hawaiian islands to communicate with each other. The ALOHA protocol developed in 1970 was the first so-called multiple access protocol, permitting geographically distributed users to share a single broadcast communication medium (a radio frequency)]; since then a number of networks have been reported. In 1973, Ethernet has been reported providing a huge growth in so-called Local Area Networks (LANs) that operated over a small distance based on the Ethernet protocol. BITNET provides and transfers file among several universities in the northeast Computer Science Network (CSNET) was formed to link together university researchers without access to ARPANET. In 1986, NSFNET was made to give access to NSF-sponsored supercomputing centers. Starting with an initial backbone speed of 56 kbps, NSFNET's backbone was operated at 1.5 Mbps by the end of the decade 1990–2000.

The changing topology based on changing transmission medium of networking is classified as the trend of networking into three generations:

   i. LAN or backbone network based on earlier cupper twisted pair, cable and electronics switches. (Network working in only electrical domain) – for example, Ethernet, token ring, token bus, wireless LAN, etc.
   ii. Second generation of network operating both electrical and optical domains where transmission is made in optical domain and switching and multiplexing/demultiplexing an amplification is made in an electrical domain. For example, FDDI, Express-net, Data Queue Dual Bus (DQDB), fiber net, etc. Next-generation DWDM optical network. (All optical)
   iii. Next-generation WDM optical network. (All optical) For example, single-hop WDM network, multi hop WDM network.

6.2   FIRST GENERATION OF NETWORK

There are four basic topologies – bus, tree, ring and star which have been used in the first generation of network. The bus and tree topologies are based on multi-point medium, where all stations/users are attached through a hardware interface known as tap directly to a linear transmission medium or bus. These topologies cover LANs [3] and metropolitan area networks (MANs) [5]. Out of these topologies, tree topology is used for packet broadcasting networks in which each station transmits data which is shared by all other stations. This means that a transmission from any stations is broadcast to and received by all other stations. Other topologies – specially tree – may be broadcasting in nature.

6.2.1   PROTOCOL ARCHITECTURES

LANs are different from other types of networks as it covers moderate-size geographic area such as a single office building, a warehouse or a campus. Protocols [1,3] for LAN explaining the issues related to transmission of blocks of data over the network. In OSI layer (described in Section1.1.1) [2], higher layer protocols are independent of the network architecture which are applied to LANs [3] and MANs [5]. The lower layer of LANs and MANs are different from lower layers of OSI model [2]. Figure 6.1 shows LAN protocols related to OSI layers which are already described in Figure 6.2. This architecture developed by IEEE 802 committee is applicable to LAN and followed by other organizations. This architecture is known as the IEEE reference model working from bottom up.

The lowest layer of IEEE 802 reference model corresponds to the physical layer which includes the following functions – Encoding/decoding of signals, preamble generation/removal (for synchronization) and bit transmission/receptions. Since the specifications of transmission medium and topology are critical in LAN design, the physical layer of the IEEE 802 model should include specifications of transmission medium and topology.

Above the physical layer, IEEE 802 reference model has MAC having the following functions;

- Governing access to the LAN transmission medium, i.e., shared access medium
- Provide interface to the next layer logical link control (LLC)
- Perform error detection and flow and error control

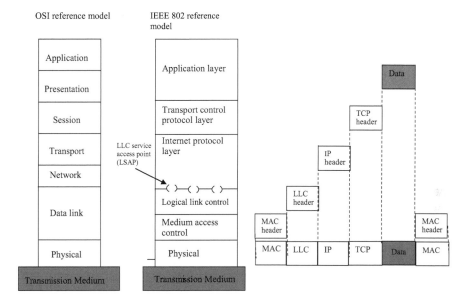

FIGURE 6.1   IEEE 802 protocol layers related with an OSI model.

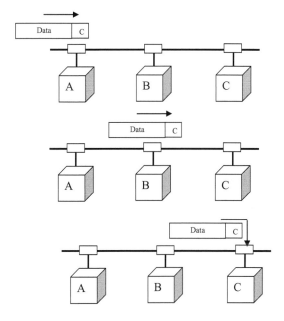

FIGURE 6.2   LAN based on bus topology and transmission of frame.

The next layer of this reference model is LLC, which includes the following functions [4]

- On transmission, making data into a frame with address and error detection fields
- On reception, disassemble frame, perform address recognition and error detection

Table 6.1 shows physical layer, MAC layer and LLC layer standards developed by the IEEE 802 committee and adopted by ISO and other upper layers shown in Figure 6.1 and are the same as those of TCP/IP standards [2] prescribed by ISO. In the figure, user data are coming from applications layer and passed down to an LLC layer via upper layers – TCP and IP layers which append control information as a header making protocol data unit (PDU). The LLC layer makes control information creating an LLC-PDU from PDU developed in the upper layers. The entire LLC-PDU is then transmitted to the MAC layer that makes control information from the front and back to the PDU forming a MAC frame. The MAC header having control information needs the operation of accessing of the medium. The table also shows MAC for the corresponding topologies using the corresponding transmission medium with data speed.

6.2.2 TOPOLOGIES

As discussed earlier, there are four basic topologies of LAN – bus, tree, ring and star. The bus is a special case of tree topology.

6.2.2.1 Bus Topology

Both bus and tree topologies are made by using multipoint medium. All stations/nodes are attached directly to the bus/transmission medium through a tap have an appropriate interface. There are two types of operations – full duplex operations which allow data to be transmitted onto the bus and received from the bus simultaneously and half duplex operation which allows data to be transmitted onto the bus and received from the bus one at a time. These operations require transmission from a station to bus in both directions. In Figure 6.2, station A wishes to transmit a frame of data to C. The frame in which header includes station C's address propagates along bus and passes to station B, which sees the address field and ignores the frame transmitting towards station C. After seeing the address field in station A, it goes and is copied there. But the transmission is also controlled by MAC. As seen in Table6.1, carrier sense medium access (CSMA)/collision detection (CD) MAC and token bus MAC are used for the bus topology [4].

Ethernet was the first initiative of bus topology based on CSMA/CD MAC and a wired LAN system [2,4], which had a 2.5 km long coaxial cable with repeaters at 500 m and capability to interconnect up to 256 machines. For 10 Mbps 10BASE5 (bus topology), 10BASE2 (bus topology) and 10BASE-T (star topology), for Fast Ethernet 100 Mbps 100BASE-TX and 100BASE-T4, for Gigabit Ethernet 1000BASE-CX and 1000BASE-T are the physical layer specifications.

TABLE 6.1

Standards Used for Physical, MAC and LLC Layers as Per IEEE 802 Reference Model [2]

LLC	IEEE 802.2 • Unacknowledged connectionless service • Connection mode service • Acknowledged connectionless service						
MAC	IEEE 802.3CSMA/CD	IEEE 802.4Token bus	IEEE 802.12Round-robin priority	IEEE 802.5Token ring	FDDI Token ring	IEEE802.6DQDB	IEEE 802.11CSMA polling
Physical	Base-band coaxial:10 Mbps Unshielded twisted pair:10, 100 Mbps Shielded twisted pair:100 Mbps Broadband coaxial:10 Mbps Optical fiber:10 Mbps	Broadband coaxial:1,5,10 Mbps Carrier band coaxial:1,5,10 Mbps Optical fiber:5, 10, 20 Mbps	Unshielded twisted pair:100 Mbps	Unshielded twisted pair:4 Mbps Shielded twisted pair:4,16 Mbps	Optical fiber:100 Mbps Unshielded twisted pair:100 Mbps	Optical fiber:100 Mbps	Wireless Infrared:1, 2 Mbps Wireless Spread spectrum:1,2 Mbps

6.2.2.2   Tree Topology

The tree topology is an extension of the bus topology in which the transmission medium is a branching cable without closed loop as shown in Figure 6.3. The tree topology begins at a point called as head end in which one or more cables start and each of them may have branches. The branches may include additional branches making it a complex layout. In tree topology, a transmission from any station propagates throughout tree branches and received by all other stations. Two problems arise in this arrangement of transmission of data–transmissions starting from any station are received by all other stations, indicating insecurity of data and no mechanism for regulation of transmission. When two stations attempt to transmit at the same time, these signals overlap and get lost, due to not having regulation, as the transmission decides to continue for a long period. To solve these problems stations should transmit data in frames of small size having unique address/identification and destination address as a header in frame. With the header, the transmission of frames is controlled by a tap in which as per destination address destination station will receive the frames. For solving the second problem a proper MAC for regulation of access is required. Of course tree is an ideal topology for broadcasting [2].

6.2.2.3   Ring Topology

Ring topology is used as a backbone based on ring structure as shown in Figure 6.4. It consists of point-to-point links in a closed loop. Each station attaches to the network with a tap. This tap is operated in three modes – listen mode, transmit mode and bypass mode. Figure 6.5 shows modes of operations of taps which contribute to the proper functioning of the ring by passing on all the data that come its way and provide an access point for attached stations to send and receive data [2,4].

In listen mode, each received bit is retransmitted with a small delay to allow to perform the required functions, which includes scan passing bit of streams–specially address of destination station, copying of each incoming bit while continuing to retransmit and modifying bit while passing, if required.

When station attached with a tap has to send the data after getting bits for the incoming link, the tap enters the transmit mode. In this state, the tap receives the data bits from the station and retransmits it to the outgoing link. During retransmissions, there may be two possibilities–bits from the same packet that may tap is still

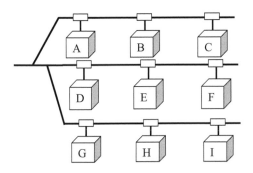

FIGURE 6.3   LAN based on tree topology.

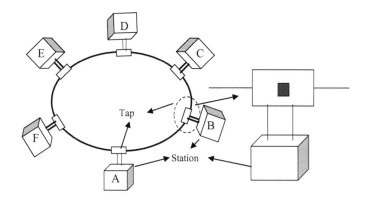

FIGURE 6.4   LAN based on ring topology.

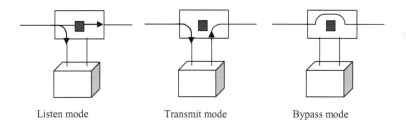

Listen mode   Transmit mode   Bypass mode

FIGURE 6.5   Different modes of tap.

in the process of sending bits from more number of packets on the ring at the same time, transmitting later.

Apart from these two modes, there is another mode of operation called bypass mode. In this mode, a bypass relay is activated so that signals pass the tap with no delay, except that from propagation in the transmission medium. This mode provides advantages to improve the performance by eliminating delay in the tap.

Figure 6.6 shows the transmission of frame from the source node/station to the destination station/node in which station C transmits frame to station A. Since the frame having destination address A originated from station C, the tap attached to

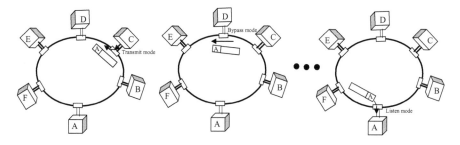

FIGURE 6.6   Transmission of frames in ring topology.

the station is in transmit mode. The taps attached to stations D, E and F are in bypass mode and the tap attached to destination station A is in listen mode. The frame having destination address A passes through the taps attached to D, E and F and finally reaches to the tap of station A, which is the destination of frame in listen mode. Since multiple stations share the ring it is required to have MAC for establishment of connection for transmission of frames. There are two types of MAC – token ring and round-robin MAC [2,3] which are discussed in the next section.

6.2.2.4 Star Topology

The stations used I star topology are directly attached to the central node called as central hub/coupler via two point-to-point links as shown in Figure 6.7. Figure 6.7a shows a single star topology consisting of one central hub. There are two approaches used for the operation of central hub. In the first approach a central hub is operated in broadcast mode. A transmission of a frame from a station (called as a source) to the central node is retransmitted to all other outgoing links by a central hub. In this case, although the frame transmission is physically based on star topology, it is logically based on bus topology. This type of transmission makes only one station at a time participate to transmit frames successively. There are disadvantages in this approach – limitation of number of users, more queue time and less security of data transmission, as frames are retransmitted to all other stations by a central hub.

In another approach, a central hub is acted as a frame switching device. The incoming frames from the source station is buffered in a central hub and then switched to the outgoing link to retransmit to the destination station. This approach provides security in transmission of frames by not allowing frames to all other stations and simultaneous transmission of frames from more source stations making less queue delay.

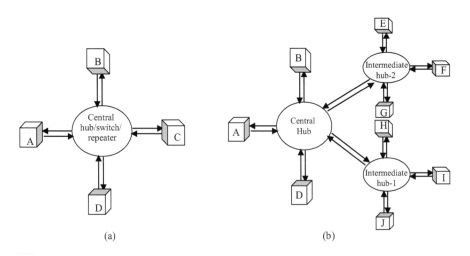

(a)   (b)

FIGURE 6.7   Star topology: (a) single star (b) multiple stars.

There is an extension of multiple star topology in which multiple hubs are present. Figure 6.7b shows multiple stars that consist of one central hub (also called as header hub) in which stations A, B and D are connected and two intermediate hubs (connected to central hub) in which other stations D, E and F are connected to intermediate hub-2 and stations H, I and J are connected to intermediate hub-1. The destination address of frames coming from source station is switched via a central hub and intermediate hubs.

6.2.2.5 Mesh Topology

Apart from regular topology the network is based on an irregular structure. In the first generation of network, networks are mostly LANs that are based on a regular topology. The mesh topology is used in wide area network and nationwide networks which connect major cities of the nation. These networks are mostly based on an optical backbone that comes under 3rd generation of network.

6.2.3 Medium Access Control

All types of networks – especially LANs and MANs having large number of stations/nodes must share the transmission capacity of the network. For efficient use of transmission capacity, it is required to have some means of control to access the transmission medium of the network. This control is named as MAC protocol, which is also used for establishing connection request for data transfer. The key requirement of MAC technique is "where" and "how" it is applied [3]. In centralized approach, the MAC has the authority to grant access to the network A station having signal to transmit must wait until it receives the permission from the controller. In distributed approach, all the stations in the network carries out a MAC function together to dynamically find out the order in which the stations transmit signal. The MACs of star topology and ring topology are mostly based on centralized scheme, whereas the bus and tree topology follow a distributed approach. The second requirement "how" is constrained by the topology with a trade-off among the factors such as cost, performance and complexity.

There are two categories of access control approaches – synchronous and asynchronous. In case of synchronous approach, a specific capacity is dedicated to a connection for a source and destination pair. This approach is used by circuit-switched, frequency division multiplexing and synchronous time division multiplexing (TDM) in case of LANs and MANs. The synchronous approach is not used normally, because the needs of the stations are unpredictable. Rather, it is preferred to allot the capacity in an asynchronous mode – more or less capacity in response of immediate demand. Asynchronous technique is divided in to three categories – round robin, reservation and contention. Table 6.2 shows standardized MAC techniques used for LANs based on different topologies.

6.2.3.1 Round Robin

In round-robin approach, each station gets the opportunity to transmit. There are two approaches used for round robin – polling and token capturing.

TABLE 6.2

MAC Techniques for LAN [2,3]

Types of MAC Technique	Bus Topology	Ring Topology	Star Topology	Switching Topology
Round robin	Token bus (IEEE 802.4) Polling (IEEE 802.11)	Token bus (IEEE 802.4)		Request/priority (IEEE 802.4)
Reservation	DQDB (IEEE 802.6)			
Contention	CSMA/CD(IEEE 802.3) CSMA(IEEE 802.11)			CSMA/CD(IEEE 802.3)

6.2.3.1.1 Polling

In case of polling, during the opportunity of transmission, the station may transmit or decline to transmit, subject to the upper bound expressed as an amount of data or time for transmission. When it finishes transmission, it relinquishes in turn and passes to the next station. The standard of polling is named as IEEE 802.11, which is used mainly in wireless LAN [2]. The central frequency is used in the range of 2.407–2.412 GHz with bandwidth 5MHz. The transmission is made by using either direct sequence spread spectrum (DSSS) signaling [8] or frequency hopping spread spectrum (FHSS) [9].

For numerous station shaving data to transmit over an extended period, the use of this technique is very efficient. If less number of stations attempt to transmit data over this extended period, then there is a considerably large overhead in passing the turn from station to station, as other stations stop transmission but simply wait for their turns. Under such circumstances, other techniques are preferable, largely depending on whether the data traffic is a stream or busy type (where steam traffic is lengthy and fairly continuous signal and busy signal is a short sporadic signal that describes an interactive terminal host traffic.

6.2.3.1.2 Token-Based MAC

The token approach is used in ring and bus topologies [2,3]. The token-based approach uses a small frame called as a token that circulates when all stations are idle.

A station having data to transmit is in queue until it detects a token passing by. It then captures the token by changing one bit in the token, transforming it from a token into a start of frame sequence for a data frame. The station then adds and transmits the remainder of the fields needed to construct a data frame. When a station snatches a token and begins to transmit a data frame, due to not having token in transmission medium, other stations that have data to transmit should wait. The steps required for the transmission of data frames in this approach are given below:

1. If a station wishing to transmit seizes a token, it begins to transmit frame and other stations having data to transmit must wait; otherwise, it must wait.
2. When the station completes the transmission of all the frames, it must release the token for the transmission of frames of the next station, wishing as per the round-robin format.

The release of token depends on the length of the frame. There are two cases of frame length – frame length is shorter than the bit length of the ring and is longer than that of the ring. If the frame length is more than that of the ring, the leading edge of the transmitted frame will return to the transmitting station before it has completed transmission, and in this case, the station may issue a token as soon as it has finished the frame sequence. If the frame is less than the bit length of the ring, after completion of transmission of a frame, it must wait until the leading edge of the frame returns, and after returning, a token will be issued. In this case some capacity of the ring may be idle, and so link utilization is poor. To tackle this problem, an early token release (ETR) approach is adapted. The ETR permits a transmitting station to make a token free as soon as frame transmission is completed, whether the frame header is sent back to the station or not. The priority is used for a token made free prior to the receipt of the previous frame header.

6.2.3.1.3 *Performance of Token Ring*

We consider a ring network having N active stations and maximum normalized propagation delay a and transmission delay of 1. To simplify the analysis, we consider that each station is always prepared to transmit a frame, allowing the development of an expression for maximum link utilization U. For a ring, there are average time of one cycle C, average time to transmit a data frame T_1 and average time to pass a token T_2. The average cycle rate is written as $1/C = 1/(T_1 + T_2)$, and the link utilization is written as [2]

$$U = \frac{T_1}{T_1 + T_2}$$

This is called as throughput normalized to the system capacity, which is a measure of fraction of time spent for transmitting data. As shown in Figure 6.8, time is normalized such that frame transmission time equals to 1 and propagation time equals to a. The propagation delay includes a delay of tap point through which a station is connected to the ring. There are two cases – $a < 1$; $a > 1$.

For $a < 1$, a station sends a frame at time t_0 and the destination station receives the leading edge of its frame at time $t_0 + a$ and completes transmission at time $t_0 + 1$, as shown in Figure 6.8. The station then emits a token which takes an average time a/N to reach the next station, where N = total number of stations connected to the ring. Thus, one cycle takes $1 + a/N$ time and transmission time 1. So link utilization $U = 1/(1 + a/N)$.

In case of $a > 1$, the transmission is slightly different. A station transmits at t_0, completes the transmission at $t_0 + 1$ and receives the leading edge of its frame at $t_0 + a$ time, at which point it emits a token which takes an average time a/N to reach the next station. The full cycle of transmission of the frame is completed with time $a + a/N$. So the link utilization is written as $U = 1/(a + a/N)$.

The same procedure is also considered for the estimation of link utilization in case of token bus, where the token passing time is a/N.

As s tends to infinity, the link utilization of the token ring is written as

$$\lim_{N \to \infty} U = 1, \quad a < 1$$

$$= 1/a, \, a > 1$$

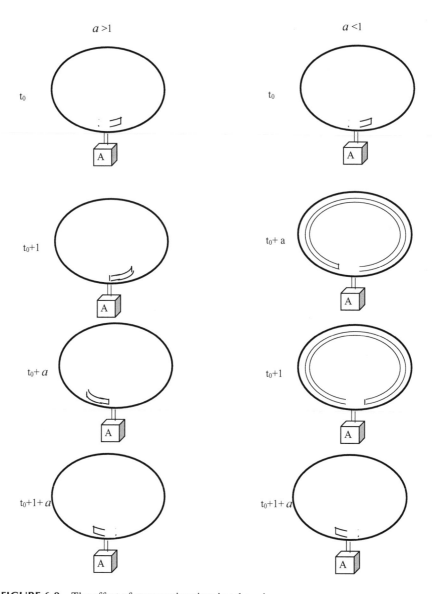

FIGURE 6.8   The effect of propagation time in token ring.

6.2.3.2   Reservation

For a stream of traffic, reservation techniques are more suitable. In this approach, time in the medium is broken into time slots, as much with synchronous TDM, where synchronous transmission is followed. A station having data to transmit make future slots reserved for an extended or even an indefinite period in a distributed or centralized manner. The reservation scheme is found in DQDB IEEE 802.6 where only error control is required. In this case synchronization of frame, flag field (01111110) is used in both start and end of frame. For error control, frame check sequence (FCS) is used before the end flag field. For addressing of frame to reach destination, address field is used just after the start flag field of frame. After making frames, these frames are accommodated in the reserved slots in the DQDB [2].

6.2.3.3   Contention

There are two types of MAC under contention category–CSMA [2] and CSMA/CD [2].

6.2.3.3.1   Carrier Sense Medium Access (CSMA)

One of the MAC techniques used in bus/tree and star topology is CSMA, which is termed as a contention technique. It is random in the sense that there is no predictable or scheduled time for any station to transmit. It means that the transmission by stations are ordered randomly. The stations have contention in the sense that stations compete for time on the medium.

First random contention MAC is ALOHA developed for packet radio networks [2]. In this technique, the stations having frame to send does the same and tries to hear an acknowledgment during maximum possible round trip propagation times. If they do not hear the same during that time, it resends the frame. If it fails to receive an acknowledgment after repeated transmission, it leaves the transmission. The receiver stations verify the correctness of the incoming frame by checking the FCS field with cyclic redundancy check (CRC) technique, and if the frame header matches the receiver's address, then the station immediately issues an acknowledgment; otherwise, the frame with error and mismatching address is invalid. Although ALOHA is simple, it pays a penalty because of collision. It is also seen that the number of collision increases rapidly with an increase of load.

To improve efficiency, the ALOHA can be modified by using slotted ALOHA in which time the channel is organized into uniform slots. The size of the slots is equal to the frame transmission time. A central clock/technique is needed to synchronize all stations. Transmission is allowed to start only at a slot boundary. If some stations do not have frames to transmit, slots belonging to these stations are idle and this leads to poor link utilization. But there is no collision of frames in this approach. Both ALOHA and slotted ALOHA provide poor link utilization [2]. So both techniques fail to take advantage of packet radio.

To reduce the number of collisions and at the same time improve link utilization, CSMA is developed [2,3]. With CSMA, a station wishing to transmit first listens to the medium to determine whether another transmission is in progress or not. There are two types of CSMA – non-persistent and persistent CSMA. In CSMA, an algorithm is needed to specify what a station should do if the medium is found busy. The algorithm of non-persistent CSMA is given below:

1. If the medium is idle, transmit; otherwise, go to step 2.
2. IF the medium is busy, wait an amount of time derived from probability distribution (retransmission delay) and repeat step 1.

The use of random delays reduces the probability of collision. When two stations attempt to transmit at the same time while another transmission is in progress and both stations wait for a different amount of time, they attempt to transmit at different times. Hence, collision may be avoided because transmissions are at two different times. But there is a problem of wasting more capacity providing poor link utilization because of the idle state of the medium. So one can use persistent CSMA in which there is no waiting time. There are two types of persistent CSMA – 1-persistent and p-persistent [2],

The algorithm of 1-persistent CSMA is given below:

1. If the medium is idle, transmit; otherwise, go to step 2.
2. If the medium is busy, continue to listen until the medium is sensed to be idle; then transmit immediately.

There are disadvantages of 1-persistent CSMA [2,3]

a. Collisions occur if two or more stations sensed the medium to transmit
b. More queue time as more collisions occur

To remove these disadvantages p - persistent CSMA is used. The algorithm of p - persistent CSMA is given below [2]:

1. If the medium is idle, transmit with p probability and a delay of one unit time with probability $(1-p)$. The time unit is the maximum propagation delay.
2. IF the medium is busy, continue to listen until the medium is sensed to be idle; then repeat step 1.
3. If the transmission is delayed by one unit time, repeat step 1.

There are difficulties even in case of p-persistent CSMA

a. Difficult to implement
b. How steps are taken when collisions occur

To tackle these problems, one should use a CD system along with medium sensing of station for detection of collisions, and after detection of collision, one should take steps for successful transmission of data frames. This leads to the development of CSMA/CD. It is also seen that in all these CSMA approaches collision can be avoided. The algorithm procedure of CSMA/CD is given below [3]:

1. If the medium is idle, transmit; otherwise, go to step 2.
2. IF the medium is busy, continue to listen until the medium is sensed to be idle; then transmit immediately.
3. If the collision is detected during transmission, transmit a brief jamming signal to assure that all stations know that there has been a collision and then cease the transmission.
4. After transmitting the jamming signal, wait a random amount of time, then attempt to transmit again, i.e., repeat from step 1.

Figure 6.9 shows the transmission procedures of frames on a bus topology by using CSMA/CD. At t_0 time station A starts to transmit a packet addressed to station D. At t_1, both B and C are made ready to transmit. Station B senses a transmission and so detains. The Station C is still not aware of station A's transmission and starts its own transmission. When A's transmission arrives at C, at t_2, C detects the collision and stops transmission. The effect of the collision propagates back to A, where it is detected sometime later as t_3 at which time A ceases transmission. A again transmits after a time interval, and other stations start transmitting frame after different time gaps.

For CSMA/CD [3], the medium makes time organized into slots of length twice the end-to-end propagation delay, i.e., $2a$. The slot time is the maximum time from the start of the transmission required to detect a collision. There are N number of active stations. If each station has a frame to send and transmits, then there is a collision taking place on the line. As a result each station restrains itself for transmitting during an available slot with probability p which is written as $1/N$.

Time on the medium has two types of intervals – a transmission interval which occupies $1/2a$ slots and a contention interval has the sequence of slots with either a collision or no transmission in each slot. The link utilization is a measure of the amount of time spent in transmission interval.

To determine the average length of contention interval we define probability A that exactly one station attempts the transmission in a slot and occupies the medium. The probability A is written as

$$A = \binom{N}{1} p^1 (1-p)^{N-1}$$

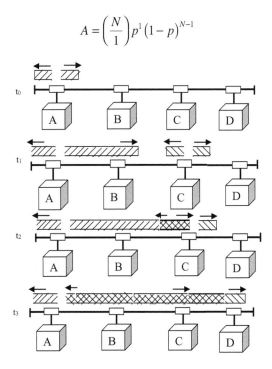

FIGURE 6.9 Operation of CSMA/CD MAC.

By substituting the value of p, the probability A can be written as

$$A = (1 - 1/N)^{N-1}$$

It is clear that to achieve maximum link utilization the probability of successful transmission of frames should be maximized. To find the link utilization, it is required to determine the mean length of contention interval. The mean value of contention interval is expressed as expectation value of slot length [2].

$$E(w) = \sum_{i=1}^{\infty} i \times \Pr \left[\begin{array}{l} \text{i slots in row with a collisions} \\ \text{or no transmission followed by a slot} \\ \text{one transmission} \end{array} \right]$$

$$= \sum_{i=1}^{\infty} i(1 - A)^i A$$

$$= \frac{1 - A}{A}$$

The link utilization U of CSMA/CD is written as

$$U = \frac{\text{transmission interval}}{\text{transmission interval} + E(w)} = \frac{1/2a}{1/2a + (1 - A)/A} = \frac{1}{1 + 2a(1 - A)/A}$$

For CSMA/CD, $N \to \alpha$, $A = \lim_{N \to \infty} (1 - 1/N)^{N-1} = 1/e$

$$Lt_{N \to \infty} U = \frac{1}{1 + 3.44a}$$

The link utilization of CSMA/CD decreases as the number of stations increases. This is because more collisions occurred in the medium, and this leads to less number of successful transmission of frames.

6.2.3.3.2 MAC Frame Format

The MAC layer accepts a block of data from LLC layer that is mentioned in the next section [3]. The MAC layer is used to carry out the functions related to medium access and for transmitting the data. MAC does these functions by using PDU, which is considered to be MAC frame. The MAC frame is slightly distinguishable from the standard High-Level Data Link Control (HDLC) frame format (already discussed in Chapter 5) due to having an LLC field which is required especially for LAN. Figure 6.10 shows the fields of the frame given below.

- **Preamble:** It is a seven-octet pattern with alternating 0s and 1s used by the receiver to establish bit synchronization.
- **Start frame delimiter (SFD):** The sequence 10101011 representing the actual start of the frame makes the receiver to locate the first bit of the rest of the frame.

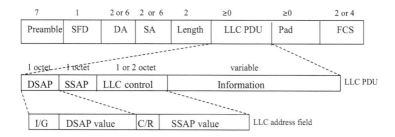

FIGURE 6.10   MAC frame format with LLC PDU.

- **Destination MAC address (DA):** It specifies the station(s) for which the frame is transmitted. It has a unique physical address, a group address or a global address. The choice of a 16 or 48 bit address length varies network to network and is same for all stations on a particular LAN.
- **Source MAC address (SA):** Source address field may be needed for the sender address of frames.
- **Length:** Length of the LLC data field.
- **LLC PDU:** LLC PDU consisting of destination service access point (DSAP), source service access point (SSA), LLC control and information bits. The DSAP and SSAP contain 7 bit addresses specifying the destination and source users of LLC. One bit of the DSAP represents whether the DSAP is an individual or group address (I/G), whereas one bit of SSAP indicates whether the PDU is a command or response (C/R) PDU. The (C/R = 0) bit indicates command, and the (C/R = 1) bit indicates response. The representation of LLC control bits is discussed in the next section. The information in variable bit represents message which may be a command or data.
- **Pad:** It is octets added to make it sure that the frame is large enough for its proper operation.
- **FCS:** It is a 16 or 32 bit FCS used for error checking of all the fields of frame except preamble, and SFD is made by a CRC scheme which is already discussed in Chapter 5.

6.2.4   LOGICAL LINK CONTROL

The LLC layer is made for sending link level PDU between two stations without the requirement of an intermediate switching node. The LLC should have the following two characteristics [2].

   i. uses multi-user access to obtain shared medium nature of link
  ii. uses some details of link access by the MAC layer.

Addressing in LLC requires the source and destination specification as LLC users. The LLC user addresses correspond to service access points (SAPs) for the user of a protocol layer. The operation and format of the LLC is the same as HDLC, which is already discussed in Chapter 5. There are three types of services provided as alternatives using LLC.

- **Acknowledged connectionless service**: It is a service that does not make the data flow or error control mechanism
- **Connection mode service**: It is similar to that of HDLC in which a logical connection is set up between two users and flow and error controls are provided.
- **Acknowledged connectionless service**: This service provides that datagram is to be acknowledged, but no prior logical connection is set up.

The LLC follows the asynchronous and balanced mode of operation of HDLC to accommodate connection mode LLC service, which is known as type 2 operation. Other mode of operation is not used in LLC. The LLC provides connectionless service using unnumbered information PDU that is type 1 operation. The acknowledged connectionless service is also made by using unnumbered PDUs, and this is known as type 1 operation.

6.2.5 Wireless LANs

For past few years, wireless LANs (W-LAN) have shown a significant impact in LAN [2,10], specially, nowadays in organizations/offices where W-LANs have become indispensable in addition to wired LANs due to its reconfiguration, relocation, ad hoc networking capability, removal of difficulty to wire in heritage building etc. as wireless LAN makes use of wireless transmission medium.

The wireless LAN is developed and standardized by IEEE 802.11 working groups. Figure 6.11 shows the standard model of wireless LAN introduced by the committee [2]. The standard model is called as an extended service set (ESS) consisting of two or more smallest building blocks. Each building block is called as a basic service set (BSS) having executed the same MAC protocol with some number stations for access to the same shared medium. The BSS may be isolated or may be connected with the backbone distribution system through an access point which is also called as bridge. The MAC is distributed or controlled by central coordination functions housed in the access point [10]. These BSSs are interconnected by a distribution system through wired LAN under an ESS which appears to be controlled by LLC.

There are three types of stations in wireless LAN based on the nature of their mobility:

- **No transition station**: A station having either stationery and moving within the direct communication range of the communicating stations of a single BSS is called as a no transition station
- **BSS transition station**: A station moving from one BSS to another BSS within an ESS is known as BSS transition station. In this case, the delivery of data to the station requires the address capability to recognize the new location of station.
- **ESS transition station**: A station moving from a BSS in one ESS to a BSS in another ESS is known as an ESS transition station.

There are three physical media mentioned in IEEE 802.11.

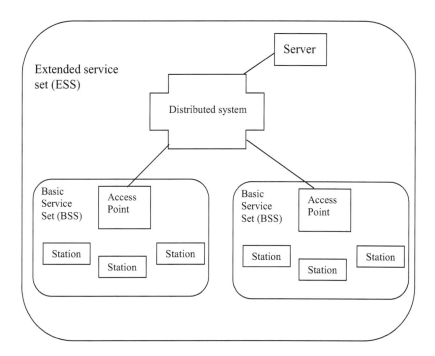

FIGURE 6.11   IEEE 802.11 architecture.

- Infrared at 1 and 2 Mbps operating at a wavelength between 850 and 950 nm.
- DSSS operating in 2.4 GHz ISM (industrial, scientific and medical) radio band. Up to 7 channel, each channel uses the data rate of 1 or 2 Mbps.
- FHSS operating in 2.4 GHz ISM band.

6.2.5.1   Medium Access Control (MAC)

The IEEE 802.11 group has two types of MAC layer – distributed access protocol and point access protocol [3]. Figure 6.12 shows the IEEE 802.11 protocol architecture consisting of three layers – Physical layer at bottom [2], distributed coordination function (DCF) layer at the middle and PCF layer at the top. In the lower level, the distributed access control uses the contention algorithm to provide access to all traffic with the DCF and at the top layer, point access protocol uses centralized MAC algorithm to provide contention-free service.

6.2.5.1.1   Distributed Access Protocol

This is a protocol for distributed system in which the decision of transmission is distributed over all the nodes and is made by using carrier sense mechanism like CSMA/CD. This carrier sense function is called DCF. It uses CSMA concept in which, before transmission by a MAC, the station tries to listen to the medium. If the medium is idle, the station transmits via MAC; otherwise, the station must wait

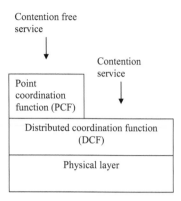

FIGURE 6.12   IEEE 802.11 protocol architecture.

until the current transmission is complete before transmission [2,3]. The DCF has no CD function because CD is not feasible in case of wireless medium. For smooth functioning of this algorithm, the DCF sets delays such as inter-frame space (IFS) used for CSMA access. There are three different IFS values:

1. the idle time of the medium during which the station tries to listen.
2. the time delay during which the station does not transmit because of the business of the medium
3. the time delay after the transmission is over.

6.2.5.1.2   Point Co-Ordination Function (PCF)

PCF is another access method used on the top of DCF. The operation consists of polling with a centralized polling master (point coordinator). The point coordinator makes use of PCF inter-frame space (PIFS) while issuing polls. Because PIFS is smaller than DCF inter-frame space (DIFS), the point coordinator can seize the medium and lock out all synchronization traffic while issuing polls and receiving response. There are two types of traffic – time-sensitive traffics controlled by point coordinator with round-robin manner and time-insensitive traffic directly controlled by DCF.

6.2.6   ASYNCHRONOUS TRANSFER MODE (ATM) LAN

The ATM LAN consists of ATM switching nodes in which multiple/integrated services are accommodated by virtual paths and channels. There are three possible types of ATM LANs [11,12]

- **Gateway to ATM LANs**: An ATM switch sets as a router and traffic concentrator for linking a premises network complex to an ATM WAN.
- **Backbone ATM switch**: Either a single ATM switch to local network of ATM switches interconnecting other LANs.
- **Workgroup ATM**: High-performance multimedia workstations and other end systems connect directly to an ATM switch.

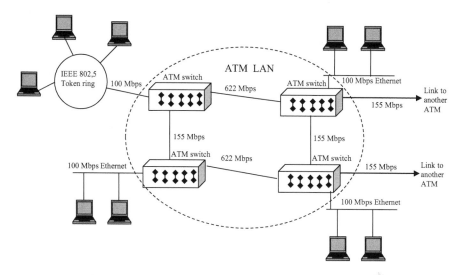

FIGURE 6.13   A backbone of ATM LAN.

Figure 6.13 shows a simple architecture of backbone of an ATM-based LAN having four ATM switches interlinked with links of data rates 155 and 622 Mbps. The ATM switch should carry out some sort of the protocol conversion form MAC protocol used on attached LAN to the ATM cell stream used on the ATM network. The ATM switches functioned as a router bridge in the network. The operation of ATM switch is already discussed in Chapter 5. This architecture is a technique for inserting a high-speed backbone into local environment. Of course, it does not address all of the needs for local communications [2,11,12].

SUMMARY

This chapter deals with data communication Networks using no optical transmission which is included as first generation of network. Initially we have mentioned background of networking-different generations. We have discussed LAN protocol architectures along with different medium access control which are also used in optical LAN. We have also discussed ATM LAN which can be connected with optical back bone based on WDM.

EXERCISES

6.1.   Distinguish cell rate and burst tolerance in ATM network. Compare committed information rat and excess burst size in frame relay networks.

6.2.   The transfer of a file containing one million characters is made from one station to another. What is the total elapsed time and effective throughput for the cases mentioned below:

a. A circuit switched star topology neglecting call set time and the data rate is considered to be 64 kbps.

b. A bus topology local network with two stations a distance D apart a data rate of B bps and a packet size of P with 80 bits overhead. Each packet is acknowledged with 88 bit packet and the propagation speed is of 200 m /microsecond. Solve for
    1. D = 1 KM, B = 1 Mbps, P = 256 bits
    2. D = 10 KM, B = 1 Mbps, P = 256 bits

c. A ring topology with a total circular length of 2D with two stations a distance D apart Acknowledgement is achieve by allowing a packet to circulate past the destination back to the source station. There are N repeaters on the ring each of which introduces a delay of one bit. Calculate for each of b(1) through b(4) for $N = 10$ and 100 bits.

6.3. At a propagation speed of 200,000,000 m/s, what is the effective length added to a ring by a bit delay at each repeater at 1000 Mbps.

6.4. A tree topology local network is considered between two buildings. If the permission can be made to a string cable between two buildings, one continuous tree network is used. Each building has an independent tree topology network and a point to point link will connect a special communication station on one network with communication station on the other network. What functions must the communication stations perform, Repeat the same for ring and star topology.

6.5. Take two stations on a broadband bus at a distance of 2 KM form each other. Let the data rate be 1Mbps, the frame length be 100 bits and the propagation velocity be 2,000,000 Km/s. Each station generates average rate of 2000 frames per second. For the ALOHA protocol, if the station start the transmission at time t, what is the probability of collision? Calculate the same for slotted ALOHA.

6.6. Show that the maximum values of the throughput S are 1/2e at offered load $= G = 0.5$ for pur ALOHA and 1/e for slotted ALOHA at $G = 1.0$ packets/s.

6.7. Take a token ring network having N stations. Make plots of the throughput S as function of propagation parameter, a for the values $N = 20$ and 200.

6.8. Take a token ring network having 30 stations. Find the throughput S for propagation parameter, a = 0.01, 0.05 and 0.1.

6.9. Take a CSMA/CD bus network having one station active out of 20 stations. Find the throughput S for message length of 1000 bits.

6.10. Take a Token bus network having one station active out of 50 stations. Find the throughput S for message length of 1000 bits.

6.11. For a Token bus network having all 50 stations active. Find the throughput S for message length of 1000 bits.

6.12. For a CSMA/CD bus network having all 50 stations active. Find the throughput S for message length of 1000 bits.

6.13. The system A consists of a single ring with 100 stations one per repeater and system B has 150 stations linked with bridge. If the probability of a link failure is P_1 a repeater failure is P_r and bridge failure is P_b. Derive P_1,

P_r and P_b. Find the system failure probability of A and complete failure of system B for $P_1 = P_r = P_b = 0.01$.

REFERENCES

1. D. Bertsekas and R. Gallager, *Data Networks*, Prentice Hall, Englewood Cliffs, NJ, 1992.
2. W. Stalling, *Data and Computer Communication*, Prentice Hall, Upper Saddle River, NJ, 1999.
3. G. E. Keiser, *Local Area Networks*, Tata McGraw-Hill, New Delhi, 1997.
4. M. Schwartz, *Computer Communication Networks Design and Analysis*, Prentice Hall, Engle wood Cliffs NJ, 1977.
5. G. Kessler and D. Train, *Metropolitan Area Networks Concepts, Standards and Services*, McGraw-Hill, New York, 1992.
6. A. Khanna and J. Zinky, The revised ARPANET routing metric, *Proceedings of SIG/comm.*, Symposium, Cambridge, MA, 1989.
7. A. Tanenbaum, *Computer Networks*, Prentice Hall, Englewood Cliff, NJ, 1988.
8. P. P. Sahu and M. Singh, "Multichannel direct sequence spread spectrum signaling using code phase shift keying," *Computer and Electrical Engineering (Elsevier)*, vol. 35, no. 1, pp. 218–226, 2009.
9. P. P. Sahu and M. Singh, "Multichannel frequency hopping spread spectrum signaling using code M-aryfrequency shift keying," *Computers & Electrical Engineering Journal (Elsevier)*, vol. 34, no. 4, pp. 338–345, 2008.
10. P. Davis and C. Mc Guffin, *Wireless Local Area Networks*, McGraw-Hill, New York, 1995.
11. H. L. Truong, W. W. Ellington, J. Y. L. Boudec, A. X. Meier and J. W. Pace, *LAN Emulation on an ATM Networks*, Prentice Hall, Englewood Cliff, NJ, 1992.
12. K. Sato, S. Ohta, and I. Tokizawa, "Broadband ATM network architecture based on virtual paths," *IEEE Transaction on Communication*, vol. 38, pp. 1212–1222, August 1990.
13. Fiber Channel Association, *Fiber Channel: Connection to the Future*, Fiber Channel Association, Austin TX, 1994.
14. A. Mills, *Understanding FDDI* Prentice Hall, Englewood Cliff, NJ, 1993.
15. P. Green, *Fiber Optic Networks*, Prentice Hall, Englewood Cliff, NJ, 1993.
16. J. Bliss and D. W. Stevenson, Special issues on *Fiber Optical for Local Communication*, Special Issue, *IEEE Journal on Selected Areas on Communication*, vol-SAC-3, November 1985.
17. J. Bliss and D. W. Stevenson, Special issues on *Fiber Optics for Local Communication*, special issue on *IEEE Journal Lightwave Technology*, vol-LT-3, June 1985.
18. M. M. Nassehi, F. A. Tobagi and M. E. Marhic, "Fiber optic configuration for local area networks," *IEEE Journal Selected Areas on Communication*, vol. SAC-3, pp. 941–949, November 1985.
19. H. K. Pung and P. A. Davics, "Fiber optic local area networks with arbitrary topology," *IEEE Proceedings*, vol. 131, pp. 77–82, April 1984.
20. A. Shah and G. Ramakrishsnan, *FDDI: A High Speed Network*, Prentice Hall, Englewood Cliff, NJ, 1994.

7 Fiber-Optic Network without WDM

In the second generation of networking, the signal is transmitted in an optical domain (preferably through optical fiber) but multiplexing, switching and amplification are performed in an electrical domain. There are four basic topologies normally used in the second generation of networking [1–8] – bus topology, ring topology and star topology. In bus topology the most common networks are Fasnet, Expressnet and dual queue data bus (DQDB) [1–3]. In case of ring topology, most popular networks are fiber distributed data interface (FDDI) and its extension version FDDI-I. In star topology, most commonly used optical networks are Fibernet-I and Fibernet-II. These fiber-optic networks are based on fiber channels [1] organized into five levels – FC-0, FC-1, FC-2, FC-3 and FC-4 [13]. Figure 7.1 shows architecture of a five-layer fiber channel in which levels FC-0–FC-4 of fiber channel hierarchy are described as follows [1–2]:

- **FC-0:** FC-0 consists of a physical media which is an optical fiber cable with laser/LED for long-distance transmissions, copper coaxial cable for short-distance communications and shield twisted pair for low speeds for short distances.
- **FC-1:** FC-1 does the function of byte synchronization and encoding. It uses 8B/10B encoding/decoding scheme providing error detection capability. In FC-1 a special code character maintains byte and word alignment.
- **FC-2:** It provides an actual transport mechanism in which framing protocol and flow control between N ports are obtained. There are three classes of service between ports.
- **FC-3:** It is a common services layer. These are port-related services and services across two or more ports in a node.
- **FC-4:** It is an upper layer protocol that supports a variety of channels such as SCSI, HIPPI, and IPI-3 SBCS and network protocols such as IEEE 802 series, ATM and Internet Protocol (IP).

7.1 BUS TOPOLOGY

In this topology there are two types of bus configuration [16,9] – dual bus and loop configuration. Fasnet is based on dual bus configuration, whereas Expressnet and DQDB are based on loop configuration (Figure 7.1).

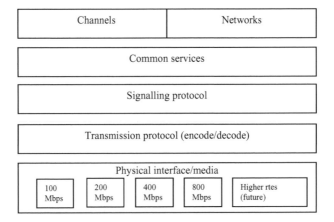

FIGURE 7.1   Fiber channel layer architecture.

7.1.1 FASNET

Figure 7.2 shows the topology of Fasnet consisting of two unidirectional fiber channels (A and B) in which the signals are propagated in the reverse direction [1,6,9]. All the users transmit the signals and receive the signals from these channels. In the figure, node-1 transmits the signal to node-2 via channel-A and node-2 transmits the signal to node-1 via channel-B, so the node uses channel-A to transmit the signal to its right-side node and uses channel-B to transmit the signal to its left-side node. Each node has four tap points and total of $4M$ tap points are required for M number of nodes. In Fasnet, the asymmetry made by the unidirectional signal transmission sets up a natural order among the users for round-robin access protocols.

There are two streams of signal flows – upstream and downstream. A node having transmission of a packet sends it in one of the channels as upstream and the recipient is downstream from the sender. The sender is treated as a head node, whereas the recipient is treated as an end user performing special functions on each channel. The head user sends clock signals to maintain bit synchronization in the system, and from the clocking information, nodes listened to the channel requires to identify

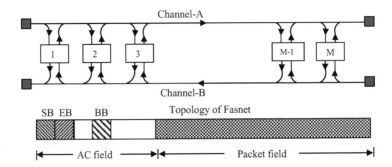

FIGURE 7.2   Fasnet architecture with slot [9].

field length slots traveling downstream. In the figure, each slot starts with an access control (AC) field and the rest of the slot is packet field. The AC field consists of a start bit (SB) indicating the start of a new round or cycle, busy bit (BB) stating that a packet has been found in the slot and end bit (EB) that is situated during the dead time between SB and BB. This bit is used by the destination node to order the head user via channel to begin a new cycle. There are two types of access protocols – gated Fasnet protocol and negotiated Fasnet protocol [18–20].

In Fasnet, a node having no transmission of packet is in IDLE state, and after arrival of a packet to be transmitted on channel-A (i.e., destined for a user to the right of this one) the node becomes WAIT state. The user can understand SB of each slot. For SB = 1, the user turns into DEFER state in which it reads and sets BB of each slot. When an empty slot is identified, the user writes its packets into it. It departs to be the IDLE or WAIT state depending on whether it has more packets to travel or not.

In negotiated Fasnet, an IDLE user is said to be ACTIVE if it has yet to transmit in the current round or DORMANT state which transmits already. A DORMANT user does not attempt to access the channel. Upon arrival of a packet to an ACTIVE idle user, the user moves immediately to the DIFFER state. So, the user becomes DORMANT in which it does not transmit another in this round.

In both cases of Fasnet, SB is placed by the head node and the end user examines all slots on channel A with decoding of the status of SB and BB. Upon identifying SB = 1, the end user locks for the first slot in which BB = 0 indicates that all users are either IDLE or DORMANT. At this time, it sets EB = 1in the next slot on channel B. The head user identifying EB = 1 goes to SB = 1 in the next slot on channel A. Thus, the overhead in starting a new round is two times of the end-to-end propagation delay plus the slot size. The additional two slots are acquired by the destination identified with BB = 0 on channel A, which lingers in the AC field of the next slot on channel B to set EB = 1 and the head user having identified EB = 1 on channel B lingers for the next slot on channel A in order to set SB = 1. It is also possible to permit the end user to set EB = 1 every time it encounters BB = 1. The average packet delay in Fasnet is written as [9]

$$D = \frac{M}{S} - \frac{1}{\lambda}$$

where M = total number of users, $S = \sum_{i=1}^{M} s_i$, where S_i denotes the expected through-put of packets form user i and λ = packet arrival rate.

7.1.2 EXPRESSNET

Figure 7.3 shows an architecture of Expressnet based on a loop configuration [6,9]. In Expressnet, there are two channels – outbound channel which sends data and an inbound channel that is solely used for reading the transmitted data. All signals sent on the outbound channel are copied again on the inbound channel, thus achieving broadcast communication among the other stations. Here, like Fasnet, the asymmetry developed by the unidirectional signal propagation sets up a natural ordering among the users needed for round-robin access protocols. A user who has a message

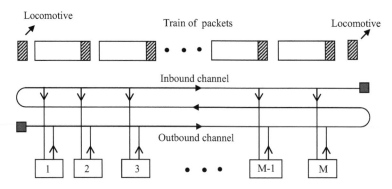

FIGURE 7.3   Expressnet architecture with flow of packets [9].

to send is said to be backlogged. If a user does not have a message to send, then it is said to be idle. A backlogged user operates as follows:

1. Linger for the next end carrier of outbound channel (The occurrence is indicated as EOC(OUT)).
2. Immediately start sending the packet and, at the same time, detect the outbound channel for activity from the upstream side.
3. If action is identified from upstream, then abandon the transmission; otherwise, complete the transmission. If it is still not done, go to step 1; otherwise, linger for the next packet.

In Expressnet [18,19], there is a user which abandons its transmission and hence the desired node is sent successfully. Moreover, a user that finished the transmission of a packet in a given round meets the event EOC(OUT) again in that round, thus guaranteeing that no user sends more than once in a given round. In the figure a train of packet generated in the outbound is distinct to be a succession of the transmissions in a given round and is entirely seen in the inbound channel by all users. The end of a train on the inbound channel indicated as EOC(IN) is identified whenever the idle time exceeds t_d (time required to identify EOC(OUT)). The EOC(IN) is permitted to visit each user in the same order. Thus, to begin a new round, the EOC(IN) is treated as a synchronizing event. The step-1 of the algorithm should be revised as

1. Linger for the first of the two events EOC(OUT) or EOC (IN). (Note that one of such events can take place at a given point of time).

To keep away from losing of synchronizing event EOT(IN) which happens if no packet is ready when it is removed from the inbound channel, all users in either backlogged or idle state send a short burst of not modulated carrier of duration t_d whenever EOT(IN) is identified. The burst is known as locomotive. If the train of packet is empty, then the end of the locomotive constitutes EOT (IN). The time gap between two consecutive trains is the transmission delay between the transmit and receive taps of a node. The time required for this gap is called as a round trip delay.

An average throughput of Expressnet is written as

$$S = \frac{\bar{n}T}{\bar{n}T + Y}$$

where $\bar{n}$ = average number of packets in a round, T = packet length which is the time required to transmit a packet and Y is an idle time of the channel between rounds represented as the inter-round overload = $2\tau^3$ (τ = end-to-end transmission delay of the signal traveling over the network). The expected average delay of a packet is considered as

$$D = \frac{M}{S} - \frac{1}{\lambda}$$

where M = total number of users and λ = packet arrival rate. The Expressnet has a smaller number of tap points than that of Fasnet, but its access protocol is more complicated in comparison to that of Fasnet. Transmission in Expressnet is asynchronous while that of Fasnet is synchronous.

7.1.3 DISTRIBUTED QUEUE DUAL BUS (DQDB)

The DQDB is emerging as one of the leading technologies for high-speed metropolitan area networks (MANs). Figure 7.4 shows DQDB based on dual bus configuration, which consists of two contra-directional buses. The principal components of the DQDB are nodes connected to these two contra-directional buses. Each bus A or B is headed by a frame generator as shown in the figure. These frame generators are responsible for continuous generation of 125 µs frames. The DQDB has a medium AC(MAC) protocol standardized as IEEE 802.6 MAN in which the frame is divided into equal size slots. The slot consists of 5 octet overhead and 48 octet payload field. The slot follows the ATM cell size as discussed earlier. The overhead called as header field consists of BB and request bit (REQ). The BB indicates whether the slot is occupied or not by segment and is used for accessing bus A, whereas REQ is used for sending requests for future segment transmission. Since a frame length is fixed, the number of slots in a frame is determined by a transmission speed C. For a given length of bus l, the number of slots N in frame is written as $N = \tau/(53 \times 8/C)$, where τ = end-to-end propagation delay. The DQDB uses DQDB state machine for transmission of segments containing payload/information.

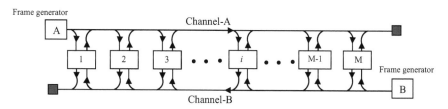

FIGURE 7.4   DQDB architecture [10].

In DQDB protocol [10], synchronous (isochronous) packets are transmitted on the basis of prior reservations. There are two buses – bus A which is treated as forward bus and bus B treated as reverse bus. The procedure for segment transmission in the forward bus is same as that in the reverse bus. Each node is either idle, when there is nothing to transmit, or otherwise it is count_down. When it is idle, the node makes count_down through a request counter (RQ_CTR) of the number of outstanding REQs from the downstream nodes. The RQ_TR increases by one for each REQ received in the reverse bus and decreases by one for each empty slot in the forward bus. When it becomes active, the node transfers the contents of RQ_CTR to a second counter named count_down (CD_CTR) and resets RQ_CTR to zero. The node then sends a request in the reverse bus by setting REQ = 1in the first slot with REQ = 0. The CD_CTR is decreased by one for every empty slot in the forward bus in the forward bus until it reaches zero. Immediately afterwards, the node transmits the segment into the first empty slot of the forward bus and signals this event to the bandwidth balancing machine (BMB). At the same time, the RQ_CTR increases by one for each new REQ received in the reverse bus for downstream nodes. The BWB machine having a threshold named as BWB_MOD counts to keep track of the number of the transmitted signals. When BWB reaches the BWB_MOD threshold, the counter is cleared and the RQ_CTR is increased by one. The value of BWB_MOD can vary from 0 to 16. The value zero means that the BWB machine is disabled. The order of the segment transmission on bus A is organized in a distributed first-come first-service (FCFS) manner by utilizing these two counters – RQ_CTR and CD_CTR. The third counter Req_B keeps track of the number of outstanding reservation requests. When Req_B is non-zero, the node sets the REQ in the next B slot if it is not already set.

DQDB network is specified by the following network parameters

- Nodes spaced equally along two buses A and B
- Capacity of each bus = 150 Mbps
- Slot size = 53 octets
- Slot duration = 2.82 μs
- Length of each bus = 147 slots = 89 km
- Bus latency = 400 μs
- Dual bus latency = 800 μs
- Total number of nodes = M = 50
- Buffer size = 10 segments

The performance of DQDB is analyzed by using M/M/1 queuing system with round-robin scheduling discipline. Each unidirectional bus is represented by a sequence of servers. Each server models signal a propagation delay between two consecutive pairs of nodes using an M/M/1 queue model which is discussed in next section.

7.2   RING TOPOLOGY: FDDI

In the second generation, the ring topology is mainly FDDI [2,3,7] which is a token ring scheme following IEEE 802.5 specification and designed for both high-speed LAN and MAN applications. In the ring of FDDI, the signal is transmitted in the optical domain but amplification, multiplexing and operation of tap are made in an electrical domain. There is a difference between FDDI and IEEE 802.5 token ring (without fiber that is required to accommodate higher data rate (100 Mbps). The MAC frame is depicted as given below.

7.2.1 MAC Frame

Figure 7.5 shows the frame format used for the implementation of an FDDI scheme. In the figure MAC format entities/fields must deal with individual bits, so the discussion that follows sometimes refers to 4 bit symbols and sometimes to bits [2]. The general frame carrying information consists of the following fields:

- **Preamble:** It is required for synchronization of frames with each station clocks and uses a field of 16 idle symbols (64 bits); Subsequent repeating stations may change the length of the field. The idle symbol is a non-data fill pattern that depends on the signal encoding the mecum.
- **Starting delimiter (SD):** It indicates the start of frame encoded as JK, where J and K are non-data symbols
- **Frame control (FC):** It is represented as a bit format CLFFZZZZ in which C indicates whether it is synchronized or not; L indicates the use of 16 bit or 48 bit address; FF indicates whether it is an LLC MAC control or a reserved frame. The remaining 4 bits (ZZZZ) indicate the type of control frame.
- **Destination address (DA):** It indicates the station in which the frame is transmitted. It may be a unique physical address, a multicast group address or a broadcast address having 16 and 48 bit address.
- **Source address (SA):** It indicates the station from which the frame is transmitted.
- **Information**: This field having variable bit pattern (>0 bit) consists of LLC data bits or information related to control operations.
- **Frame check sequence (FCS):** It has 32 bits estimated from information bits by using cyclic redundancy check (CRC) used for error detection in the information bits.
- **Ending delimiter (ED):** It contains a non-data symbol marking the end of the frame except FS field.

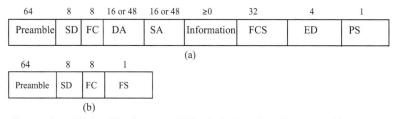

(a)

(b)

SD = start frame delimiter, FC = frame control, DA = destination address, SA = source address, FCS = frame check sequence, ED = ending delimiter and FS = frame status

FIGURE 7.5 FDDI frame format: (a) General and (b) token frame format.

- **Frame status:** It has error detected (E), address recognized (A) and frame copied (F). Each indicator is represented by symbol R for reset or false and S for set or true.

The token frame format has the following fields.

- **Preamble:** It is the same as that in general frame format
- **SD:** It is the same as that in general frame format
- **FC:** It has a bit pattern of 10000000 or 11000000 indicating that it is a token
- **ED:** It is a non-data symbol marking the end of the frame.

7.2.2   MAC Protocol of FDDI

The basic FDDI MAC protocol is fundamentally the same as IEEE 802.5. Following are the two main differences between these protocols [8].

1. In FDDI, a station waiting for a token seizes the token by failing to repeat the token transmission as soon as the token frame is recognized. After the token is completely captured, the station starts transmitting one or more data frames. In case of a normal IEEE 802.5 operation, flipping to convert a token to the start of a data frame is made but it is impractical because of high data rate in FDDI.
2. In FDDI, a station has been transmitting data frames and releases a new token as soon as it completes data frame transmission, even if it has not begun to receive its own transmission. This is the same method as early token release option of 802.5. Because of high data rate in FDDI, it is insufficient to require the station to wait for its frame to return as in IEEE 802.5.

Figure 7.6 shows a token ring operation in FDDI [20] in which single and double-frame transmission from A to C are presented in diagram (a) and (b) respectively. After seizing the token, station A transmits frame F_1 having destination station C immediately. In case of (b), transmission of frames F_1 and F_2 is followed by a token T. After receiving two frames F_1 and F_2, it releases token by station C. A similar procedure is also followed in case of (a). The priority scheme cannot be included in FDDI MAC operation, and the use of a reservation field is not effective. Specifically, the FDDI standard is intended to provide for greater control over the capacity of the network than 802.5 to meet the requirements for high-speed LAN and also accommodates a mixture of stream and bursty traffic.

Protection has been made by using an extra ring along with primary ring to provide a backup path if failure occurs in the FDDI ring. There are two types of failures against which protection can be made – node failure and link failure. Figure 7.7a shows a double-ring FDDI having protections against link and node failures in Figure 7.7b. The signals are protected against link failure by connecting two rings near to the failure whereas the signals are protected against the node failure by connecting two rings in the nodes adjacent to failure nodes.

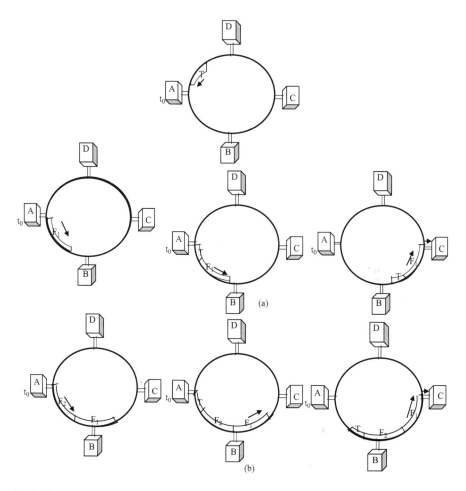

FIGURE 7.6 Token ring operation in FDDI: (a) single-frame transmission and (b) double-frame transmission.

7.3 STAR TOPOLOGY

In star topology for the second generation of network, there are passive and active star topologies. The operation of passive star is not controlled externally, whereas in case of active star, the operation is controlled externally. Fibernet and solar net are based on passive star and Fibernet-II is based on active star topology [11,12]. In star network, there are different configurations – star repeater network, passive reflective star network and passive transmissive star network. In case of star repeater network, the fibers are run from each station to the central point as shown in Figure 7.8a. In the repeater station, the optical signal is converted into electrical, amplified in electrical level and then converted into optical level transmitted to the receiver of the node. Figure 7.8b shows a reflective star coupler configuration in which single fiber is used

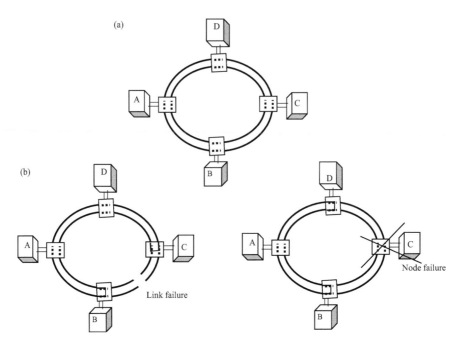

FIGURE 7.7   (a) Double-ring FDDI (b) its protection against link and node failure.

for routing the signal from the node to star and from star to the node. But there are difficulties to bidirectionality in the same fiber. Fibernet is based on the transmissive star which is discussed below.

7.3.1   FIBERNET

Figure 7.9 shows the architecture of Fibernet consisting of a 7-port transmissive star coupler in which the input fibers are passed through a tapered capillary tube to form a closely packed hexagonal array. The assembly is epoxy, and then polished. A similar array of output fibers is coupled to the first array using a clad mixing rod whose numerical aperture matches that of the fibers and three components are cemented with index matching epoxy. The number of ports is determined by the number of layers k of hexagonally closed fibers enclosed by the capillary tube. In general, the number of fibers in k layers is $N_k = k^3 - (k-1)^3$. There are two types of prototypes reported and demonstrated so far with number layers $k = 2$ and $k = 3$ which provides the number of ports 7 and 19 respectively. The insertion loss of a 7-port transmissive star coupler is −7.4 dB averaged through all ports, whereas that for a 19-port star coupler is −10 dB. The coupling coefficients between all the ports (other than the two axial ports) are within 2 dB, whereas that for the axis-to-axis coupling coefficient is about 5 dB, higher than that for average.

Fibernet is an Ethernet with fiber channel [18,19] in which packet switching is distributed in two senses – the packets are transmitted by a station/node only after determining that no previous station transmission is in progress. During the transmission of its own packet, a station monitors its transmission so that no collision

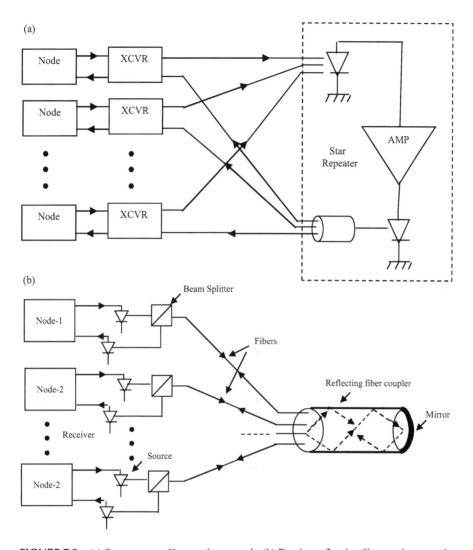

FIGURE 7.8 (a) Star repeater fiber-optic network. (b) Passive reflective fiber-optic network.

takes between its own packets and other's packets. The efficiency E is related to the packet length P in bits, transmission capacity C in bits per second and time in seconds to propagate a packet from one end to the other and is formulated as

$$E = \frac{P/C}{P/C + 1.7\,T}$$

The Fibernet experiment has been performed 150 Mbps pseudorandom data over ½ km distance with 19 port star coupler with zero error detected in a test sequence of 2×10^{11} pulses. The pulse dispersion the rough long fiber link confirmed that its bandwidth which is in excess of 300 MHz. The bit error rate (BER) is ~10^{-9}.

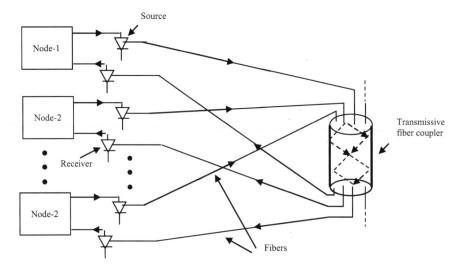

FIGURE 7.9   Transmissive star fiber optic network (Fibernet).

The disadvantage of Fibernet is limitation in the number and length of fiber cables. The collision takes place in Fibernet, but it can be dealt with the use of CSMA/CD medium AC. Time delay violation may occur during transmission.

7.3.2 FIBERNET-II

To remove limitation of number of fiber ports and length of cables, another configuration of Fibernet known as Fibernet-II is reported [12]. Figure 7.10 shows the configuration of Fibernet-II based on active star in which the transmission is controlled externally by control module. There are two fiber cables in which outbound optical fiber cable is connected to transmitter modules and inbound optical fiber cable is connected to the all receiver module. The transceiver is connected to the stations with standards Ethernet cables. Ethernet compatibility interface is at the junction between transceiver and transceiver cable. When a station transmits on an idle network, the active star repeater detects the optical signal on inbound optical fiber, electrically regenerates it and optical transmits it all the stations via outbound fibers. If two or more stations transmits simultaneously, the collision is taken place and detected in the repeater and signaled to the transceiver using 1 MHz square wave modulated optical signal indicating collision occurrence. This optical signal is easily distinguished form valid data by using simple timing circuit in Fibernet-II.

The Fibernet II star repeater is designed for convenient modular growth. Figure 7.11 shows a star repeater consisting of a receiver module, transmitter module and control module which are connected to dual coaxial cables X and R that are terminated at each end with 50 ohm resistors. The receiver and transmitter circuits are as same as those used in other networks. The electrical outputs of independent optical receivers in the receiver module drive in the back plane of R with identical voltage labels specified for Ethernet coaxial cables. The receiver output currents

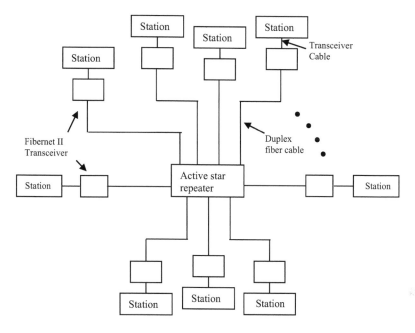

FIGURE 7.10 Schematic diagram of Fibernet II based on active star.

are summed on the R which behaves as Ethernet cable. The control module monitors data on the R bus, compares it with a data reference voltage to reset the signal levels and passes it via the digital switch to the X bus. The transmitter module then transmits this data signal on the outgoing fibers. Collision detection occurs in the control module and is accomplished by monitoring the average voltage on the R bus. This voltage is averaged for a few bit times and compared with the reference voltage. When the collision occurs, the average voltage is approximately doubled. If the collision in indicated, the collision presence signal of 1 MHz square wave modulated signal (generated by digital switch in control module and substituted the data signal) is then forwarded by the transmitter module to all the receivers. The MAC of the Fibernet II follows the rules of CSMA/CD. The control module monitors data on R and X bus.

The optical cables used in Fibernet II have two graded index fibers, each with a core diameter of 100 μm and numerical aperture of ~0.29. The wavelength of the optical signal is 850 nm.

The optical power budget for Fibernet II is given below.

Minimum optical power budget of transmitter module = 100 μW (−10 dBm)
Average input at receiver = −28 dB
Attenuation for fiber cable = −6 dB/km
Margin of aging = −6 dB
Insertion loss = −2 dB
Fiber length = 1.25 km
Network diameter = 2.5 km

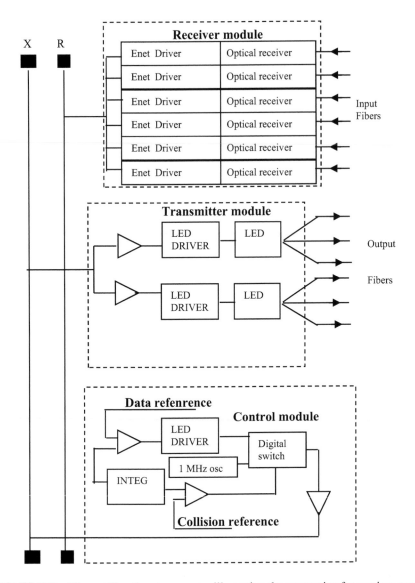

FIGURE 7.11 Fibernet II active star repeater illustrating the connection for receiver, transmitter and control modules.

7.4 WAVELENGTH ROUTED NETWORKS WITHOUT WDM

Wavelength Routed Networks have the following merits over Broadcast-and-Select Networks.

    i. Wavelength reuse
    ii. No power splitting loss
    iii. Good scalability

In this network, the end nodes are equipped with wavelength crossconnect ((WXC) for switching one wavelength of any output fiber to the required input fiber) and transceivers tunable to required wavelengths. Message is sent from one node to another node using a wavelength continuous route called *lightpath*, without requiring any optical–electronics–optical conversion and buffering. Intermediate nodes route the lightpath using WXCs. The architecture of WXC is shown in Figure 1.8. A lightpath is an all-optical communication path between two nodes. It is identified as a physical path, and the wavelength on this path can be realized as a high-bandwidth pipe capable of transmitting several gigabits per second. The requirement that the same wavelength must be used on all the links of selected route is called *wavelength continuity constraint*. One fiber cannot contain two lightpaths of the same wavelength. This is called *distinct wavelength assignment constraint*. However, two lightpaths can use the same wavelength if they use disjoint sets of links. This property is called *wavelength reuse*. Packet switching can be supported in WDM optical communication by *virtual topology* (a set of lightpaths or optical layer) is imposed over a *physical topology* (not broadcast system) by setting the OXCs in each interval nodes. At each interval node, the packet is converted to electronic form and retransmitted on another wavelength.

Existing internet backbone network consists of high-capacity IP routers [13] connected by point-to-point fiber links as shown in Figure 7.12. These links are realized by SONET or ATM-over-SONET technology [13,14]. The backbone routers use IP-over-ATM or IP-over-ATM-over-SONET [14] to route traffic in the backbone network (as shown in Figure 7.12). Most of the SONET works at OC-3 to OC-12 and now demand goes for OC-48 or more. It is impractical and uneconomical to upgrade every time the bit rate handling capacity of SONET according to the requirement

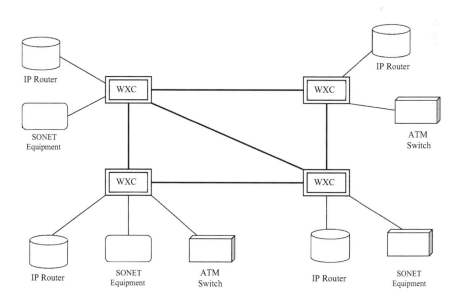

FIGURE 7.12 Wavelength routed backbone network.

of increasing traffic demand; however, a cost-effective solution to this problem is implementation of WDM technology in backbone networks. This scheme has been displayed in. The electronics processor at nodes can be an IP router, ATM switch or a SONET system to provide the electronics packet switching. DWM transport network can be decomposed into three layers.

1. **Physical media layer:** It is the lowermost level, which contains the physical fiber connections, that also deals with the hardware tools and the techniques for reliable signal transmission (transmission, amplification, reception, BER minimization, etc.).
2. **Optical layer:** It contains a set of lightpaths. It serves the upper layer (concurrent clients) with client-independent services of variable bit rates. Lightpaths can carry SONET data, IP packets/datagrams or ATM cells. It also maintains the strategy for rerouting lightpaths in case of node or physical fiber failure. ITU-T has decomposed this layer into three categories: optical channel layer, optical multiplex section layer and optical transmission section layer. This layer is just above the physical layer.
3. **Client layer:** It is the top layer and maintains the topology for the transmission of different types of data like SONET, ATM cells or IP packets.

SUMMARY

This chapter discusses the second generation of networking in which the signal will be transmitted in an optical domain (preferably through optical fiber) but multiplexing, switching, etc. will be performed in an electrical domain. There are four basic topologies normally used in the second generation of networking [1,8] – bus topology, ring topology and star topology discussed in this chapter. In bus topology we have mentioned the most common networks namely Fasnet, Expressnet and DQDB [1,3]. In case of ring topology, we have discussed the most popular networks such as FDDI and its extension version FDDI-I. In star topology, Fibernet-I and Fibernet-II are mentioned. We have also discussed wavelength routed network without WDM.

EXERCISES

7.1. Find out the average packet delay in Fasnet of 10 number of users and average throughput of each user = 70% and packet arrival rate = 1000 packets per second.
7.2. Find out the average packet delay in Expressnet of 12 number of users and average throughput of each user = 70% and packet arrival rate = 800 packets per second.
7.3. Find out the number of transmitters and receivers required for FASNET having 20 number of users.
7.4. Determine the number of transmitters and receivers required for Expressnet having 20 number of users.
7.5. Determine the number of transmitters and receivers required for DQDB having 20 number of users.

7.6. Design an FDDI network having a coverage area 5 square kilometer and 10 number of users and find the average packet delay of the same.

7.7. Design Fibernet II network having a coverage area 5 square kilometer and 10 number of users and find the average packet delay of the same.

REFERENCES

1. Fiber Channel Association, *Fiber Channel: Connection to the Future*, Fiber Channel Association, Austin, TX, 1994.

2. A. Mills, *Understanding FDDI*, Prentice Hall, Englewood Cliff, NJ, 1993.

3. P. Green, *Fiber Optic Networks*, Prentice Hall, Englewood Cliff, NJ, 1993.

4. J. Bliss and D. W. Stevenson, Special issues on *Fiber Optical for Local Communication*, special issue on *IEEE Journal Selected Areas on Communication*, vol. SAC-3, November 1985.

5. J. Bliss and D. W. Stevenson, Special issues on *Fiber Optics for Local Communication*, special issue on *IEEE Journal Lightwave Technology*, vol-LT-3, June 1985.

6. M. M. Nassehi, F. A. Tobagi, and M. E. Marhic, "Fiber optic configuration for local area networks," *IEEE Journal Selected Areas on Communication*, vol. SAC-3, pp. 941–949, November 1985.

7. H. K. Pung and P. A. Davics, "Fiber optic local area networks with arbitrary topology," *IEEE Proceedings*, vol. 131, pp. 77–82, April 1984.

8. A. Shah and G. Ramakrishsnan, *FDDI: A High Speed Network*, Prentice Hall, Englewood Cliff, NJ, 1994.

9. F. Tobagi and M. Fine, "Performance of unidirectional broadcast local area networks: Expressnet and Fasnet", *IEEE Journal of Selected Areas of Communication*, vol. 1, pp. 913–926, 1983.

10. B. Mukherjee and S. Banerjee, "Alternative strategies for improving the fairness in and an analytical model of the DQDB network," *IEEE Transaction on Computers*, vol. 42, 151–167, 1993.

11. E. Rawson and R. Metcalfe, "Fibernet: multimode optical fibers for local computer networks," *IEEE Transactions on Communications*, vol. 26, pp. 983–990, 1978.

12. R. Schimdt, E. Rawson, R. Noorton, S. Jackson, and M. Bailey, "Fibernet II: a fiber optic Ethernet", *IEEE Journal on Selected Areas of Communication*, vol. 1, 701–711, 2003.

13. K. Kitayama, N. Wada, and H. Sotobayashi, "Architectural considerations for photonic IP router based upon optical code correlation," *IEEE Journal of Lightwave Technology*, vol. 18, 1834–1840, 2000.

14. J. Manchester, J. Anderson, B. Doshi, and S. Dravida, "IP over SONET", *IEEE Communication Magazine*, vol. 36, pp. 136–142, 1998.

8 Single-Hop and Multihop WDM Optical Networks

There are three generations of optical network identified based on physical level structure. The first generation includes networks before fiber-optic technology [1–5], which is already discussed in Chapter 6. The second generation exhibits network with optical fiber technology [6–9] in traditional architecture without all-optical devices such as WDM and optical switches. These second-generation networks are also discussed in Chapter 6. Although some improvement in performances such as higher data rates, lower bit error rates (BER) and reduced electromagnetic coupling are made by using optical fiber in second generation, the limitation of this generation in data rates is due to the use of electro to optic and optic to electro converter during amplification switching and multiplexing. These limitations are removed by using an all-optical concept in which once the information enters the network, it may remain in the optical domain until it is delivered to the destination (avoiding any electrical bottleneck. This all-optical concept is used in third generation. This chapter begins with a discussion on a local lightwave network employing WDM constructed based on a passive-star coupler (PSC). These networks are of two types – single-hop and multihop. In single-hop network only one via is used for the establishment of connection, whereas in multihop networks, single or more than one via nodes may be used for the establishment of the connection request.

8.1 SINGLE-HOP NETWORKS

A local area network (LAN) is made by keeping capabilities of optical technology in mind so that WDM and tunable optical transceivers (transmitters or receivers) are used. The vast optical bandwidth of a fiber is engraved up into smaller-capacity channels, each of which can operate at peak electronic processing speeds (i.e., over a small wavelength range) of 40 Gbps [10–13]. By tuning its transmitter(s) to one or more wavelength channels, a node can transmit into those channel(s). Similarly, a node can tune its receiver(s) to receive from the appropriate channels. The system can be configured as a single-hop broadcast-and-select network in which all of the inputs from various nodes are combined in a WDM PSC and the mixed optical information is broadcast to all outputs.

Figure 8.1 shows single-hop broad and select (BSC) WDM network in which the transmitted is tuned to any of the wavelengths $\lambda_1, \lambda_2, \ldots, \lambda_N$ for transmission of signals to the corresponding receiver which can receive the transmitted wavelength. For broadcasting of signal to all the receivers the same signal has to be transmitted with all wavelengths.

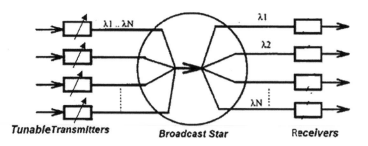

FIGURE 8.1   Singe-hop broadcast-and-select WDM network.

An $N \times N$ star coupler can be used as a WDM passive star for single-hop net-work consisting of an $N \times 1$ combiner followed by a $1 \times N$ splitter 1; thus, the signal strength incident from any input can be (approximately) equally divided among all of the N outputs. The passive property of the optical star coupler is important for network reliability, since no power is needed to operate the coupler; also, it allows information relaying without the bottleneck of electrooptic conversion

Figure 8.2 shows a $D \times D$ single-hop metro network architecture using arrayed-waveguide grating (AWG) [11–13]. The AWG allows for spatial wavelength reuse, i.e., the wavelengths of the network are simultaneously employed at each AWG input port without having collisions at the AWG output ports. This spatial wavelength reuse is made to enhance the network performance for a fixed set of wavelengths with respect to BSC-based single-hop networks (Figure 8.1). The scalability is diffi-cult to get in an AWG-based single-hop WDM network because the number of trans-ceivers required at each node is equal to the total number of nodes. The problem is resolved by installing optical couplers or splitters between the AWG and the nodes. In the figure, a cyclic AWG has D input ports and D output ports. At each AWG

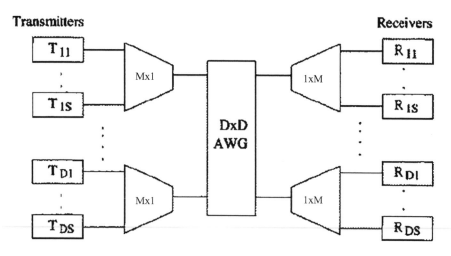

FIGURE 8.2   An AWG-based single-hop metro network.

input port, an $M \times 1$ combiner combines the transmissions from the transmitters of S attached nodes. At each AWG output port, a $1 \times M$ splitter has the signal to M individual fibers attached to the receivers of the nodes. Thus, this architecture needs total number of transceivers $N = M \times D$. There are other topologies such as bus or tree, apart from star topology. In case of bus topology, there is an additional attempt capability that a node, before/during its transmission, senses the activity on the bus from upstream transmissions.

The input lasers (transmitters) or the output filters (receivers) or both are made tunable to open up a multitude of networking possibilities. In a single-hop network, a dynamic coordination between nodes are needed. For transmission of a packet transmission, one of the transmitters of the sending node and one of the receivers of the destination node are tuned to the same wavelength during full transmission session. In the single-hop environment, the tuning time of transmitters and receivers should be tuned to different channels quickly so that packets may be sent or received in quick succession. The main problem in configuration is the large tuning time in comparison to packet transmission times. On the other hand smaller tunable range of these transceivers limits the number of channels.

8.1.1 CHARACTERISTICS OF A BASIC SINGLE-HOP WDM STAR NETWORK

Single-hop WDM networks based on $N \times N$ star topology [7] having all nodes connected to a central hub node is shown in Figure 7.3 as an all-optical architecture for LAN applications. In a single-hop WDM network, routing and signaling are not required, due to its minimum hop of one and inherent transparency. Because of their simple operation and reduced network cost, single-hop WDM networks have been used widely as an all-optical architecture for LAN and MAN applications. For a single-hop system to be efficient, the bandwidth allocation among the contending nodes should be made dynamically. There are two types of allocations – pre-transmission coordination and no pre-transmission coordination. The pre-transmission coordination systems make use of a shared control channel (CC) through which nodes arbitrate their transmission requirements, and the actual data transfers take place through a number of data channels. Idle nodes may be required to monitor the CC. Before data packet transmission or data packet reception, a node must tune its transmitter or its receiver, respectively, to the proper data channel. Generally, no such CC exists in systems that do not require any pre-transmission coordination, and arbitration of transmission rights is performed either in a pre-assigned fashion or through contention-based data transmissions on regular data channels (e.g., requiring nodes to either transmit on or receive from pre-determined channels).

As a result, for a large user population whose size may be time-varying, deterministic scheduling is required to monitor signal transmission. In this technique, at least one channel is required for control signaling, so that signal containing data is transmitted without interruption. But in this approach, the main problem is that if signal has more data to transmit, other users have to wait for more time to transmit the signal to destination. This leads to a limitation in the number of users in the network. Because of the above fact, pre-transmission coordination may be the preferred choice in which priority or shared transmission can be made.

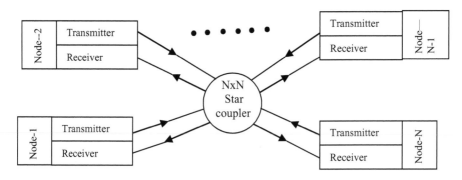

FIGURE 8.3 A basic architecture of single hop WDM network.

For shared transmission, an alternative classification of WDM systems can be developed based on whether the nodal transceivers are tunable or not. In Figure 8.3, a node can have one of the following four structures of transceiver:

1. Fixed Transmitter–Fixed Receiver (*FT–FR*)
2. Tunable Transmitter–Fixed Receiver (*TT–FR*)
3. Fixed Transmitter–Tunable Receiver (*FT–TR*)
4. Tunable Transmitter–Tunable Receiver (*TT–TR*)

The *FT–FR* structure is generally suitable for constructing multihop systems in which no dynamic system reconfiguration may be necessary. Although a single-hop *FT–FR* system with a small number of nodes is employed [14] (next section) it may not require any coordination in CC selection between two communicating parties. But for efficient system both *FT–TR* and *TT–TR* structures are used. If each nodal transmitter of the node is assigned a different channel under the *FT–FR* or *FT–TR* structures, then no channel collisions will take place and simple medium access protocols are used, but the maximum number of nodes is limited by the number of available channels. Single-hop system having the *TT–TR* structure is one of the most flexible structure in accommodating a scalable user population, but the main difficulty is to deal with the channel-switching overhead of the transceivers.

In addition, for some applications, a node may require multiple transmitters or receivers. The following single-hop systems can be used:

1. $FT^i TT^j - FR^m TR^n$ no pre-transmission coordination
2. $CC - FT^i TT^j - FR^m TR^n$ CC-based system

where a node has i number of fixed transmitters, j number of tunable transmitters, m number of fixed receivers and n number of tunable receivers. In this classification, the default values of i, j, m and n, if not specified, will be unity. Thus, LAMBDANET [14] is a *FT–FR*~system, since each of the nodes in the system needs one fixed transmitter amid an array of M fixed receivers. The *TT* and *TR* portions of the classification are suppressed since the system requires no tunable transmitter or receiver [15].

Most experimental WDM network prototypes are in single-hop category, and any CC is not employed for pre-transmission coordination. From the central hub node point of view, single-hop WDM networks are typically either based on a central PSC or a central AWG [15]. Some experimental WDM systems such as ACTS'S SONATA [16], Stanford's HORNET [17] and STARNET [18], and IBM's Rainbow [19] are also reported. The work in this field was begun by the British Telecom Research Lab (BTRL). The AT&T Bell Labs demonstrated first with a channel spacing of the order ~1 nm [20–21]. The Heinrich Hertz Institute (HHI) reported the first broadcast star demonstration of a video distribution using coherent lightwave technology. Then afterwards, a number of works have been mentioned – two 45-Mbps channels and employing tunable receivers, two 600-Mbps channels and two 1.2-Gbps channels employ tunable transmitters. An experimental system employing six 200-Mbps channels, spaced by 2.2 GHz, is reported in Ref. [22]. The work in Ref. [22] has a 16-channel system based on tunable receivers, where each channel has a capacity of 622 Mbps. Figure 8.4 mentions different single-hop WDM systems with no pre-transmissions and pre-transmissions.

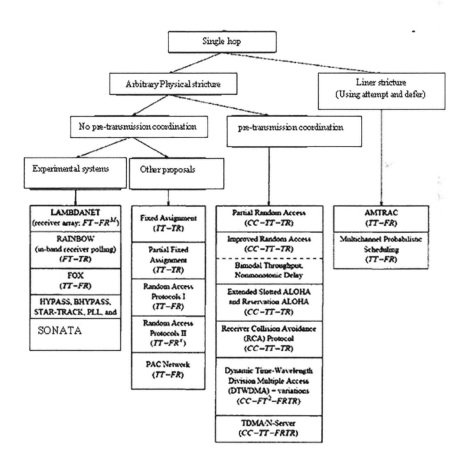

FIGURE 8.4   Single-hop WDM network reported by different researchers.

8.2   DIFFERENT SINGLE-HOP OPTICAL NETWORKS

There are many existing optical network-based single-hop communication –
switchless optical network of Advanced Transport Architecture (SONATA),
LAMBDANET, Rainbow, Fiber-Optic Crossconnect (FOX), STARNET, etc. In this
section these existing single-hop networks are discussed.

8.2.1   SONATA

Single-hop optical network without switches has been demonstrated as SONATA
supported by ACTS in Europe [16]. The main purpose of SONATA is to form a
single-layer network platform for end-to-end optical connections between a large
number of terminals. Figure 8.5 shows a SONATA architecture in which wavelength
nimbleness at terminals was made by the network structure to a single node providing
passive routing functions and actively controlled wavelength conversion. The optical physical layer of a "switchless" network is made by integrating wavelength-agile
transmitters, wavelength-agile burst-mode receivers, a passive wavelength-routing
node, a wavelength converter and optical fiber amplifiers in gated loops. Burst-mode
transmission at 622 Mbps is employed. By means of BER measurements, the effects
of optical amplifier noise, crosstalk and wavelength conversion were evaluated, and
the limits of the optical layer were also analyzed.

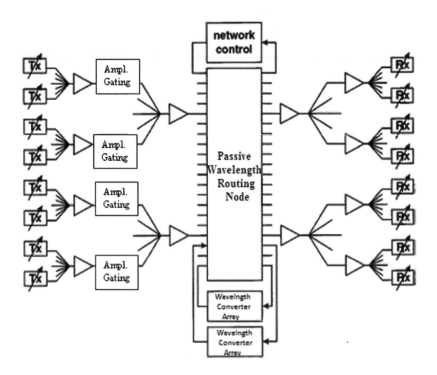

FIGURE 8.5   SONATA architecture [16].

Another configuration is made by considering the integration of wavelength-agile transmitters, wavelength-agile burst-mode receivers, passive wavelength routing node, a network controller and external services with a burst-mode transmission of speed 155 Mbps. The terminals are placed in different rooms and are not able to receive the data from two external sources – client service and cell generator/ analyzer instrument. This configuration presents the feasibility of connection setup and medium-access procedures using the same optical network used for the data.

8.2.2   LAMBDANET

The LAMBDANET [16] consists of an *FT–FR* with M nodes where each transmitter is equipped with a laser transmitting at a fixed wavelength. It is employed for a broadcasting network. Using a broadcast star at the center of the network, each of the wavelengths in the network is broadcast to every receiving node. It is made by using an array of M receivers at each node in the network, employing a grating demultiplexer to separate the different optical channels. In this network, 18 wavelengths are used with 2 Gbps over 57.5 km. Here, the array of M receivers was designed and realized in optoelectronic integrated circuits (OEIC) [16].

8.2.3   RAINBOW

In IBM's Rainbow [19], circuit-switched metropolitan area network (MAN) backbone consists of 32 IBM PS/2s as network nodes communicating with each other at data rates of 200 Mbps. The network structure is a broadcast-star topology having the lasers and filters centrally adjacent to the star coupler. The laser source is a fixed wavelength source, but the Fabry–Perot etalon filters in the receiver are tunable with sub-millisecond switching times, since it uses an *FT–TR* system shown in Figure 8.6. The Rainbow follows an in-band receiver polling mechanism in which each idle receiver is required to continuously scan the various channels to determine whether a transmitter wants to communicate with it. The transmitting node continuously transmits a setup request (a packet containing the destination node's address), and has its own receiver tuned to the intended destination's transmitting channel to listen for an acknowledgment from the destination for circuit establishment. The destination node, after getting the connection request, will send an acknowledgment on its transmitter channel for establishing the connection. Because of getting its long connection-acknowledgment delay, this technique is inappropriate for packet-switched traffic. But it can be suitable for circuit-switched applications with long holding times. Under the in-band polling protocol, a timeout mechanism is needed for nodes after sending a setup request; otherwise, there is the possibility of a deadlock. There are two types of Rainbow prototypes – Rainbow-I and Rainbow-II [23].

FIGURE 8.6   *FT-TR* system.

The Rainbow-I prototype is set up at Telecom '91 in Geneva. It is a broadcast-star architecture with each node having a fixed transmitter and a tunable receiver (*FT–TR*). The *FT–TR* system considers an in-band polling protocol.

Rainbow-II is an optical MAN having 32 nodes. Each node has a data speed of 1 Gbps over a distance of 10–20 km. The same optical hardware and medium access control protocol are used as that of Rainbow-I. Rainbow-II is to provide a connectivity to host computers using standard interfaces such as the standard high-performance parallel interface (HIPPI) while overcoming distance limitations, to deliver a throughput of 1 Gbps to the application layer, and to employ for the applications of the bandwidth of Gbps. The Rainbow-II has an experimental test bed at the Los Alamos National Laboratory (LANL), where performance measurements and experimentation with gigabit application are currently being conducted [23].

Apart from the above two prototypes, there is another prototype Rainbow-III which supports 100 packet-switched nodes in which each operates at 1 Gbps and uses the same protocol as that of Rainbow-I.

8.2.3.1   Rainbow Protocol

The original Rainbow-I WDM local optical network prototype supports up to 32 stations (or nodes) connected in a star topology over a range of 25 km. Figure 8.7 shows a typical architecture of Rainbow and a covers larger area than that provided by a LAN (typically a few km). So this network can be used as a MAN. Data can be transmitted on each WDM channel at a rate of up to 300 Mbps. The Rainbow-II network is a follow-up to Rainbow-I, also supporting 32 nodes, and employing the same optical hardware and multiple access protocol as Rainbow-I. Thus, this network protocol is equally applicable to both Rainbow-I and Rainbow-II. In the Rainbow architecture, each node (or station) is equipped with a single fixed transmitter, which is tuned to its own unique wavelength channel, and a single tunable Fabry–Perot filter, which can be tuned to any wavelength (*FT–TR*). Tuning time of a particular channel is ~25 ms. The tunable receiver scans across all the channels, looking for connection requests or acknowledgments from other stations. Rainbow's protocol follows a circuit-switched principle. The large filter tuning time leads to a high connection setup time. The equilibrium point analysis (EPA) technique is a means of analyzing complex systems by assuming that the system is always at an equilibrium point [47]. This technique has been successfully used to analyze a number of communication systems, e.g., satellite systems, and has been found to provide accurate results.

The signaling protocol is given below:

Each station is assigned its own unique channel on which its transmitter is fixed. Upon the arrival of a message at Station A and destined for Station B, Station A first tunes its receiver to Channel B so that it takes Station B's acknowledgment signal. Station A then begins to send a continuous request signal on Channel A. This request signal has a periodically repeated message having the identities of both the requesting station and the intended destination. If Station B's receiver is continuously scanning across all channels, it comes across the request on Channel A, the receiver will stop on that channel and Station B's transmitter will send out an acknowledgment on Channel B. Station A's receiver is tuned to Channel XB to receive the acknowledgment and then Station B's receiver is tuned to Channel XA. Station A's transmitter

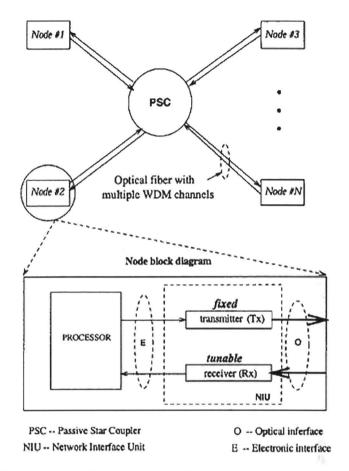

FIGURE 8.7   Passive star single hop network: Rainbow.

will then begin transmitting the message on Channel XA. This establishes a full duplex connection. Upon completion of the transmission, both stations resume scanning for requests.

With this protocol, there is the possibility of deadlock. If two stations begin sending connection setup requests to each other nearly simultaneously, they will both have to wait until the other sends an acknowledgment, but since both stations are waiting for each other, acknowledgments will never be sent. To avoid this problem, the Rainbow protocol also includes a timeout mechanism. If an acknowledgment is not received within a certain timeout period measured from the message arrival instant, the connection is blocked and the station is under scanning mode.

8.2.3.2   Model of Rainbow

The medium access control of Rainbow model is based on round-robin (or polling) systems which are used to analyze Rainbow model [19]). In Rainbow protocol, although a station's receiver is performing a round-robin operation, the operation may be interrupted by the station's transmitter. The modeling challenge is to relate the transmitter and receiver operations at a station in a simple manner with the following assumptions:

- N stations.
- Each station has a single buffer to store a message, and any arrival to a non-empty buffer is blocked. A message departs from the buffer after it is completely transmitted.
- The sending station, upon arrival of a message, tunes its receiver to the channel of the target station prior to sending the connection setup request.
- Stations control the channels in a round-robin fashion in the sequence:
  $$1, 2, ..., N, 1, 2, ...$$
- Time slot of length of 1 µs. This was chosen to provide a fine level of granularity in the system's model.
- The τ time slot is taken to tune a receiver to any particular channel.
- Messages arrival at each station follows the Bernoulli process with parameter σ, where σ is the probability that a message can arrive a station with an empty slot.
- Message lengths are geometrically distributed with the average message length being $1/p$ slots.
- The propagation delay between each station and the PSC is R slots. The signal propagation delay in fiber is approximately 5 µs/km, the value of R can be quite large ($R = 50$ slots for a station to star distance of 10 km).
- The timeout duration is represented by ϕ (in time slots).
- Transmission time is negligible for connection request and acknowledgment.

The state diagram for the model is shown in Figure 8.8. A station can be in any state and remains in that state for a geometrically distributed amount of time if it transmits with TR or it remains in any state for a fixed amount of time (one slot) if it is in any other state. A station departs from the state TR with probability p at the end of a time slot and remains in state TR with probability $1 - p$. The states are defined as follows:

- $TU_1, TU_2, ..., TU_\tau$ are states during which a station's receiver is under scanning across the channels for connection requests. It takes τ time slots to tune to a particular station. An arrival occurs with probability σ. From the state TU_τ if there is no arrival, the station either finds a connection request on the channel to which it has just completed tuning with some probability M and proceeds to send an acknowledgment, find a request with probability $1 - M$ and proceeds to tune its receiver to the next channel.
- After an arrival occurs, the station starts tuning its receiver to the channel of the destination. This process has τ time slots.
- $RQ_1, RQ_2, ..., RQ_{2R+\phi}$: Upon sending a request, it takes a propagation delay of R time slots for the request to reach the destination as shown in Figures 8.8 and 8.9. An acknowledgment will be received after a propagation delay of $2R$ time slots. The station continues to send the request signal

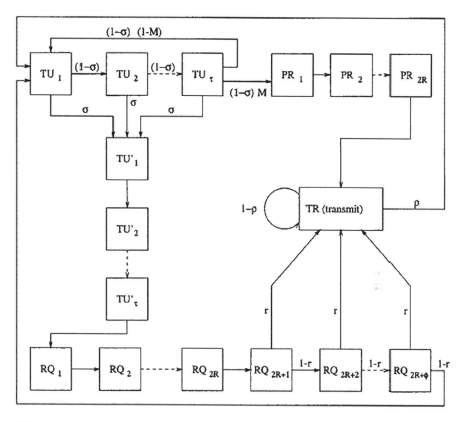

FIGURE 8.8   State diagram of Rainbow [13].

for a duration of ϕ time slots or until it receives an acknowledgment, where ϕ is the timeout duration. After sending a request for a duration of ϕ slots, the station must remain idle for an additional $2R$ time slots of propagation delay to get an acknowledgment. This ensures that all acknowledgments will result in a connection. If no acknowledgment is received, the current message is "timed out" and considered "lost", and the station returns to a scanning mode. The probability of getting an acknowledgment is denoted by r, which is the same for each of the states RQ_{2R+1} to $RQ_{2R+\phi}$, since the system has less memory, and an acknowledgment can be sent at any time by the acknowledging station. (The parameter τ will be related to the probability M later in this analysis.) As soon as an acknowledgment is obtained, the station, without delay, starts transmission of its message and goes into the transmission state TR.

• $PR_1, PR_2, \ldots, PR_{2R}$: The station goes into these states if it finds a connection request while scanning. After receiving the request, the station sends an acknowledgment to the requesting station. The acknowledgment takes R time slots of propagation delay to reach the station requesting the connection, after which the requesting station starts its transmission. It takes

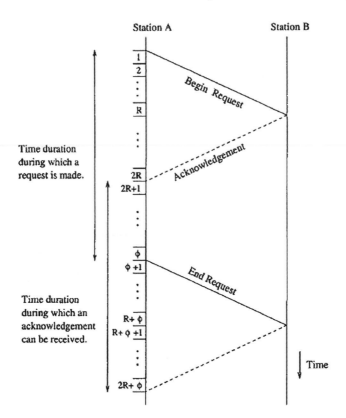

FIGURE 8.9 Timing diagram for connection set up.

another R slot of propagation delay for the message to reach at the destination station, after which the station goes into the transmission state, TR, to receive the message. A connection will always be established if an acknowledgment has been sent.

- TR (transmission): A station either transmits or receives a message. A station is kept in this state for a duration of more than one time unit and relieves with probability p at the end of a slot. After message transmission or reception, the station is kept under the scanning operation.
- Here N_{TU_i} is to be the expected number of stations in state TU_i; N_{TU} is to be the expected number of stations in state TU_i^j; N_{RQ_i} is the expected number of stations in state RQ_i; N_{PR_i} is the expected number of stations in state PR_i and N_{TR} is to be the expected number of stations in the transmission state. The system is written as a Markov chain model with the state space vector given below:

$$N = \begin{cases} N_{TU_1}, N_{TU_2}, \ldots, N_{TU_\tau}, N_{TU_1^j}, N_{TU_2^j}, \ldots, N_{TU_\tau^j}, N_{RQ_1^j}, N_{RQ_2^j}, \ldots, N_{RQ_{2R+\varphi}^j}, \\ N_{PR_1}, N_{PR_2}, \ldots, N_{PR_{2R}}, N_{TR} \end{cases} \quad (8.1)$$

The performance analysis of Rainbow model [19,23] is difficult to be performed by using Markov analysis techniques because of the very large state space. Therefore, the analysis of the system can be carried out at an equilibrium point using EPA as given below.

The assumption in EPA is as follows:

- The system is assumed to be always operating at an equilibrium point. This is an assumption, since the system actually stays around the equilibrium point.
- At an equilibrium point, the expected number of stations entering each state is equal to the number of stations departing from each state in each time slot.

We obtain flow state equations for a set of K equations with K unknowns, where K is the number of states. The expected number of stations for each state is written in terms of the expected number of stations in state TU_1. The flow equations can be written as follows. The N's, representing the random variables, is taken as the average values.

$$N_{TU_i} = (1-\sigma)^{i-1} N_{TU_i} \quad \text{for } i = 2,3,\ldots\tau \tag{8.2}$$

$$N_{PR_1} = N_{PR_{21}} = \cdots = N_{PR_{2R}} = (1-\sigma)^\tau M \times N_{TU_i} \tag{8.3}$$

$$N_{TU_1'} = N_{TU_2'} = \cdots = N_{TU_\tau'} = N_{RQ_1}$$
$$= N_{RQ_2} = \cdots = N_{RQ_{2R}} = \left[1-(1-\sigma)^\tau\right] N_{TU_i} \tag{8.4}$$

$$N_{RQ_{2R+j}} = (1-r)^{j-1} [1-(1-\varpi)^\tau] N_{TU_i} \quad \text{for } j = 1,2,\ldots\phi$$

$$\rho \times N_{TR} = N_{PR_{2R}} + \sum_{j=1}^{\phi} \tau \times N_{RQ_{2R+j}}$$

$$= \left[(1-\sigma)^\tau M + \left\{1-(1-r)^\phi\right\} \times \left\{1-(1-\sigma)^\tau\right\}\right] \times N_{TU_i} \tag{8.5}$$

For the unknown variables N_{TU_1}, M is the probability that a request is established by a scanning station for a time τ. There is another probability that another station is in states RQ_{R+1} to $RQ_{R+\phi}$, and that the request is intended for the scanning station. The probability M is written as

$$M = \frac{1}{N-1} \times \frac{1}{N} \left(\sum_{i=R+1}^{R+\phi} N_{RQ_i}\right) \tag{8.6}$$

Putting the value of N_{RQ_i} the M is obtained as

$$M = \frac{1}{N-1} \times \frac{1}{N} \times \left[1 - (1-\sigma)^{\tau}\right] \times \left\{R + \frac{1}{\tau}\left(1 - (1-r)^{\phi-R}\right)\right\} N_{TU_i}$$

The rate of transmission into the active state from the request state is the rate of transmission into the active state from the state PR_{2R}. A station only begins transmission if the destination station will be receiving the transmission.

$$N_{PR_{2R}} = \sum_{i=R+1}^{R+\phi} r \times N_{RQ2R+} \tag{8.7}$$

where

$$(1-\sigma)^{\tau} M = \left[1 - (1-\sigma)^{\tau}\right]\left[1 - (1-r)^{\phi}\right]$$

In the steady state, the sum of stations in each state is equal to the total number of stations in the system and N is written as

$$N = \left[1 - (1-\sigma)^{\tau}\right] \times \left\{\frac{1}{\sigma} + \tau + 2R + \left(\frac{\tau+\rho}{\tau\rho}\right)\left[1 - (1-r)^{\phi}\right]\right\} N_{TU_i}$$

$$= \left[(2R + 1/\rho)(1-\sigma)^{\tau} M\right] N_{TU_i} \tag{8.8}$$

where variables r, M and TU_i are determined by using the above equation providing a steady-state solution to the entire system. The performance is made in terms of throughput, time delay and timeout probability. The normalized throughput is written as the ratio of expected fraction of stations for transmission and total number of stations in the system

$$S = \frac{N_{TR}}{N} \tag{8.9}$$

The time delay is a time from a message's arrival to the system until the message completes its transmission. It has the time for tuning the destination station's channel, the propagation delay for the request and acknowledgment signals, the time until an acknowledgment is received and the message transmission time. It is given by

$$D = \tau + 2R + \sum_{j=1}^{\phi} j \times (1-r)^{k-1} \times r + \frac{1}{\rho}$$

The timeout probability is the probability that a station will be time out after making in the request mode and is written as

$$P_{TO} = (1-r)^{\phi}$$

The blocking probability is the probability of an arrival to be blocked. The blocking probability is as follows:

$$P_B = 1 - \frac{\sum_{i=1}^{\tau} N_{TU_i}}{N} \tag{8.10}$$

For low arrival rates, most of the stations are under scanning for requests, and only a small number of stations are under request. The stations requesting connections has a high probability of being acknowledged due to having an increase in timeout duration. A deadlock occurs if any station being requested is also requesting a connection. This results in a higher probability of deadlock than in the analytical model for low arrival rates.

8.2.4 FIBER-OPTIC CROSSCONNECT (FOX)-BASED SINGLE-HOP NETWORK

In a single-hop optical network, Fiber-Optic Crossconnect (FOX) [14] was employed with fast tunable lasers in a parallel processing environment (with fixed receivers) (i.e., a *TT–FR* system). The architecture has two star networks – one for signals traveling from the processors to the memory banks, and another for information flowing in the reverse direction [23, 24]. The binary exponential back-off algorithm was used for dealing with the contentions due to having slow memory access. This algorithm achieves sufficiently lower contention. Since the transmitters are capable to tune with a tuning time of less than a few tens of nanosecond, it has more efficiency with packet transmission times of the range 100 ns to 1 μs, because tuning time is less than the packet transmission time.

8.2.5 STARNET

STARNET is a WDM LAN having the passive-star topology [18]. It provides supports to all of its nodes by using two virtual sub-network structures or a high-speed reconfigurable packet-switched data sub-network structure or a moderate-speed fixed-tuned packet-switched control sub-network structure. The STARNET node consists of a single fixed-wavelength transmitter employing a combined modulation technique to simultaneously send data on both sub-networks on the same transmitter carrier wavelength and two receivers – a main receiver operating at a high speed of 2.5 Gbps and an auxiliary receiver operating at a moderate speed of 125 Mbps equivalent to that of a fiber distributed data interface (FDDI) network. The main receiver is tuned to any node's transmitting wavelength depending on the prevailing traffic conditions.

8.2.6 OTHER EXPERIMENTAL SINGLE-HOP SYSTEMS

Thunder and Lightning network is another single-hop network that gives 30-Gbps ATM structure using optical transmission and electronic switching. Electronic switch having 7.5-GHz Galium Arsenide (GaAs) circuits was used to obtain clock recovery, synchronization, routing and packet buffering and to facilitate the transition to manufacture.

HYPASS [44] is an extension of FOX-based single-hop system in which both transmitters and receivers were tunable (i.e., *TT–TR* system) contributing vastly improved throughputs. Other experimental single-hop systems are BHYPASS, STAR-TRACK, passive optical loop (POL) and broadcast video distribution systems used to provide broadcasting services.

8.3 COORDINATION PROTOCOL FOR A SINGLE-HOP SYSTEM

There are two types of coordination protocols for single-hop system – non pretransmission and pre-transmission coordination protocols.

8.3.1 NON PRE-TRANSMISSION COORDINATION

There is a non pre-transmission coordination protocols that do not require any pretransmission coordination. These protocols are based on fixed assignment of the channel bandwidth.

8.3.1.1 Fixed Assignment

There is a one-hop communication based on a fixed assignment technique using time-division multiplexing (TDM) expanded over a multichannel environment [13]. Here each node has one tunable transmitter and receiver called as *TT–TR* systems. The tuning times of transceivers are to be zero for N available channels. Time is divided into cycles, and it is predetermined at what point in a cycle and over what channel a pair of nodes is allowed to communicate. The allocation matrix is generalized for an arbitrary number of nodes M and an arbitrary number of channels N [13]. The allocation matrix accommodates tuning times (in integral number of slots) through a staggered approach [48]. There are three nodes (numbered $n = 1$, 2, 3) and two channels (numbered $t = 0$ and I), and a channel allocation matrix which indicates a periodic assignment of the channel bandwidth can be formulated as shown in Table 8.1 and, in which, $t = 3n$ where $n = 0, 1, 2, 3, \ldots$.

An entry (i, j) for channel k in slot i means that node i has exclusive, permission to transmit a packet to node j over channel k during slot i.

This scheme has the usual limitations of any fixed assignment technique, i.e., it is insensitive to the dynamic bandwidth requirements of the network nodes, and it is not easily scalable in terms of the number of nodes. Also, the packet delay at light loads can be high [18].

TABLE 8.1
Slot Allocation in Fixed Assignment

Channel No.	t	$t + 1$	$t + 2$
0	(1,2)	(1,3)	(2,1)
1	(2,3)	(3,1)	(3,2)

Another fixed assignment approach can be used as a versatile time–wavelength assignment algorithm in which node i consists of t_i number of transmitters and r_i number of receivers which are all tunable over all available channels [25]. The algorithm is designed such that, for a traffic demand matrix, it will consider to be the tuning times as minimum as possible in the schedule or reduction of a packet delay.

To accommodate arbitrary switching times and non-uniform traffic loads it requires the establishment of a periodic TDM frame structure consisting of a transmission subframe and switching subframe during which all of the necessary switching functions of nodal transmitters/receivers are performed. Another way to accommodate switching times and non-uniform traffic loads is to distribute the nodal switching requirements all over the frame, and so it is less restrictive and more efficient.

The scheduling of an arbitrary traffic matrix with a tunable transmitter and a fixed-tuned receiver are considered at each node. For off-line scheduling, the effect of tuning delay is found to be small even if the tuning time is as large as the packet transmission time, and the expected schedule completion time is obtained. The average packet delay is insensitive to the tuning time under a near-optimal schedule in which transmitter tunes just-in-time to the appropriate channel just before its packet transmission. In a traffic demand matrix, the algorithm estimates a proper time-wavelength schedule. A receiver collision occurs when a collision-free data packet transmission cannot be picked up by the intended destination, since the destination's receiver may be tuned to some other channel for receiving data from some other source.

8.3.1.2   Partial Fixed Assignment Protocols

In partial fixed assignment based on Destination Allocation (DA) the channel allocation procedures are less restrictive [26]. In the DA protocol, the number of node pairs which can communicate over a slot is increased from the earlier value of N (the number of channels) to M (the number of nodes). During a slot, a destination is still essential to get from a fixed channel, but more than one source can transmit to it in this slot. In this case, the receiver collisions are avoided, but the possibility of channel collision is introduced. The three-node, two-channel case slot allocation for channel is shown in Table 8.2.

In case of Source Allocation (SA) protocol, the control of access to the channels is reduced. Now, N ($N \leq M$) source nodes transmit, each over a different channel. For a node to each of the remaining ($M - 1$) nodes, there is a possibility of receiver collisions. A periodic slot allocation policy for the three-node, two-channel is shown in Table 8.3.

In case of Allocation Free (AF) protocol, all source–destination can transmit on any channel over any slot duration [26].

TABLE 8.2
Slot Allocation in Partial Fixed Assignment

Channel No.	T	$t + 1$
0	(1.2), (3,2)	(1,3)
1	(2,3)	(2,1), (3,1)

TABLE 8.3
Periodic Slot Allocation

Channel No.	t	$t+1$	$t+2$
0	(1.2), (1,3)	(1.2), (1,3)	(2,1), (2,3)
1	(2,1), (2,3)	(2,1), (3,2)	(3,1), (3,2)

8.3.1.3 Random Access Protocol I

The random access protocol is another non pre-transmission coordination in which each node consists of a *TT–FR* system. A node will receive signals from the channel through the node's address. There are two slotted-ALOHA (Additive Links On-line Hawaii Area) protocols reported for random access [27]. In the first protocol [27], time is slotted on all the channels, and these slots are synchronized across all channels, and the slot length is equal to a packet's transmission time. In the second protocol, each packet is considered to be of L mini-slots, and time across all channels is synchronized over mini-slots. The slotting across the entire packet length was performed better than mini-slotting, since the latter scheme increases the susceptibility period of a data packet transmission. Also, the maximum throughput on each data channel is found to be l/e, which is the value for the single-channel case.

8.3.1.4 Random Access Protocol II

A slotted-ALOHA and a random TDM protocol are used for random access protocol II [13]. Both these protocols consider a limited tuning range and zero tuning time. Both protocols are based on slotted architectures. The node consists of *TT–FR* system. For node $iT(i)$ and $R(i)$ are considered to be the set of wavelengths over which node i can transmit and receive, respectively. Under the slotted-ALOHA scheme, if node i wants to transmit to node j, it arbitrarily selects a channel from the set $T(i) \cap R(j)$ and transmits its packet on the selected channel with probability $p(i)$. In case of random time-division multiple access (TDMA), the scheme operates under the assumption that all network nodes, even though they are distributed and capable of making the same random number to perform the arbitration decision in a slot. This can be done by taking all nodes with the same random number generator starting with the same seed. Thus, for every slot, and each channel at a time, the distributed nodes generate the same random number, which indicates the identity of the node with the corresponding transmission right. Analytical Markov chain models for the slotted ALOHA and random TDMA schemes are formulated to determine the systems' delay and throughput performances.

8.3.1.5 The PAC Optical Network

In case of Packet Against Collision (PAC) protocol, the node has a *TT–FR* system. Here the packet collisions are removed by employing PAC switches at each node's interface with the network's star coupler. The node's transmitter access is allowed to a channel (through the PAC switch) only if the channel is available. The packets simultaneously accessing the same channel are avoided. The collision avoidance is met only if it is under a multichannel environment.

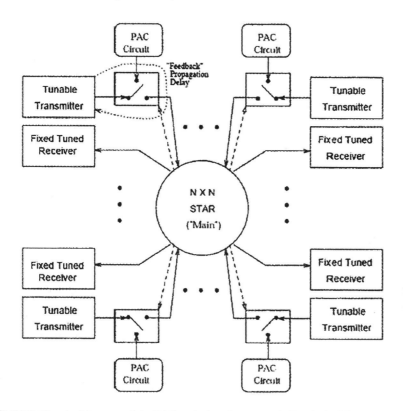

FIGURE 8.10 Architecture of the PAC optical packet network (the dashed lines are used to detect energy on the various chan from the "main" star) [13].

The PAC circuit does carrier sensing by using an n-bit burst. The carrier burst is switched through a second $N \times N$ control star coupler combined with a fraction of all the packets coming out of the main star plus all carrier bursts. Figure 8.10 shows PAC performing a flow of carrier burst using $N \times N$ control star. Here, the resulting electrical signal operates the optical switch connecting the input to the network. When two or more nodes access the channel simultaneously, all of them detect the carrier, and their access to the network is blocked.

8.3.2 Pre-transmission Coordination Protocols

8.3.2.1 Partial Random Access Protocols

For efficient single-hop communication, each node is required to have a single tunable transmitter and a receiver, and a CC to control tunability i.e., a CC–TT–TR system [28]. The following assumptions are taken:

- The tuning times should be zero
- Transceivers should be tunable over the entire wavelength range under consideration.

Three typical protocols are used for partial random access providing a high-speed lightwave LAN with their performance capabilities. A first protocol is termed as ALOHA/ALOHA, where the first term represents the protocol used for the CC, and the second term indicates the protocol used for the chosen WDM data channels. With the ALOHA/ALOHA protocol, the control signals and data packets are transmitted at random over the control and randomly chose WDM data channel without acknowledgment between the packets.

A second protocol is termed as ALOHA/Carrier Sense Multiple Access (CSMA), wherein an idle WDM data channel is sensed before sending the control packet on the CC while concurrently jamming the sensed idle WDM data channel. Immediately thereafter, the data packets are sent over through the idle WDM data channel.

A third protocol is based on a CSMA/N-Server Switch protocol having all idle transceivers monitoring the CC to use the control packets maintaining a list of idle receivers and WDM data channels. When a transceiver becomes active, it transmits its control packet over the CC under the condition if it is found idle. Here time is normalized with the duration of a control packet transmission fixed with a size of one unit. There are N data channels and length of data packets are L units [28]. A control packet consists of three types of information – the source address, the destination address, and a data channel wavelength number, which is assigned randomly by the source and with which the corresponding data packets are transmitted. Under the ALOHA/ALOHA protocol, a node sends a control packet over the CC with a randomly selected time, after which it immediately transmits the data packet on data channel I ($1 < i < N$, which was specified in its control packet as shown in Figure 8.11).

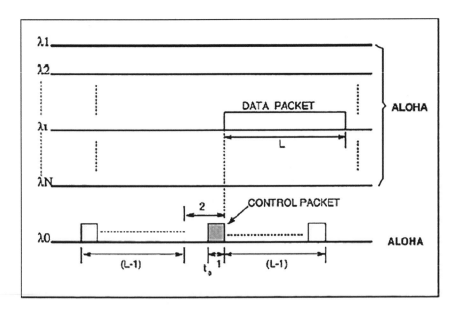

FIGURE 8.11   ALOHA/ALOHA protocol [13].

Any other control packet sent during the tagged packet's transmission collide with the tagged packet and the packets are destroyed. Since different nodes can be at different distances from the hub, these times are specified relative to the activity seen at the hub. There is another successful control packet transmission over the period to $-L$ to $+L$, and the data channel chosen by that control packet is also the same. The difference with other protocols is that the other protocols [13] ignores the possibility of receiver's collisions. The slotted-ALOHA/ALOHA protocol is similar, except that access to the CC via the slotted-ALOHA protocol. Other schemes [45] are ALOHA/CSMA, CSMA/ALOHA and CSMA/N-server protocols. However, the main difficulty of the CSMA-based schemes is carrier sensing in near-immediate feedback of practical feature applied in high-speed systems even for short distances.

8.3.2.2   Improved Random Access Protocols

The improved random access protocol is a modification of partial random access protocols [29]. The realistic protocols do not need any carrier sensing as the channel propagation delay in a high-speed environment of wide area network is more than the packet transmission time. Here slotted-ALOHA for the CC and ALOHA and N-server mechanism are employed in improved random access protocols. Another improvement in the protocols is that a node waits its access to a data channel until its transmission on the CC has been successful [11].

The probability of receiver collisions is reduced for large population systems. A receiver collision happens when a source sends to a destination without any channel collision. It is observed [11] that the slotted-ALOHA/delayed-ALOHA protocol [13] has a bimodal throughput characteristic. If there is a large number of data channels, then the CC's bandwidth is under dimensioned and CC is blocked. Under the slotted-ALOHA/delayed-ALOHA protocol with L-slot data packets, the number of data channels is written as [29]

$$N = \left(\frac{2L-1}{e} \right)$$

The slotted-ALOHA/delayed-ALOHA protocol's performance is reduced because of the receiver collisions for finite population systems. Because of that, throughput is also reduced. The difficulty for receiver collision detection is the simplicity of the systems having the availability of only one tunable receiver per node for tracking of both CC and the data channel activities.

8.3.2.3   Receiver Collision Avoidance (RCA) Protocol

To find receiver's collision, some intelligence is required for the receivers. The receiver collision's problem is solved at the data link layer. The protocol modified with intelligence is named as Receiver Collision Avoidance (RCA) protocol [30] operating under the same basic system having TT–TR per node and a contention-based CC. The protocol accommodates the transceiver tuning times of duration T slots. All nodes are assumed to be D slots away from the hub and $N = L$, but these conditions can be generalized.

8.3.2.3.1 Channel Selection

Before sending a control signal, the source chooses the channel employed to transmit the corresponding data packet. In order to remove data channel collision, the RCA protocol uses a simple and fixed data channel assignment policy. Each control slot is denoted as 1, 2, 3, ... N, periodically. Specifically, each control slot is assigned with a fixed wavelength if the corresponding control packet is successfully sent in that slot. So this assignment makes the corresponding data channel transmission collision-free.

Each node has a Node Activity List (NAL) keeping the information on the CC during the most recent $2T + L$ slots. Each entry indicates the slot number and a status. For the status representing a successful receipt of control packet, the corresponding NAL entry has the source address, the destination address and the wavelength selected, which are copied from the corresponding control packet. The packet transmission follows certain rules. If a packet is generated in transmitter i for the destination receiver j, the transmitter i will send out a control packet only with the following condition that node is NAL does not contain any entry with either node i or node j as a packet destination. The control packet thus transmitted will be received back at node i after $2D$ slots, during which node is receiver must also be tuned to the CC. Based on the NAL updated by node is receiver, if a successful control packet to node i (without receiver collision) is received during the $2T + L$ slots prior to the return of the control packet, then a receiver collision is detected and the current transmission procedure has to be rejected and restarted. Otherwise, transmitter i starts to tune its transmitter to the selected channel at time $t + 2D + 1$, and the tuning takes T slots, after which L slots are used for data packet transmission, which is followed by another T-slot duration during which the transmitter tunes back to the CC. The Packet Reception procedure is straightforward for receiving the packets.

8.3.2.4 Reservation Protocols

In case of reservation protocol for a single-hop system, a single transmitter and a single receiver per node are required. The system's performance is improved by including nodes having multiple transceivers. The dynamic time–wavelength division multiple access (DT-WDMA) protocol [31] requires that each node be equipped with two transmitters and two receivers – one transmitter and one receiver at each node are always tuned to the CC, and each node has exclusive transmission rights on a data channel on which its other transmitter is always tuned to, and the second receiver at each node is tunable over an entire wavelength range, i.e., this is a CC–FT^2–$FRTR$ system.

Figure 8.12 shows on the protocol's operation [31]. There are N nodes for a system requiring $N + 1$ channels – N for data transmission and the $(N + 1)$th for CC. It is divided into the slots synchronized over all channels at the passive star (hub). A slot on the CC has N mini-slots – one for each of the N nodes. Each mini-slot has a source address field, a destination field and an 'additional field by which the source node can send the signal with the priority of the packet for queued transmission. The control information has to be sent collision-free, and after transmitting in a control

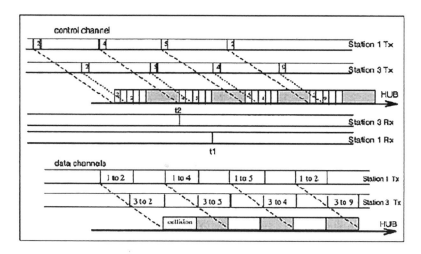

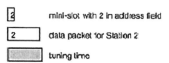
mini-slot with 2 in address field

data packet for Station 2

tuning time

t_1 and t_2 are instants when Stations 1 and 3, respectively, learn about the status of their first control packet transmissions (to station 2)

FIGURE 8.12 Dynamic time-wavelength division multiple access (a Reservation protocol) [13] t_1 and t_2 are instants when Stations 1 and 3, respectively, learn about the status of their first control packet transmissions (to station 2).

mini-slot, the node transmits the data packet in the following slot over its own dedicated data channel. By monitoring the CC over a slot, a node determines whether it is to receive any data over the following data slot. If a receiver finds that more than one node is transmitting data to it over the next data slot, it checks the priority fields of the corresponding mini-slots and selects the one with highest priority. To receive the data packet, the node tunes its receiver to the source node's dedicated transmission channel. Collision occurs when two or more nodes transmit data packets to the same destination over a data slot duration, in which only one of these transmission slots is successfully received. Also, this mechanism has an embedded acknowledgment feature since all other nodes can know about successful data packet transmissions by following the same distributed arbitration protocol. In addition, the mechanism supports arbitrary propagation delays between various nodes and the passive hub. The main limitation of the system is its scalability property, as it requires that each node's transmitter has its own dedicated data channel.

8.4 MULTIHOP OPTICAL NETWORK

In a single-hop system, there are limitations – (a) non-availability of faster tunable transmitter and receiver, (b) less number of wavelength channels due to less tunable range of transmitter and receiver, (c) limitation in the number of nodes and

(d) limitation in coverage area. To resolve these limitations in a single-hop system, it is required to use more number of hops for transmission of signal from source to destination in which less tunability in transmitter and receiver is needed [32]. Although there is a requirement for tuning in transmitter and receiver in a multihop system, this tuning is relatively static in comparison to a single-hop system. It is unlikely that there will be a direct path between every node pair (in which case each node in a N-node network must be equipped with $(N - 1)$ number of fixed-tuned transceivers) so that, in general, traffic (i.e. a packet) from a source to a destination may have to hop through some intermediate nodes. These systems have an operational feature of routing and performance parameter such as minimal average packet delay, minimal number of hops and balanced link flows.

Figure 8.13 shows a multihop WDM network consisting of N number of nodes in which each node transmits two wavelengths and receives two wavelengths [32]. In the figure, the physical topology is based on a star topology in which both transmitter and receiver are tuned to two wavelengths for transmission and receiving of signal respectively. The transmission of signal wavelength is from one node to the other node with one or more via nodes or direct via WDM star coupler. For example in the figure, node-1 transmits the signal directly to node-2 via a WDM star coupler with wavelength channel λ_3, whereas for transmission of signal from node-2 to node-1 first the signal is transmitted to node-3 with wavelength λ_4 and then node-3 transmits the signal to node-1 by using wavelength channel λ_2, i.e., transmission from node-2 to node is made via node-3

The transceiver tuning times have a little impact on multihop systems, since the multihop virtual topology is essentially a static one. There are two other important issues to be considered in a multihop system.

First, in the structure, distance between nodes must be small so that the average packet delay must be minimal or the maximum flow on any link in the virtual structure must be minimal. Two nodes are at a hop distance of h if the shortest path between them requires h hops. In a multihop structure, each such hop means travel to the star and back. The maximum hop distance between any two nodes is referred to as the structure's diameter.

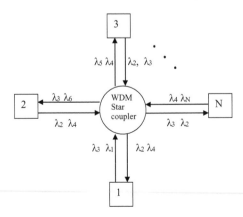

FIGURE 8.13 A N-node multi-hop WDM network.

Second, the nodal processing time has other issues to be considered because of nodal processing complexity. Consequently, simple routing mechanisms must be employed. A routing-related sub-problem is the buffering strategies at the intermediate/via nodes. In this direction, the approaches use deflection routing under which a packet, instead of being buffered at an intermediate node, may be intentionally misrouted but still reach its destination with a slightly longer path.

Multihop network [13,32] can be based on either irregular or regular topology. Irregular multihop structures generally address the optimality criterion directly, but the routing complexity can be large since this structure lacks any structural connectivity pattern. The routing complexity with wavelength assignments of irregular cum mesh topology is discussed in the following chapters.

The routing schemes of regular structures are simple because of their structured node-connectivity pattern. However, their regularity in structure has also imposed on constrains for the set of solutions in addressing the optimality problem, and the number of nodes in a complete regular structure usually forms a special set of integers, rather than an arbitrary integer. In this chapter, regular structures have been discussed. These structures include perfect shuffle (called ShuffleNet), de Bruijn graph, toroid (Manhattan Street Network, MSN), hypercube, linear dual bus and a virtual tree. The characteristics of alternative routing strategies have also mentioned corresponding to these structures such as deflection routing in ShuffleNet [13].

The offered loads by the various nodes are not normally symmetric and varying, which are dealt with special-purpose networking equipment such as servers and gateways. Regular structures are generally used for uniform loading patterns, while irregular structures are made for arbitrary and non-uniform workloads. The performance effect of non-uniform traffic and the corresponding adaptive routing schemes to control congestion are important issues for regular structures as nodes of irregular structure may have non-symmetric and non-uniform traffic. Another issue that should be taken in a multihop network for selection channel is whether to employ dedicated channel or shared channel (SC). Under the case of dedicated channels, each virtual link employs a dedicated wavelength channel. Since internodes traffic may be bursty, the traffic on an arbitrary link should be bursty. For this case, some of the links' utilizations may be low in case of dedicated channels. Consequently, the SC mechanism advocates the use of two or more virtual links to share the same channel in order to improve link utilization.

8.4.1 Optimal Virtual Topologies Using Optimization

There are two types of parameters – link data flow and time delay which are considered for optimization studies based on maximum link flow [32], followed by optimizations based on minimization of the mean network-wide packet delay [33].

8.4.1.1 Link Flow

A network consisting of an arbitrary number of nodes N has been taken for flow-based optimization. The nodes are indexed 1, 2, ..., N, and each node consists of T transmitters and T receivers. There are $N \times T$ channels since each transmitter and receiver transmits and receives on its own unique channel. The capacity of each WDM channel is C bps. The following parameters are considered:

- λ_{sd} is the traffic matrix, where λ_{sd} is the traffic flow for source node s and destination node d for $s, d = 1, 2... N$.
- f_{ij} is the data flow in link ij (where i and j denote ith and jth node respectively)
- f_{ij}^{sd} is the fraction of the λ_{sd} traffic flowing through link ij
- Z_{ij} be the number of directed channels from node i to node j.
- $C_{ij} = Z_{ij} \times C =$ capacity of link ij.
- $\dfrac{f_{ij}}{C_{ij}} =$ fraction of the (i, j)-link capacity

In an arbitrary topology, the link has maximum utilization and is given by

$$\max_{ij} \left\{ \frac{f_{ij}}{C_{ij}} \right\}$$

The flow for wavelength assignment (FWA) is a mixed integer optimization problem with a min–max objective function subject to a set of linear constraints [13]. The main characteristic of this problem formulation allows the traffic matrix to scale up by the maximum amount before its most heavily loaded link saturates. In the first part, the connection attempts to connect nodes with more traffic between them in one hop and the connectivity can be solved by a special version of the simplex algorithm. The second part is a routing problem formulated as a multicommodity flow problem with a nonlinear, non-differentiable, convex objective function, and it is solved by using the flow-deviation method [13]. Iterative improvement is used by a number of least-utilized branches (K) two at a time. A branch-exchange operation is carried out by swapping the transmitters (or receivers) of the two least-utilized branches for solving the routing problem on the new connectivity diagram or accepting the swap if the new topology leads to a lower network-wide maximum link utilization.

8.4.1.2   Delay-Based Optimization

For an optimal virtual topology, an alternative method is to minimize the mean network-wide packet delay. The packet delay consists of two components – the propagation delays encountered by the packet as it hops from the source through intermediate nodes to the destination and queuing delay at the intermediate nodes. In a high-speed environment for large channel capacity C and moderate link utilizations the queuing delay is ignored in comparison to the propagation delay component [13]. Thus, this optimization also requires the distance matrix d_{ij}, where d_{ij} is the distance from node i to node j per the underlying physical topology. The mean network-wide packet delay is written as [13]

$$T_D = \sum_{i=1}^{N} \sum_{j=1}^{N} \frac{f_{ij} d_{ij}}{v \cdot \lambda} + T_Q \tag{8.11}$$

where v = velocity of light in fiber, f_{ij} is the data flow through link ith node and jth node, $\lambda = \sum_{s=1}^{N} \sum_{d=1}^{N} \lambda_{sd}$ = total offered load to the network, and T_Q is the nodal queuing delay component [13]. For optimization, traffic matrix and distance matrix are to be made. The design variables are virtual topology and link flows. The constraints are flow conservation and nodal connectivity (including the number of transmitters and receivers per node).

In Ref. [33], optimization algorithms such as simulated annealing and genetic algorithm have been used to solve both dedicated channels and SCs (where TDMA is employed for channel sharing).

8.4.2 REGULAR STRUCTURES

Regular topologies for multihop lightwave networks include ShuffleNet, de Bruijn graph, Toroid and Hypercube [34].

8.4.2.1 ShuffleNet

A (p,k) ShuffleNet has $N = kp^4$ nodes arranged in k columns of p nodes each (where p,k = 1, 2, 3, ..., p = number of physical connections from each node and k = number of columns in ShuffleNet) [34], and the kth column is wrapped around to the first in a cylindrical fashion. The nodal connectivity between adjacent columns is a p-shuffle. This interconnection pattern is distinctly represented as numbering of nodes in a column from top to bottom as 0 through $p_k - 1$, and directing p arcs from node i to nodes $j, j + 1, ..., j + p - 1$ in the next column, where $j = (i \bmod p)$. This structure has the mean hop distance between any two randomly chosen nodes. From any "tagged" node in any column (say the first column), p nodes can be reached in one hop, another p^2 nodes in two hops, until all remaining $(p_k - 1)$ nodes in the first column stayed. In the second pass, all nodes visited in the first pass can now be stayed, although multiple (shortest-path) routes are there for doing so. Figure 8.14 shows a $(2,2)$ ShuffleNet with total number of nodes $N = 14$. In the $(2,2)$ ShuffleNet in node 6 can be reached from node 0 either via the path 0–5–3–6 or the path 0–4–1–6, both of which are "shortest paths".

The number of channels for N user (p, k) ShuffleNet is written as [34]

$$W = \frac{N \cdot T \cdot R}{p}$$

where T and R denote the number of transmitters and receivers per user. For the ShuffleNet spanning tree is used for assigning fixed routes generated for any user. It is obtained in Table 8.4.

The expected number of hops between two users is written as

$$E[\text{numbers of hops}] = \frac{kp^k (p-1)(3k-1) - 2k \left(p^{k-1} - 1\right)}{2(p-1)\left(kp^k - 1\right)}$$

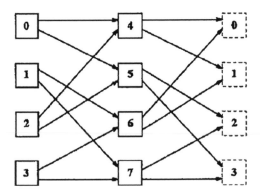

FIGURE 8.14   A (2,2) Shufflenet having 14 nodes [13].

TABLE 8.4

Number of Hopes in Sufflenet

Number of Hops (h)	Number of Users h Hops from Source
1	p
2	p^k
.	.
.	.
$k-1$	p^{k-1}
k	p^k-1
$k+1$	p^k-p
$k+2$	p^k-p^2
.	.
.	.
$2k-1$	$p^k - p^{k-1}$

If the routing algorithm balances all traffic loads, then the channel efficiency is written as

$$\eta = \frac{1}{E[\text{number of hops}]}$$

$$= \frac{2(p-1)\left(kp^k-1\right)}{kp^k(p-1)(3k-1)-2k\left(p^{k-1}-1\right)}$$

For large p, the channel efficiency is written as [35]

$$\eta = \frac{2}{(3k-1)}$$

As k increases in the ShuffleNet, the number of wavelength channels which is required for transmission of packet increases. For reduction of number of wavelength channels, it needs a routing procedure for a single transmitter and single receiver per user ShuffleNet with SCs. This routing procedure uses lower lost single transmitter and receiver per user with TDMA sharing of channels.

Thus, the number of nodes which are h hops away from a "tagged" node can be written as

$$n_h = p^h, \quad h = 1,2,...k-1$$

$$= p^k - p^{h-k}, \quad h = k,k+1,k+2,...,2k-1 \tag{8.11}$$

Then, the average number of hops between any two randomly selected nodes can be formulated as

$$\bar{h} = \frac{kp^k(p-1)(3k-1)-2k(p^k-1)}{2(p-1)(kp^k-1)} \tag{8.12}$$

The ShuffleNet's diameter, defined as the maximum hop distance between any two nodes is written as $2k-1$. Due to multihopping, a fraction of a link's capacity is actually used for direct traffic between the two specific nodes connected by a link, whereas the remaining link capacity is used for forwarding of multihop traffic. In a symmetric (p, k) ShuffleNet, the routing algorithm uniformly loads all the links, and the above utilization of any link is given by

$$U = 1/\bar{h}.$$

Since the network has $N_L = kp^{k+l}$ links, the total network capacity is obtained as

$$C = \frac{kp^{k+1}}{\bar{h}} \tag{8.13}$$

The per-user throughput is formulated as [46]

$$C/N = p/\bar{h}.$$

The per-user throughput depends on different (p, k) combinations and increase with a small value of k and a large value of p.

There are a number of approaches that are used for the routing in ShuffleNet [34,35]. A very simple approach for addressing "fixed routing" is given below:

A node in a (p, k) ShuffleNet is specified by the address (c, r) where $c \in 0$ 0, 1, ... , $k-1$ is the node's column coordinate labeled 0 through $k-1$ from left to right as shown in Figure 8.14 and $r \in 0, 1, 2, ... , p^k-1$ is the node's row coordinate labeled 0 through p^k-1 from the top to bottom. Thus, we may write $r = r_{k-1} \cdot r_{k-2} \cdots r_2 \cdot r_1 \cdot r_0$ [49]. This addressing scheme with the p-shuffle interconnection pattern has capability that from any node (c, r) (where $r = r_{k-1} \cdot r_{k-2} \cdots r_2 \cdot r_1 \cdot r_0$), the row addresses of all the nodes accessible in the next column have the same first $k - 1$ p-ary digits specified by $r_{k-2} \cdot r_{k-3} \cdots r_2 \cdot r_1 \cdot r_0$, and these are different in only the least-significant digit. For routing purpose, the destination address (c_d, r_d) is also included in every packet. When such a packet arrives at an arbitrary node (c_a, r_a), then, it is deleted from the network if $(c_d, r_d) = (c_a, r_a)$, which indicates reaching the packet at its destination. The node (c_a, r_a) evaluates the column distance X.

$$X = k + c_d - c_a \quad c_d \neq c_a$$
$$= k \quad c_d = c_a \tag{8.14}$$

Out of p nodes in the next column, (c_a, r_a) forwards to the current packet and finds the packet whose least-significant digit is written as r_{X-1}^d. In particular, the packet is sent to the node with the identity $[(c_a + 1 \bmod k, r_{k-2} \cdot r_{k-3} \cdots r_1 \cdot r_0 \cdot r_{x-1}^d]$. The routing scheme chooses the single shortest path between nodes (c_a, r_a) and (c_d, r_d) if the number of hops between them is k or less; otherwise, it considers one among several possible shortest paths.

The transfer function is written as [49]

$$T(D) = D + D^5 + 15D^9 + 225 \, D^{17} \cdots$$

where the above transfer function denotes 1 single-hop path, 1 five-hop path, 15 nine-hop path, 225 thirteen-hop path and so on.

8.4.2.2 de Bruijn Graph

One of most popular structures for multihop network is based on de Bruijn graph (Δ, D) [36,37]. A de Bruijn graph $(\Delta > 2, D > 2)$ is a graph with the nodes $\{0, 1, 2, \ldots, \Delta - 1\}^D$ with an edge from node $a_1 a_2 \cdots a_D$ to node $b_1 b_2 \cdots b_D$ if $b_i = a_{i+1}$, where $a_i, b_i \in (0, 1, 2, \ldots, \Delta - 1)$ and $1 \leq I \leq D - 1$. Each node has in-degree and out-degree Δ, some of the nodes have self-loops, and the total number of nodes in the graph equals $N = \Delta^D$. An (2,3) de Bruijn graph is presented in Figure 8.15a.

A link of node A–node B is indicated by $(D + 1)$ Δ-ary digits, the first D of which shows node A, and the last D digits represent node B. In a similar fashion, any path of length k is written in terms of $D + k$ digits. For the shortest path from node $A = (a_1, a_2, \ldots, a_D)$ to node $B = (b_1, b_2, \ldots, b_D)$, we have taken the last several digits of A and the first several digits of B to obtain a perfect match over the largest possible number of digits. If this match is of size k digits, i.e., $(b_1 b_2 \cdots b_{D-k})$ and $(a_{k+1} a_{k+2} \cdots a_D)$, then the k-hop shortest path from node A to node B is given by $(a_1 a_2 \cdots a_D b_{D-k+1} b_{D-k+2} \cdots b_D)$. An upper bound on the average hop distance between two arbitrary nodes in a de Bruijn graph is given by [49]

$$\overline{h_{de}} \leq D \frac{N}{N-1} - \frac{1}{\Delta - 1} \tag{8.15}$$

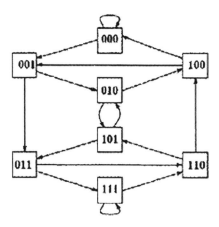

FIGURE 8.15 (a) A de Bruijn (2,3) graph [13].

(Continued)

For a large Δ, this bound is theoretically lower bound on the mean hop distance in an arbitrary directed graph with N nodes and maximum out-degree $\Delta \geq 2$ [49]. For the same mean hop distances in (Δ, D) de Bruijn graphs and (p, k), the topologies based on de Bruijn graphs accommodate a larger number of nodes than that for ShuffleNets [37] due to the fact that the diameter (the maximum hop distance) in a ShuffleNet is very large (it equals $2k - 1$ in a (p, k) ShuffleNet). An undesirable characteristic of the de Bruijn graph is that, even if the offered traffic to the network is fully symmetric, the link loadings can be unbalanced. This is due to the inherent asymmetry in the structure, e.g., in the (2,3) de Bruijn graph, the self-loops on nodes "000" and "111" carry no traffic (and hence are wasted), and the link "1000" only carries traffic destined to node "000" while link "1001" carries all remaining traffic generated by or forwarded through node "100". As a result of the link-load asymmetry, the maximum throughput supportable by a de Bruijn graph is lower than that supportable by an equivalent ShuffleNet structure with the same number of nodes and the same nodal degree. A simplified delay analysis using M/M/1 queuing models for links and independence assumptions shows that for uniform loading, the average packet delay in a de Bruijn graph is lower than that in an equivalent ShuffleNet. Using a longest-path routing scheme load-balanced routing is obtained in the de Bruijn graph and throughputs higher than ShuffleNets [37].

8.4.2.3 Torus (MSN)

An $N \times M$ MSN is one of the regular mesh structures [38] having its opposite sides linked to form a torus. Unidirectional communication links join its nodes into N rows and M columns, with adjacent row links and column links alternating in direction. Figure 8.15 shows a 4×6 MSN having four rows and six columns. A locally adaptive deflection routing algorithm is used for the routing of MSN in Ref. [39]. Another advantage of the MSN is highly modular and easily growable. In these architectures optical deflection switches are used in MSNs [39].

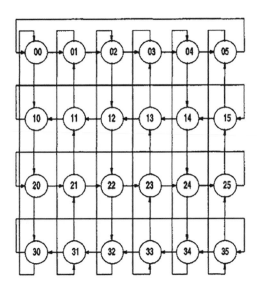

FIGURE 8.15 (CONTINUED)   **(b)** A 4×6 Manhattan street network with unidirectional links [13].

8.4.2.4 Hypercube

The hypercube interconnection pattern has been already used for multiprocessor architectures. It is sued as a virtual topology for multihop lightwave networks [40]. A p-dimensional binary hypercube has $N = 2p$ nodes, each of which has p neighbors. A node needs p transmitters and p receivers using one transmitter–receiver pair to communicate directly and bidirectional with each of its p neighbors. Any node i with an arbitrary binary address has its neighbors – whose binary address differs from node is address in exactly one bit position. Figure 8.16 shows the simplest form of the hypercube interconnection pattern having eight nodes. It is a binary hypercube whose nodes are located by binary indexes [41].

The merits of this structure are its small diameter ($\log_2 N$) and short average hop distance ($N \log_2 N)/(2(N - 1)$). Its disadvantage is that the nodal degree increases logarithmically with N.

The radix in the nodal address notation is written in generalized form using arbitrary integers, and generalized hypercube structure uses a mixed radix system to represent the node addresses. Total number of nodes for p neighbors hypercube, $\prod_{i=1}^{P} n_i$, where the n_i are the number of nodes of ith neighbor. A node's address P ($0 \le P \le N - 1$) is represented by the p-tupple $(m_p m_{p-1} \cdots m_1)$, where $0 \le m_i \le n_i - 1$. The P is written as $P = \sum_{i=1}^{P} w_i m_i$, where $m_i = \prod_{j=1}^{P} n_j$. The generalized hypercube has similar merits and demerits as its binary version.

8.4.2.5 GEMNET

An attractive approach to interconnect computing equipment (nodes) in a high-speed packet-switched network is to use a regular interconnection graph. The graph has the following properties

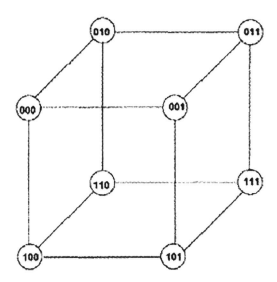

FIGURE 8.16 A hypercube network having eight nodes [13].

- small nodal degree
- low network cost
- simple routing
- fast packet processing
- small diameter
- growth capability

The graph has a scalable property where nodes can be added and deleted at all times with a modularity of unity. A new network structure is named as a Generalized shuffle exchange-based Multihop Network architecture (GEMNET) [42]. GEMNET is a logical (virtual), multihop topology for constructing lightwave networks using WDM and also a regular multihop network architecture, representing a family of network structures which includes the well-studied ShuffleNet [34] and de Bruijn [37] together. Figure 8.17 shows a simple ten node GEMNET in which two wavelengths are transmitted and two wavelengths are received by each node of the network. By using wavelength-routing switches, wide-area, multihop optical networks are made with this topology. Its logical virtual topology of the network is presented in Figure 8.18.

A typical (K, M, P) GEMNET consists of $K \times M$ nodes – each of degree P are arranged in a cylinder of K columns and M nodes per column so that nodes in adjacent columns are arranged according to a generalization of the shuffle-exchange connectivity pattern using directed links. The generalization permits any number of nodes in a column as opposed to the constraint of p_K nodes in a column. The logical topology in Figure 8.18 is a (2,5,2) GEMNET. In GEMNET, there is no restriction on the number of nodes as opposed to the cases in ShuffleNet [34] and de Bruijn graph [37] which can support only $K \cdot P^K$ and p^D nodes, respectively, where $K, D = 1, 2, 3, \ldots$

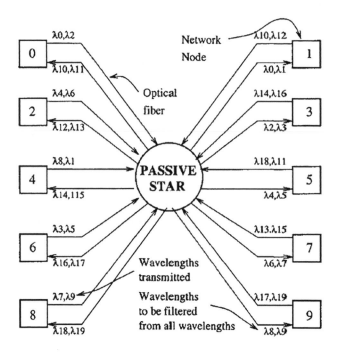

FIGURE 8.17   A GEMNET having ten nodes [13].

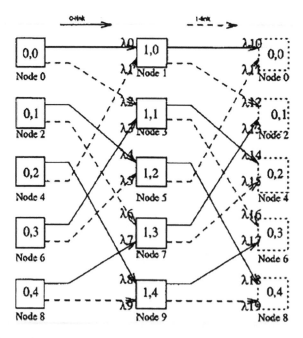

FIGURE 8.18   Logical connection virtual topology of GEMNET having ten nodes shown in Figure 8.17.

and $P = 2, 3, 4, \ldots$. The GEMNET shows arbitrary-sized networks in a regular graph; conversely, for any network size, at least two GEMNET configurations exist – one with $K = 1$, and other with $M = 1$. GEMNET is also scalable in units of K, if the nodes are equipped with either tunable transmitters or receivers.

In GEMNET architecture, its construction and routing algorithms are considered for balancing its link loads, mean as well as bounds on its hop distance, and algorithms to add nodes and delete nodes from an existing GEMNET.

8.4.2.5.1　Interconnection Structure of GEMNET Architecture

In GEMNET Architecture [42], there are N number of nodes that are evenly divisible. In GEMNET, the $N = K \times M$ nodes are arranged in K columns ($K \geq 1$) and M rows ($M \geq 1$) with each node having degree P. Node a ($a = 0, 1, 2, \ldots, N - 1$) is located at the intersection of column c ($c = 0, 1, 2, \ldots, K - 1$) and row r ($r = 0, 1, 2, \ldots, M - 1$), or simply location (c, r), where $c = (a \bmod K)$ and $r = [a/K]$, where $[.]$ represents the largest integer smaller than or equal to the argument. The P links emanating out of a node are referred to as i-links, where $i = 0, 1, 2, \ldots, P - 1$. For a given number of nodes N, there are as many (K, M, P) GEMNETs as there are divisors for N. Specifically, when $K = 1$ or $M = 1$, we can accommodate any-sized network. However, $M = 1$ results in a ring with P parallel paths between consecutive nodes. Due to the cylindrical nature of GEMNET, the nodes in this column will be finally covered in an additional $K - 1$ hops. Thus, $D = [\log_p M] + K - 1$, where $[\log_p M]$ represents the largest integer but smaller than or equal to the argument.

8.4.2.5.2　GEMNET Routing

Each connection in GEMNET [42] has to be specified by a source node and destination node using (c_s, r_s) and (c_d, r_d). We consider the column distance ∂ as the minimum hop distance, where the source node touches a node (not necessarily the destination) in the destination node's column. When $c_d \geq c_s$, we have $\partial = c_d - c_s$, because $(c_d - c_s)$ forward hops from any node in column c_s will cover a node in column c_d. When $c_d < c_s$, ∂ is given by $S = (c_d + K) - c_s$ because, after forwarding c_d by K (i.e., $c_d + K$) i.e. when $c_d \geq c_s$. Thus, ∂ is formulated as $\partial = [(K + c_d) - c_s]_{\bmod K}$. The hop distance from source node (c_s, r_s) to destination node (c_d, r_d) is made by the smallest integer h of the form $(\partial + jK)$, $j = 0, 1, 2, \ldots$, satisfying the expression

$$R = \left[M + r_d - \left(r_s \times p^h \right)_{\bmod M} \right]_{\bmod M} < p^h \qquad (8.16)$$

where R is the route code specifying a shortest route from the source node to the destination node, expressed as a sequence of h base-P digits. In general, if $R = [\alpha_1, \alpha_2, \alpha_3, \ldots, \alpha_h]_{base\ p}$, then the node about to send the packet on its jth hop will route the packet to its α_jth outgoing link. The maximum iterations required to solve for R is $[D/K]$, where D is the diameter of the network. When $K = 1$, the number of iterations is maximum and time complexity in computing R is $O(\log_p N)$. ∂ is the minimum number of hops required to reach a node in column c_d from a node in column c_s. The $[r_s \times p^h]_{\bmod M}$ is the row index of the node in column c_d reachable from the source node in h hops by following the all-0-link path. Then, $(h - 1)$ 0-links followed by a 1-link leads to the node with row index $[r_s \times p^h + 1]_{\bmod M}$ in column c_d, and so on.

However, on the hth hop, a maximum of P^h nodes can be covered. The node reached on the hth hop from the source node by following the all-$(P-1)$-link path will be $[r_s \times p^h + (p^h - 1)]_{mod M}$. Thus, if R is less than p^h (which means that the destination node falls somewhere between the all-0-link path and the all-$(P-1)$-link path), then the destination node is reachable in h hops and its route code is given by R. Often, the p^h nodes covered on the hth hop could be greater than the number of nodes in that column. This means that multiple shortest paths may exist to some nodes in that column. If $(R + jM) < P^h$ for $j = 1, 2, 3, \ldots$, then $(R + jM)$ is also a routing code with pathlength h for any j that satisfies this inequality. Thus, if the shortest path from node a to node b is h hops, the number of shortest paths is given by

$$Y = \frac{p^h - R}{M} \qquad (8.17)$$

Hence, for a given N, the number of alternate shortest paths increases as M decreases. The larger the number of shortest paths, the more opportunity there is to route a packet along a less congested path and the greater is the network's ability to route a packet along a minimum length path when a link or node failure occurs.

8.4.2.5.3 Routing Algorithms for Balancing Link Loading in GEMNET [42]

In any other multihop network, the balancing of the traffic on different links is required. If a link's utilization is heavy, the corresponding link queue becomes long and the delay for packets traversing that link also becomes high. If multiple shortest paths exist, then traffic can be routed over the shortest path for balancing the link flows. Two alternative routing schemes, called "partially balanced" and "random", are found to perform well in comparison to the "unbalanced" scheme. Both of these schemes first estimate the base-R code using the equation (8.16) as under the "unbalanced" scheme. Then, if only one shortest path to the destination exists, the base-R value is used to route the message. If multiple shortest paths exist, the "partially balanced" scheme selects the route code R_1 if $R_1 = R + \left[(C_d \bmod P) \times M \right] < p^h$. When multiple paths exist, this approach spreads the traffic across different links based on the destination node number. However, if the number of shortest paths exceeds P, this approach limits its selection to the first P such paths. The random scheme simply computes the number of alternate shortest paths and assigns the route code $R_2 = R + (M \times Z)$, where Z is a uniformly distributed random integer in the range $[0, Y - 1]$ and Y is the number of shortest paths given by equation (8.17).

In a GEMNET, beginning at a node, all of the nodes are reachable on a certain hop count belonging to a specific column. These characteristics can make some fairly tight bounds on the minimum and maximum average hop distances. These bounds and their maximum difference are analyzed below [42]. If the set of nodes that reached on hop i were to overlap all of the previously visited nodes, then the number of new nodes that reached would be minimized to determine the maximum average hop distance. This pattern of visiting nodes would put an upper bound on an average hop distance of the network. Except for hops $D - K + 1$ through D, the number of new nodes covered on hop i is equal to P^i when $i<K$, and $P^i - P^{k-1}$ when $i \geq K$. For hop $D - K + 1$, fewer than P^i new nodes are covered and for hops $D - K + 2$

through D, also fewer than P^i new nodes are covered. The C, the last column to cover p^i new nodes on hop i, equals to

$$C = D - K \text{ if } D - K \le K - 1$$
$$= K - 1 \text{ otherwise} \tag{8.18}$$

We consider the function

$$\partial(x, y) = 1 \text{ if } x < y,$$
$$= \text{ otherwise}$$

Then, the maximum average hop distance bound for GEMNET is given by

$$\overline{h_{\max}} = \frac{1}{M \times K - 1} \left\{ \sum_{i=0}^{C} iP^i + \partial(C+1, D-K) \sum_{i=C+1}^{D-K} i\left(P^i - P^{i-K}\right) \right.$$

$$\left. + \sum_{i=D-K+1} i\left[M - \partial(K,i)P^{i-K}\right] \right\} \tag{8.19}$$

When $C = D - K$, all nodes in the network are symmetric, equation (8.19) provides the actual average hop distance. To determine the minimum average hop distance, we have to visit new nodes on hop i (i.e., nodes which have not been visited before) until it forces some of the nodes reached to be previously visited nodes. This would put a lower bound on the average hop distance. So, the maximum number of nodes P^i nodes can possibly cover on hop i, until either hop $D - K$ or $D - K - 1$. The L is the last column in which P^i new nodes could be covered. As described previously, the two possible values for L are $D - K$ and $D - K - 1$. If the total number of nodes reached in a column on the $(D - K)$th hop, assuming P^i nodes covered on each hop is $\le M$, then $L = D - K$, otherwise $L = D - K - 1$.

The minimum average hop distance can be written as [42]

$$\overline{h_{\min}} = \sum_{i=0}^{L} iP^i + \sum_{i=L+1}^{L+K} i\left(M - P^{i \bmod K} \sum_{j=0}^{\left[\frac{i}{K}\right]-1} P^{Kj} \right)$$

The largest potential difference between the maximum and minimum average hop distances occurs when $P = 2$, $K = 1$ and $N(= M \times K)$ is large. When $P = 2$ and $K = 1$, the difference between the max and min hop bounds $\left(\overline{h_{\max}} - \overline{h_{\min}}\right)$ is obtained as [42]

$$\Delta = \frac{N \cdot D - 2^D - 1}{N-1} - \frac{\left[(L+1)P^{L+1} - P^{L+2} + 2\right] + N(L+1) - \left(2^{L+1} - 1\right)(L+1)}{N-1}$$

where the first and the second terms correspond to the max and min bounds, respectively.

8.4.2.5.4   Scalability of a GEMNET

The scaling of a GEMNET is an important issue. Specifically, it shows the ways to grow up a GEMNET. The approaches to scaling a GEMNET is implemented on a single passive star, because a topology reconfiguration under this implementation can be easily performed by retuning some of the transmitters or receivers at the various nodes. Since a one-column GEMNET can accommodate any number of nodes, there is always at least one interesting GEMNET (besides the one with P parallel rings corresponding to $M = 1$). The easiest way to grow a GEMNET is to add one node at the bottom of each of its column. Thus, a GEMNET can grow by K nodes at a time (i.e., with modularity K). The location for adding K nodes revolves to be in the bottom row because the interconnection pattern is interrupted at the farthest point from the topmost row. Adding nodes closer to the top would have caused the interconnection pattern to be interrupted earlier. An example of growing a (1,6,2) GEMNET by one node to a (1,7,2) GEMNET is shown in Figure 8.18. Approximately the next M/P nodes perform one retuning, the next M/P nodes perform two retunings and so on. When adding K nodes to an N-node GEMNET in the fashion described above, we get

$$\text{No of retuning} = \sum \left[\frac{N}{P}\right] \times (i-1) \geq \frac{N \times (P-1)}{2}$$

Since the total number of links equals to NP, for a $P = 2$ GEMNET, this means that approximately one-fourth of the total number of transmitters or receivers in the network need to be retuned, while for a $P = 3$ GEMNET, approximately one-third of the transmitters or receivers need to be retuned.

8.5   SC MULTIHOP SYSTEMS

To increase link utilization, sharing of wavelength channel is needed in multihop network. The number of channels, w, in the network is constraint by technology and is usually less than the number of nodes N. A general method using channel sharing is considered to construct practical multihop networks under this limitation. Channel sharing may be achieved through TDM.

8.5.1   CHANNEL SHARING IN SHUFFLE NET

For channel sharing of ShuffleNet [43], (p,k) ShuffleNets with $p = 2$ is chosen. There are k columns of nodes in a (p,k) ShuffleNet. All nodes in the same row share their transmissions on p channels. Although node i ($0 \leq i \leq pk-1$) in an arbitrary column transmits to the nodes $j, j + 1 \cdots j + p - 1$ in the next column [where, $j = (i \bmod p^k - 1)_p$] via p distinct channels, all k of the nth transmitters ($n = 0, 1, ..., p$) from the ith nodes in all columns must share the same channel. Due to channel sharing, a packet arrives its destination in fewer hops, on an average, i.e., channel sharing reduces with the hop distance. For $p = 2$, an upper bound on the expected number of hops is obtained to be [43]

$$\overline{h_1} = \frac{1}{2^k}\left\{2 + (k-1)2^k\right\}$$

The mean hop distance in the above equation is for self-hopping also. The channel sharing in generalized ShuffleNet uses a different SC allocation mechanism [13].

8.5.2 CHANNEL SHARING IN GEMNET

Here we discuss the channel-sharing approach and its effect on the nodal degree in GEMNET [42]. In a (K, M, P) GEMNET, N nodes of degree P set in the network with K columns and M nodes per column in adjacent columns are arranged with the shuffle exchange connectivity pattern using directed links. Without channel sharing, the number of channels required in a (K, M, P) GEMNET is equal to $N \times P = K \times M \times P$, where nodal degree $P = N/w$. Channel sharing through TDM provides a higher logical nodal degree (viz., ability to reach more nodes directly), even though with a lower capacity on each of the logical links. We consider a (K, M, P) SC-GEMNET having K columns and M rows, where M is an integral multiple, n, with P (i.e., $M = nP$) and $K = N/M = N/(nP)$. Figure 8.19 shows several 12-node ($N = 12$) SC-GEMNETs with different values of w. Figure 8.19a shows the trivial 12-channel case. As each node has only one fixed transmitter and one fixed receiver, the only SC-GEMNET is

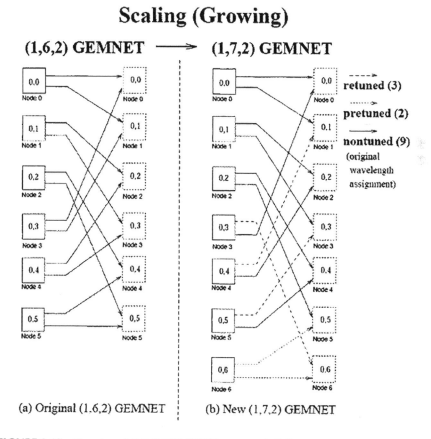

FIGURE 8.19   Growing of (1,6,2) GEMNET by one node [13].

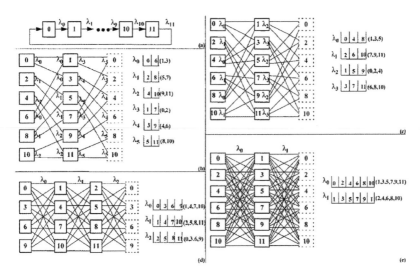

FIGURE 8.20 Twelve node SC-GEMNET with corresponding timing diagram for various values of (a) $w = 12$, (b) $w = 6$, (c) $w = 4$, (d) $w = 3$, (e) $w = 2$ [13].

constructed is a ring (i.e., $K = 12$, $M = 1$). Figure 8.20 also shows the other four cases which allow for "perfect sharing", i.e., $w = 6, 4, 3, 2$. In each of these figures, both a logical topology and the corresponding TDM frames on the various channels are provided in the figure. In each of these figures, both a logical topology and the corresponding TDM frames on the various channels are provided in the figure. In the TDM frames, the numbers in the slots indicate the transmitting nodes and the numbers in parentheses indicate the nodes having their receivers tuned to the channel. For each value of w, multiple SC-GEMNETS may be present. For the $w = 4$ case, it is also possible to construct a $K = 4$, $M = 2$ SC-GEMNET. The maximum average hop distance of the (K, M, P)-SC-GEMNET with diameter $D = [\log_{kp} M] + K - 1$ is obtained as

$$\overline{h_{max}} = \frac{MK\left(D + \dfrac{1}{2} - K/2\right) - K\left(\dfrac{P^{D-K+1} - 1}{P - 1}\right)}{M \times K - 1}$$

In determining a lower bound, the average hop distance is obtained as

$$\overline{h_{min}} = \sum_{i=L+1}^{L+K} iP^i + \sum i\left(M - P^{i \bmod K} \sum_{j=0}^{\left[\frac{1}{K}\right]-1} P^{Kj}\right)$$

$$L = D + K \quad \text{if } P^{(D-K)\bmod K} \times \left[\left(P^K\right)^{Z+1} - 1\right]/\left(P^K - 1\right) \le M$$

$$= D + K - 1 \quad \text{otherwise}$$

where

$$Z = \left[(D - K)/K_j \right]$$

In general, two-column GEMNETs (i.e., $I < = 2$) have the best performance in terms of the average hop distance. It is interesting to consider what happens when the available number of channels w is not an exact divisor of N. Such a system can be treated in two ways: (1) $w' < w$ channels may be used such that N/w' nodes share a channel with all nodes getting an equal share of the bandwidth, and $w - w'$ channels are wasted; (2) alternatively, all of the available channels may be used by employing different TDM frame lengths on different channels, i.e., with unequal sharing. For example, consider a system with $N = 0$. If $w = 50$, each channel is shared by two nodes, and the TDM frame length in each channel is two slots–one slot allocated to each of the two nodes that transmit on that channel. If $w = 51$, there are two possibilities. First, only $w' = 50$ channels may be used, and the 51st channel may be wasted resulting in a loss of efficiency of $1/51 = 2\%$. Alternatively, 98 nodes may share 49 channels–2 nodes per channel, and the last two nodes may use the remaining two channels exclusively. The unfairness problem becomes more pronounced if $w = 99$.

SUMMARY

This chapter discusses a local lightwave network employing WDM based on star topology. These networks can be broadly categorized into one of the two types – single-hop and multihop star WDM network. Various approaches for single-hop network design is mentioned in the chapter. We have mentioned reviews of some experimental WDM systems such as ACTS's SONATA [16], FOX single-hop network [14], Stanford's HORNET [17] and STARNET [18], and IBM's Rainbow [19]. We have discussed the coordination protocols used in a single-hop system.

In multihop system, we have started with the basic operation of multihop network. Then optimal virtual topologies using optimization is mentioned. We have mentioned operation of existing multihop networks such as Shufflenet, de Bruijn Graph, Torus (MSN), Hypercube and GEMNET. Finally, we have discussed channel sharing in Shufflenet and GEMNETSystems.

EXERCISES

8.1. Compare the physical topologies: star, bus, and tree with respect to: (a) number of simultaneous connections, (b) scalability and (c) delay.

8.2. Consider a PSC using 2×2 couplers (Figure 8.1). (a) Determine the number of 2×2 couplers needed for this design, and prove that each output gets $1/N^2$ of the input power. (b) The second approach is the butterfly arrangement (also called a multistage interconnection network) mentioned in the text. Determine the number of 2×2 couplers needed for this design and prove that each output gets $1/N$ of the input power. (c) What are the tradeoffs between the two designs.

8.3. State the comparison between the maximum power loss in an N node PSC and the maximum power loss in an N node linear bus. For the transmitted power is P, consider only losses from splitting and coupling.

8.4. Compare pre-transmission coordination protocols and protocols without pre-transmission coordination.

8.5. Compare single-hop WDM systems employing fixed transceivers to single-hop WDM systems employing tunable transceivers.

8.6. A single-hop *TT–TR* network with ten nodes connected via a PSC uses a fixed channel assignment, and each transmitter and receiver is capable of tuning to three non-interfering channels. How many TDM slots are required? Show that there is no interference assuming negligible tuning times.

8.7. In a single-hop network with four nodes and three channels, (a) give a schedule for the fixed channel assignment technique and (b) a schedule for the DA protocol.

8.8. A partial fixed-assignment protocol is used in a passive star network. There are 12 nodes in the network and four channels available. What is the shortest time slot scheme (i.e., the scheme with the shortest period), avoiding receiver collision, but not necessarily channel collision. When might such a scheme be desirable, and when might it not?

8.9. A fixed-assignment protocol is designed for a system with N nodes and W channels. The time slot duration is T. Find a lower bound on expected packet delay in the system. Find the lower bound on a partial fixed-assignment protocol. By what factor do these values differ?

8.10. Suppose we have a system with four nodes. Each node is equipped with a single tunable transmitter amid two fixed receivers. Assign channels to each of the transmitters $T(i)$ and receivers $R(i)$ for the random TDM protocol. What must we assume when assigning channels?

8.11. Consider a packet network in which the packet arrival process is a Poisson process with rate A. In this network, users can start packet transmissions at times $T, 2T, 3T, \ldots$, where T is the length of a packet. All the packets that arrive during the time interval $[(n-1)T, nT)$ are transmitted during the time "slot" $[nT, (n+1)T)$. When a single packet is transmitted in a time slot, that packet is successfully received by the receiver. When two or more packets are transmitted in the same slot, a collision occurs and all the packets are lost. (a) Find the probability P_k that k packets are transmitted in a time slot. (b) The *throughput S* of this network is defined as the fraction of time slots which result in successful packet transfer. Find S as a function of X and T. (c) Find the value of X that maximizes the throughput and the resulting throughput.

8.12. Consider the following WDM protocol. The network has a large number of nodes and W channels. Each node has W fixed-tuned receivers, one per channel and a tunable transmitter. The time is slotted, and packets arrive according to a Poisson process with rate G packets/slot. When a node has a packet to transmit, it selects a wavelength randomly and transmits the packet in the first time slot on that wavelength. (a) Find the probability that a wavelength is successfully utilized in a given time slot. (b) Find the traffic rate that maximizes the throughput per channel and the resulting throughput. (c) Find the average number of packets successfully carried by this network per time slot values of N (number of nodes), W (number of channels), L, (tuning time for a receiver) and traffic matrix a_{ij}, where i, $j = 1, 2, \ldots, N$, find a lower bound on the optimal schedule length.

8.13. In the ALOHA/ALOHA protocol, calculate the throughput as a function of the normalized offered load and calculate the maximum throughput. Assume that there are N stations, and packets arrive to each station according to a Poisson process with rate A. Traffic is uniformly distributed across all stations. (Normalized offered load is defined as the rate of arrivals per packet transmission time.)

8.14. In slotted-ALOHA/ALOHA, what is the length of the vulnerable period for the control packet and data packet? Assume control packets to be one time unit in length and data packets to be L time units in length.

8.15. Why are CSMA protocols usually not very attractive for a single-hop WDM optical network?

8.16. Describe the procedure for receiving a packet in the RCA protocol.

8.17. In the RCA protocol, we set the size of the NAL to $2T + L$ slots. Explain why.

8.18. For LAMBDANET, approximately how much of the low-loss region of bandwidth was utilized?

8.19. Consider the linear bus with attempt and defer nodes. Suppose the bus has five nodes over 5 km, and we require the received power at each node to be at least 30% of the transmitted power. How many amplifiers are required? Assume that amplifiers provide a gain of 25 dB.

8.20. Which kind of traffic is more well-suited to the Rainbow protocol–packet-switched traffic or circuit-switched traffic? Why?

8.21. Suppose station A is trying to set up a connection with station B. Draw simple state diagrams for stations A and B that illustrate this process.

8.22. Why is the timeout mechanism necessary in the Rainbow protocol? Show by example how, in the absence of a timeout mechanism, the system can become deadlocked. Give an alternate method of avoiding deadlocks in the Rainbow protocol.

8.23. Consider a Rainbow network with 100 nodes, an average scan time of 100 ps for each channel (including tuning time as well as signal detection time) and a round-robin channel scanning algorithm on each receiver. What is the expected time required to broadcast a packet of 1000 bytes from one node to all of the other nodes'? Propagation time, as well as acknowledgment transmission time, is negligible. Also assume that the network has no traffic when the broadcast occurs. Bandwidth per channel is 250 Mbps.

8.24. In an eight-node (2,2) ShuffleNet, calculate the average delay for a uniform traffic matrix (the arrival rate for each source–destination pair is X packets per second), assuming that the service time for a packet at a node is exponentially distributed with mean $1/p$ seconds. Consider shortest-path routing. Describe how a packet gets routed from node 0 to node 7 in a (2,2) ShuffleNet.

8.25. Show how a packet gets routed from node *0* to node 6 in a (2,3) de Bruijn graph. What is the maximum hop distance in this graph?

8.26. Draw a (3,2) de Bruijn graph. Compute the average hop distance.

8.27. Compare and contrast the following topologies and calculate the average hop distance for each:
   a. (2,2) ShuffleNet
   b. (2,3) de Bruijn graph
   c. eight-node binary hypercube
   d. eight-node 4×2 MSN.

8.28. Find the average hop distance for a 4×4 MSN.

8.29. Derive the average hop distance and diameter for a *p*-dimensional binary hypercube.

8.30. Suppose we wish to build a multihop network in which each node has a nodal degree of two, and the diameter of the network is three. We consider only ShuffleNet and MSN. Give the possible logical topologies and compare (a) the maximum number of nodes that can be supported and (b) the average hop distance.

8.31. A de Bruijn graph topology is chosen by a major internet network service provider. The network must contain at least one million nodes. (a) If a binary ($A = 2$) de Bruijn graph is used, what is the minimum number of bits required to represent the label of a node? (b) How many self-loops does the graph contain?

8.32. In a (2,3) de Bruijn graph. Determine the total number of paths from node 0 to node 5 that have at most eight hops.

8.33. Find the transfer function between points *A* and *B* in the following graph. How many paths of distance ten hops are there from *A* to *B*? How many paths of distance 15 hops are there from *A* to *B*?

8.34. Draw a two-dimensional radix-3 hypercube. Is the resulting graph isomorphic to a 3×3 torus? Will a two-dimensional radix-4 hypercube be isomorphic to a 4×4 torus? What is the number of edges in a *d*-dimensional radix-*r* hypercube?

8.35. Draw a generalized hypercube with 12 nodes. Explain how channel sharing can be used in this hypercube network.

8.36. In a six-dimensional binary hypercube, how would you represent the root of this hypercube?

8.37. List the advantages of channel sharing.

8.38. Show that a $(1, 2^N, 2)$ GEMNET is equivalent (having a one-to-one correspondence between all nodes and links) to a $(2, N)$ de Bruijn graph.

8.39. Show that a (K, M, P) GEMNET is equivalent (having a one-to-one correspondence between all nodes and links) to a ShuffleNet.

8.40. Given nine nodes, find a GEMNET topology for which $K > 1$ and $M > 1$.

8.41. Under what conditions do we obtain the maximum and minimum hop distances in a GEMNET?

8.42. In a (3,7,2) GEMNET. Find the route code *R* from source node (0,4) to destination node (2,1).

8.43. Find the number of shortest paths in a (3,27,3) GEMNET between nodes $s = (1,25)$ and $d = (0,10)$.

8.44. Compute the diameter and average hop distance for a 3×3 GEMNET. Compute the base route code from node (1,0) to node (1,2). Find all shortest paths from node (1,O) to node (1,2).

8.45. In a network of six nodes, construct a SC-GEMNET logical topology for the following cases by taking a *FT–FR* system.
   a. Number of channels = 6
   b. Number of channels = 3
   c. Number of channels = 2

8.46. In a (2,4,2) SC-GEMNET with $w = 3$, show a possible TDM frame structure.

8.47. Consider a (3,4,2) SC-GEMNET with unicast traffic. Packets arrive at each node according to a Poisson process with rate $X = 2$ to 5 pkt/s. The distance between each node and the star coupler is 10 km. There are six wavelengths used in the system. Slot length is equal to1 ps. Find the average packet delay. (Approximate the delay on each wavelength channel as a *M/M/*1 queuing system.)

REFERENCES

1. D. Bertsekas and R. Gallager, *Data Networks*, Prentice Hall, Englewood Cliffs, NJ, 1992.
2. W. Stalling, *Data and Computer Communication*, Prentice Hall International, New York, 1999.
3. G. E. Keiser, *Local Area Networks*, Tata McGraw-Hill, New Delhi, India, 1997.
4. M. Schwartz, *Computer Communication Networks Design and Analysis*, Prentice Hall, Englewood Cliffs, NJ, 1977.
5. G. Kessler and D. Train, *Metropolitan Area Networks: Concepts, Standards and Services*, McGraw-Hill, New York, 1992.
6. Fiber Channel Association, *Fiber Channel: Connection to the Future*, Fiber channel Association, Austin, TX, 1994.
7. P. E. Green, *Fiber Optic Networks*, Prentice Hall, Englewood Cliffs, NJ, 1993.
8. E. Rawson and R. Metcalfe, "Fibernet: Multimode optical fibers for local computer networks," *IEEE Transactions on Communications*, vol. 26, pp. 983–990, 1978.
9. R. Schimdt, E. Rawson, R. Noorton, S. Jackson and M. Bailey, "Fibernet II: A fiber optic Ethernet," *IEEE Journal on Selected Areas of Communication*, vol. 1, pp. 701–711, 2003.
10. F. Jia and B. Mukherjee, "The receiver collision avoidance (RCA) protocol for a single-hop WDM lightwave network," *IEEE/OSA Journal of Lightwave Technology*, vol. 11, pp. 1053–1065, May/June 1993.
11. F. Jia and B. Mukherjee, "A high-capacity, packet switched, single-hop local lightwave network," *Proceedings, IEEE GLOBECOM '93*, Houston, TX, pp. 1110–1114, Dec. 1993.
12. F. Jia, B. Mukherjee, and J. Iness, "Scheduling variable-length messages in a single-hop multichannel local lightwave network," *IEEE/ACM Transactions on Networking*, vol. 3, pp. 477–488, August 1995.
13. B. Mukherjee, *Optical WDM Networks*, Springer-Verlag, 2006.
14. M. S. Goodman, J. L. Gimlett, H. Kobrinski, M. P. Vecchi, and R. M. Bulley, "The LAMBDANET multiwavelength network: Architecture, applications, and demonstrations," *IEEE Journal on Selected Areas in Communications*, vol. 8, pp. 995–1004, August 1990.

15. H. Yang, M. Herzog, M. Maier, and M. Reisslein, "Metro WDM networks: Performance comparison of slotted ring and AWG star networks," *IEEE Journal on Selected Areas in Communications*, vol. 22, no. 8, pp. 1460–1473, October 2004.

16. A. Bianco, E. Leonardi, M. Mellia, and F. Neri, "Network controller design for SONATA: A large-scale all-optical WDM network," *Proceedings, IEEE International Conference on Communications (ICC) '01*, Helsinki, Finland, pp. 482–488, June 2001.

17. K. V. Shrikhande, et al., "HORNET: A packet-over-WDM multiple access metropolitan area ring network," *IEEE Journal on Selected Areas in Communications*, vol. 18, no. 10, pp. 2004–2016, Oct. 2000.

18. T. K. Chiang, et al., "Implementation of STARNET: A WDM computer communications network," *IEEE Journal on Selected Areas in Communications*, vol. 14, no. 5, pp. 824–839, June 1996.

19. J. P. Jue, M. S. Borella, and B. Mukherjee, "Performance analysis of the RAINBOW WDM optical network prototype," *IEEE Journal on Selected Areas in Communications*, vol. 14, pp. 945–951, June 1996.

20. N. A. Olsson, et al., "68.3 km transmission with 1.37 Tbit km/s capacity using wavelength division multiplexing of ten single-frequency lasers at 1.5 pm," *Electronic Letters*, vol. 21, pp. 105–106, 1985.

21. E.-J. Bachus, R.-P. Braun, C. Casper, and E. Grossman, "Ten channel coherent optical fiber transmission," *Electronic Letters*, vol. 22, pp. 1002–1003, 1986.

22. H. Toba, K. Oda, K. Nosu, and N. Takato, "Factors affecting the design of optical FDM information distribution systems," *IEEE Journal on Selected Areas in Communications*, vol. 6, pp. 965–972, August 1990.

23. E. Hall, J. Kravitz, R. Ramaswami, M. Halvorson, S. Tenbrink, and R. Thomsen, "The Rainbow-I1 gigabit optical network," *IEEE Journal on Selected Areas in Communications*, vol. 14, pp. 814–823, June 1996.

24. A. Mokhtar and M. Azizoglu, "Multiacess in all-optical networks with wavelength and code concurrency," *Fiber and Integrated Optics*, vol. 14, no. 1, pp. 37–51, 1995.

25. A. Ganz and Y. Gao, "A time-wavelength assignment algorithm for a WDM star network," *Proceedings, IEEE INFOCOM '92*, Florence, Italy, pp. 2144–2150, May 1992.

26. I. Chlamtac and A. Ganz, "Channel allocation protocols in frequency-time controlled high speed networks," *IEEE Transactions on Communications*, vol. 36, pp. 430–440, April 1988.

27. P. W. Dowd, "Random access protocols for high speed interprocessor communication based on an optical passive star topology," *IEEE/OSA Journal of Lightwave Technology*, vol. 9, pp. 799–808, June 1991.

28. I. M. I. Habbab, M. Kavehrad, and C.-E. W. Sundberg, "Protocols for very high speed optical fiber local area networks using a passive star topology," *IEEE/OSA Journal of Lightwave Technology*, vol. 5, pp. 1782–1794, December 1987.

29. N. Mehravari, "Performance and protocol improvements for very high speed optical fiber local area networks using a passive star topology," *IEEE/OSA Journal of Lightwave Technology*, vol. 8, pp. 520–530, April 1990.

30. F. Jia and B. Mukherjee, "Performance analysis of a generalized receiver collision avoidance (RCA) protocol for single-hop WDM lightwave networks," *Proceedings, SPIE '92*, Boston, MA, vol. 1784, pp. 229–240, September 1992.

31. M.-S. Chen, N. R. Dono, and R. Ramaswami, "A media access protocol for packet-switched wavelength division multi-access metropolitan area networks," *IEEE Journal on Selected Areas in Communications*, vol. 8, pp. 1048–1057, August 1990.

32. S. B. Tridandapani and B. Mukherjee, "Channel sharing in multi-hop WDM lightwave networks: Realization, and performance of multicast traffic," IEEE Journal on Selected Areas in Communications, vol. 15, pp. 488–500, April 1997.

32. J.-F. P. Labourdette and A. S. Acarnpora, "Partially reconfigurable multihop lightwave networks," *Proceedings, IEEE Globecom '90*, San Diego, CA, pp. 34–40, December 1990.

33. J. A. Bannister, L. Fratta, and M. Gerla, "Topological design of the wavelength-division optical network," *Proceedings, IEEE INFOCOM '90*, San Francisco, CA, pp. 1005–1013, June 1990.

34. M. G. Hluchyj and M. J. Karol, "ShuffleNet: An application of generalized perfect shuffles to rnultihoplightwave networks," *IEEE/OSA Journal of Lightwave Technology*, vol. 9, pp. 1386–1397, October 1991.

35. M. J. Karol and S. Z. Shaikh, "A simple adaptive routing scheme for congestion control in Shuffle Net multihop lightwave networks," *IEEE Journal on Selected Areas in Communications*, vol. 9, pp. 1040–1051, September 1991.

36. K. Sivarajan and R. Ramaswami, "Multihop networks based on de Bruijn graphs," *Proceedings, IEEE INFOCOM '91*, Bal Harbour, FL, pp. 1001–1011, April 1991.

37. K. Sivarajan and R. Ramaswami, "Lightwave networks based on de Bruijn graphs," *IEEE/ACM Transactions on Networking*, vol. 2, no. 1, pp. 70–79, 1994.

38. D. Banerjee, B. Mukherjee, and S. Ramamurthy, "The multidimensional torus: Analysis of average hop distance and application as a multihoplightwave network," *Proceedings, IEEE International Conference on Communications (ICC '94)*, New Orleans, pp. 1675–1680, May 1994.

39. N. F. Maxemchuk, "Routing in the Manhattan street network," *IEEE Transactions on Communications*, vol. 35, pp. 503–512, May 1987.

40. P. W. Dowd, "Wavelength division multiple access channel hypercube processor interconnection," *IEEE Transactions on Computers*, vol. 41, no. 10, pp. 1223–1241, October 1992.

41. B. Li and A. Ganz, "Virtual topologies for WDM star LANs: The regular structure approach," *Proceedings, IEEE INFOCOM '92*, Florence, Italy, pp. 2134–2143, May 1992.

42. J. Iness, S. Banerjee, and B. Mukherjee, "GEMNET: A generalized, shuftle- exchange-based, regular, scalable, and modular multihop network based on WDM lightwave technology," *IEEE/ACM Transactions on Networking*, vol. 3, no. 4, pp. 470–476, August 1995.

43. A. S. Acampora, "A multichannel multihop local lightwave network," *Proceedings, IEEE Globecom '87*, Tokyo, Japan, vol. 3, pp. 1459–1467, Nov. 1987.

44. E. Arthurs, M. S. Goodman, H. Kobrinski, and M. P. Vecchi, "Hypass: An optoelectronic hybrid packet-switching system," *IEEE Journal on Selected Areas in Communications*, vol. 6, pp. 1500–1510, December 1988.

45. I. M. I. Habbab, M. Kavehrad, and C.-E. W. Sundberg, "Protocols for very high speed optical fiber local area networks using a passive star topology," *IEEE/OSA Journal of Lightwave Technology*, vol. 5, pp. 1782–1794, December 1987.

46. A. S. Acampora and M. J. Karol, "An overview of lightwave packet networks," *IEEE Network Magazine*, vol. 3, pp. 29–41, January 1989.

47. A. Fukuda and S. Tasaka, "The equilibrium point analysis—A unified analytic tool for packet broadcast networks," *Proceedings, IEEE GLOBECOM '83*, San Diego, CA, pp. 33.4.1–33.4.8, November 1983.

48. K. Bogineni, K. M. Sivalingham, and P. W. Dowd, "Low complexity multiple access protocols for wavelength-division multiplexed photonic networks," *IEEE Journal on Selected Areas in Communications*, vol. 11, pp. 590–604, May 1993.

49. K. Sivarajan and R. Ramaswami, "Light wave networks based on de Bruijn graphs," *IEEE/ACM Transactions on Networking*, vol. 2, no. 1, pp. 70–79, 1994.

9 Optical Access Architecture

Before discussing optical access architecture, it is required to know about the basic access techniques used for networks such as local area network (LAN) [1], wide area network (WAN) [2] and metropolitan area network (MAN) [3]. One property of a LAN is that its backbone having a shared transmission link permits all attached users to simultaneously attempt to obtain the access for transmission. Thus, two or more stations transmit simultaneously providing their signals to interfere and transmission becomes disrupted. To deal these conflicts, a number of access protocols are already proposed for controlling access to network backbone. These all fall into the broad category of asynchronous time-division multiplexing (TDM), since this mechanism is fit for handling the bursty nature of network traffic. The asynchronous TDM mechanism provides contention methods (or random-access schemes) and deterministic (or controlled) techniques [2].

The random-access schemes are ALOHA (a ground-based radio packet broadcast network developed in Hawaii), carrier sense multiple access (CSMA), CSMA with collision detection (CSMA/CD) and register-insertion protocols. ALOHA is one first broadcasting network [1–3] in which its broadcast concept is used in the local networking environment, which may be compared with other random-access broadcast schemes adapted in LANs. Controlled access to a LAN is carried out either in a centralized or a distributed fashion. A popular centralized technique is based on polling, in which node is to access the channel at any one time. Two commonly used distributed controlled-access schemes are token passing and the slotted-ring method, which is used as token-passing access in ring networks. Different medium access control is already used in LAN in Chapter 6. The performances of these techniques are analyzed mathematically, and these applications are extended in WANs in this chapter before discussing about accessing in optical networks.

9.1 PERFORMANCE MEASURES AND NOTATION OF ACCESS ARCHITECTURE

The performance of access of the network is characterized by its average packet arrival rate λ and its average length of service time $1/\mu$ (μ being the average service rate). In this chapter, access methods are analyzed in terms of average packet delays versus the channel throughput, considering the M/G/1 queue wherein packets arrive according to a Poisson distribution, a general service distribution is used, and there is one server (buffer plus processor). The M/G/1 queue is discussed in detail in Chapter 11. The throughput of a network quantifies in bits per second of the successful traffic being transmitted between stations. Since packets are serviced under noises and distortion during transmission from station to station, it is customary to only count the error-free bits (error-free packets) during throughput analysis. The throughput S is written in a dimensionless normalized case as [1]

$$S = \frac{\lambda K}{R} \tag{9.1}$$

where K = packet length in bits and R = channel transmission rate in bits per second. The throughput is written in terms of the offered load G, which represents the traffic demand in the network. Since some packets are to be retransmitted because of error detected in packets, the total offered traffic λ_T is the summation of the packets reached successfully on first transmission attempt and repetitions of those that were damaged or detected error in transit. The normalized offered load G is thus written as [1]

$$G = \frac{\lambda_T K}{R} \tag{9.2}$$

where $\lambda_T \geq \lambda$. Analogous to the throughput, G is a dimensionless number which is varied from 0 to ∞ since λ_T is an arbitrarily large value. Maximum throughput in a particular type of access scheme is capacity of channel under that access scheme, and is determined by maximizing S with respect to G. Channel utilization is represented by efficiency ρ Channel utilization is the average fraction of time that a channel is busy in which all packets are transmitted without error and there is no overhead, then throughput become utilization also called as a channel efficiency. The D and H are the number of data bits and overhead bits, respectively, in a packet and ρ is written as

$$\rho = \frac{SD}{D + H} \tag{9.3}$$

The average transfer delay T is the time interval from the generation of the last bit of a packet at the information source until the reception of this bit at the destination. The time delay consists of the following: queuing delay at the source's interface buffer before the packet is processed for transmission, processing delays as the protocol interpreter managing the transmission of the packet, propagation time as the time required for a packet through the network and processing delays at store-and-forward nodes internal to the network, waiting time at the buffer associated with the destination station and processing delays at the destination station.

9.1.1 RANDOM-ACCESS METHODS

In random access contention methods are considered in networks having many users to transmit the messages (typically in the form of packets) to other stations through a common channel. In a random access such as ALOHA [1,2], any particular station transmits, and all stations are contended for time on the network where there is no control mechanism for determining turn of user to transmit (hence the term contention). When more than one user wants to send the messages at the same time, the messages suffer collision and are to be retransmitted later. This scheme is completely asynchronous when there is no coordination among users, and it provides poor efficiency. To improve the efficiency, a scheme known as slotted ALOHA is used, which doubles the capacity of the ALOHA system [1]. Here the channel has discrete uniform time slots, which is the time interval used for one frame transmission time. All users are then synchronized by means of a central clock, so that transmission of the first user begins at the start of a time slot and stops at the end of a time slot. Thus, partial overlapping of a colliding message is removed.

The efficiency is improved through a scheme in which the user listens to the channel to see if it is free before transmitting a message packet. If the user senses the channel, the packet is transmitted; if the channel is sensed to be busy, the packet transmission is delayed to a later time. Through this scheme, which is called CSME or listens before talking, the chance of packet collision is decreased. Although the CSMA scheme provides an improvement over ALOHA methods, packet collisions are still present. To further improve channel utilization by sending out a message and immediately cease transmission before it is completed when a collision with another simultaneously transmitting user is detected. This scheme is called CSMA/CD [4,5].

9.1.1.1 ALOHA

First, we consider ALOHA to analyze, which is the simplest possible broadcast protocol [1,6]. The original ALOHA method is pure ALOHA.

9.1.1.1.1 Pure ALOHA

Here users send the data immediately if they have data to send. If a transmission was successful, then a sender waits for an acknowledgment from the receiver for a time period equal to one propagation time. If no acknowledgment is received, the packet will be resent because of the collisions between packets sent within a packer transmission time t_p from different users, as is indicated in Figure 9.1a. All packets have the same length requiring one-time unit t_p (called a slot) for transmission. An attempt is taken by a user to send packet A starting at time t_0. If another user makes packet B between t_0 and $t_0 + t_p$, the end of packet B will suffer collision with the beginning of packet A. This occurs because, owing to long propagation delays, the sender of packet A did not know that packet B was already under way when the transmission of packet A was started. Similarly, if another user attempts to transmit

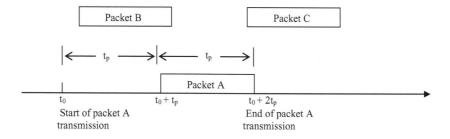

FIGURE 9.1   (a) Vulnerable period during which collision occurs in ALOHA.

(*Continued*)

between $t_0 + t_p$ and $t_0 + 2t_p$ (packet C), the beginning of packet C may collide with the end of packet A. Thus, of two packets overlapping by even the slightest amount in the vulnerable period shown in Figure 9.1a, both packets will be damaged requiring retransmission. The G is the total traffic entering the channel from an infinite population of users, where G also represents the numbers of packet transmissions that are attempted in a time period t_p.

To find the throughput, we consider the probability P_k of k transmission attempted per packet time, which follows a Poisson distribution with a mean G per packet time. So we can write P_k as

$$P_k = \frac{G^k e^{-G}}{k!} \tag{9.4}$$

The average throughput S is just the offered load G times the probability of a transmission being successful. Thus

$$S = GP_0 \tag{9.5}$$

where P_0 = probability that a packet not suffering a collision (that is the probability of no other traffic being made during a period having two packet times long). The probability of zero packets being made in an interval of two packet times long is

$$P_0 = e^{-2G} \tag{9.6}$$

So, we can write

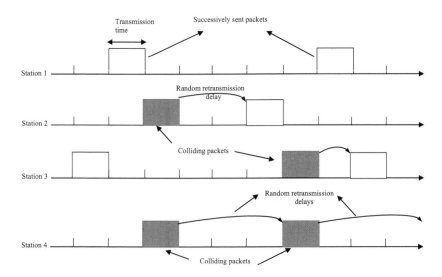

FIGURE 9.1 (CONTINUED) (b) Transmission attempts and random retransmission delays in slotted ALOHA.

(Continued)

$$S = Ge^{-2G} \tag{9.7}$$

The equation (9.7) shows the average throughput S depending directly on the offered load.

9.1.1.2   Slotted ALOHA

To enhance the efficiency of ALOHA, a slotted ALOHA scheme was introduced. All users use the time slots to have a user transmitting a packet synchronized exactly with the next possible channel slot. Consequently, during the vulnerable period, this packet collides with the other data having reduced to one packet time period versus two for pure ALOHA [1]. The transmission attempts and random retransmission delays for colliding packets are shown in Figure 9.1b for a four-user network.

As the defenseless period is decreased by half, the probability of no other traffic occurring during the same time period as a particular packet sent is $P_0 = e^{-G}$. This in turn leads to a throughput

$$S = Ge^{-G} \tag{9.8}$$

As shown in Figure 9.1b, the maximum efficiency of the slotted ALOHA system occurs at $G = 1$, where $S = 1/e$ or about 0.368, which is twice that of pure ALOHA.

We begin with the relative-frequency concept of probability for station i the steady state average throughput S_i and the offered traffic G_i are

S_i is the probability that station i successively transmits the number of slots used by station i for successful transmission/total number of slots. The G_i is the probability that station i attempts transmission of number of slots in which Station i attempts to transmit/total number of slots.

For M users, G_i is the probability that the user i transmits a packet in any given slot, where $i = 1, 2, ..., M$. Total average channel traffic is

$$G = \sum_{i=1}^{M} G_i \quad \text{packets per slot} \tag{9.9}$$

The average total throughput is

$$S = \sum_{i=1}^{M} S_i \quad \text{packets per slot} \tag{9.10}$$

The probability S_i that user i has a successful transmission in a particular time slot is merely equal to the probability G_i that user i sends a packet multiplied by the probability that none of the $M-1$ users sends packets; thus

$$S_i = G_i \prod_{i=1} (1 - G_i) \tag{9.11}$$

where the probability that none of the $M-1$ users sends packets $= \prod_{i=1} (1 - G_i)$

We consider that all M users are identical. Then each user has a throughput $S_i = S/M$ packets per slot and a total transmission rate of $G_i = G/M$ packets per slot, where G and S are given by equations (9.9) and (9.10), respectively [1,6]. Substituting these expressions into equation (9.11), it yields

$$S = G\left(1 - \frac{G^{M-1}}{M}\right) \qquad (9.12)$$

As $M \to \infty$, we have $S = Ge^{-G}$, which is the derived average throughput for an infinite population slotted ALOHA system.

9.1.2 CARRIER SENSE MULTIPLE ACCESS (CSMA)

In CSMA, all the stations in the network are aware of the transmission of packets by any station in the network within a fraction of the packet transmission time. This is known as listen-before-talk (LBT) method, where a station wants to attempt the transmission by keeping away from collisions by first listening to the medium to determine whether another transmission is in progress. When the channel is sensed to be idle, a station can take one of the three different approaches (depending on the network design) to insert a packet onto the channel. There are three protocols in this direction – non-persistent CSMA, l-persistent CSMA and p-persistent CSMA. When a station knows that a transmission is unsuccessful, in each protocol, the rescheduling of the packet transmission is the same. In these rescheduling the packet is sent again according to a randomly distributed retransmission delay. Before we analyze the CSMA protocols, the following are the assumptions [7]:

1. The station does not send and receive at the same time.
2. The channel state is sensed instantaneously.
3. The message errors resulting from random noise are negligible compared with the errors caused by overlapping packets).
4. Any fractional overlap of two packets are made destructive so that both retransmit.
5. The propagation delay is same between all source–destination pairs and is smaller than the packet transmission time.
6. The generation of packets (both new ones and retransmitted ones) from an infinite source of users obeys a Poisson distribution. Each user makes traffic at an infinitesimally small rate so that the average channel traffic sums to G packet time T_p.

9.1.2.1 Non-Persistent CSMA

The steps of non-persistent CSMA scheme follow as [2]

Step-1: If the medium is idle, transmit
Step-2: If the medium is busy, wait for a random amount of time and re-sense channel and repeat

Here the basic equation for the throughput S is expressed in terms of the offered traffic rate G and the parameter a which is defined as

$$\alpha = \frac{\text{Propagation delay}}{\text{Packet transmission time}}$$

The parameter "a" indicates a vulnerable period during which a transmitted packet has a collision, since the propagation delay is much smaller than the packet transmission time. The non-persistent CSMA protocol [1] shown in Figure 9.1c has both successful and unsuccessful attempts at transmitting packets. The activity on the channel is fragmented into busy periods during which transmission attempts are made and idle periods during which no station transmits. We consider the packet transmitted at a time $t = 0$ when the channel was sensed to be idle, and this packet originates at station 1 taking a time interval a for all other stations on the channel to become aware that station 1 is transmitting. If no other station transmits between time $t = 0$ and $t = a$, then the transmission attempt will be successful due to having a as the propagation time. At the same time for $t > a$ all other stations sense the channels to be busy. The busy period for successful transmission is then $1 + a$, which is the propagation delay plus the packet transmission time (in normalized time units). The busy period is followed by an idle period during which no station transmits. Another station has a packet to send during the time interval a. The station will transmit this packet after sensing the channel to be idle (the packet from station 1 has not arrived yet) and will thereby cause a collision. As shown in Figure 9.1c, Y be the arrival time of the last packet which collides with the one sent by station I, so that $0 \leq Y \leq a$. The transmission of all packets arriving in the time interval Y will be completed after $Y + 1$ time units. Since the station transmits

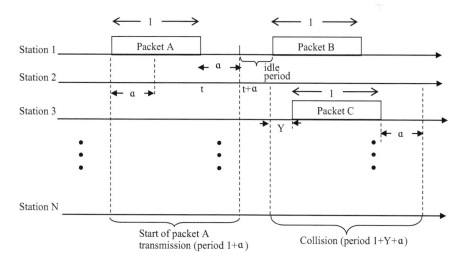

FIGURE 9.1 (CONTINUED) (c) Successful transmission and unsuccessful transmission for non-persistent CSMA.

during the time period a, the channel finally is sensed unused at time $Y + 1 + a$. Thus, the time duration $Y + 1 + a$ is the busy period for an unsuccessful transmission attempt. It is important to note that there can be at most one successful transmission during a busy period.

The channel throughput S is obtained as

$$S = \frac{E[U]}{E[B] + E[I]} \tag{9.13}$$

where $E[B]$ = expected duration of the busy period, $E[I]$ = expected length of the idle period and $E[U]$ = average time during the cycle that the channel is used without collisions.

This equation considers that all cycles are statistically the same. Under steady-state conditions, the throughput is the ratio of the average successful packet transmission time for a cycle to the total cycle time. To determine $E[U]$ we note that, for a successful transmission during a busy period, the probability that no station transmits during the first a time units of the period

$$E[U] = e^{-aG} \tag{9.14}$$

The idle period is the time interval between the end of a busy period and the next arrival to the network. At the end of a busy period, the packet arrivals follow a Poisson distribution, and the average duration of an idle period is given by

$$E[I] = \frac{1}{G} \tag{9.15}$$

For the busy periods, the time duration B of a transmission attempt is given by the random length

$$B = 1 + a + Y \tag{9.16}$$

When $Y = 0$ for successful transmissions, the average value of B then is

$$E[B] = 1 + E[Y] + a \tag{9.17}$$

Since Y is the only random quantity on the right-hand side of equation (9.16), equation (9.17) indicates that the last packet collided with the packets transmitted from station 1 on the average $E[Y]$ time units after the busy period begins, staying one time unit in transmission, and finally having the channel a time units later. The probability density function Y is a probability that no packet arrival occurs in an interval of length $(a - y)$. We can write

$$f(y) = Ge^{-G(a-y)} \quad \text{for } 0 \leq y \leq a \tag{9.18}$$

The $E(Y)$ is written as

$$E[Y] = \int_0^a yf(y)dy$$

(9.19)

$$= a - \frac{1}{G}\left(1 - e^{-aG}\right)$$

The expected duration of the busy period $E[B]$ is then written as

$$E[B] = 1 + 2a - \frac{1}{G}\left(1 - e^{-aG}\right)$$

(9.20)

So the average throughput is obtained for non-persistent CSMA as

$$S = \frac{Ge^{-aG}}{G(1 + 2a) + e^{-aG}}$$

(9.21)

The propagation time tends to be very small, i.e., $a \to 0$, S can be written as

$$\lim_{a \to 0} S = \frac{G}{1 + G}$$

(9.22)

A throughput of 1 can be theoretically achieved for an infinitely large offered channel load G.

9.1.2.2 Slotted Non-Persistent CSMA

The collision occurred in non-persistent CSMA method [1] is decreased in slotted non-persistent CSMA in which the function is same as that of the slotted ALOHA scheme. Here the time axis is fragmented into intervals of length τ (called as time slot) as shown in Figure 9.2d. All stations are synchronized to begin transmission only at the beginning of a slot. All packets are to be of same length of t_p (i.e., a transmission time t_p). When a packet reaches during a time slot, the station senses the state of the channel at the beginning of the next slot and then either transmits if the channel is idle or defers to a later time slot if there is traffic present. Here packets from stations 1, 2 and n arrive during the first time slot. If the next slot is empty, all three stations transmits as shown in the figure. All stations, including the ones at which packets arrive in the busy period (shown as packets a, b, c and d in Figure 9.2), sense the channel to be busy. These stations thus postpone their packets for transmission at a later time. As can be seen in Figure 9.2, the length of each busy period is exactly $(t_p + \tau)$ or, in normalized time units $(1 + a)$, where $a = \tau/t_p$. Analogous to the unslotted CSMA case, an idle period consists of an integer than the transmission time following a busy period. This is shown in Figure 9.2d. This packet gets sent in the next slot, thereby ending a cycle. The analysis for the slotted non-persistent CSMA is same as that of the unslotted case, where $E[U]$ is the average time during a cycle. For normalized time units, $E[U]$ is written as

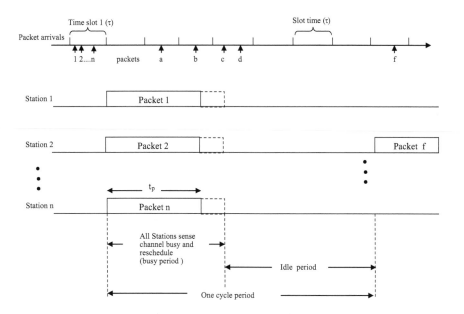

FIGURE 9.2 Packets arrivals in busy and idle periods in slotted non-persistent CSMA.

$$E[U] = P_s \tag{9.23}$$

where the conditional probability P_s is given by

$$
P_s = \frac{\text{Probability that one packet arrives in slot } a}{\text{Some arrival occurs}}
$$

$$
= \frac{P\{\text{one packet arrives in slot } a\}}{P\{\text{some arrival occurs}\}}
\tag{9.24}
$$

Using Poisson arrival statistics, we then have

$$P\{\text{one packet arrives in slot } a\} = aGe^{-aG} \tag{9.25}$$

$$P\{\text{some arrival occurs}\} = 1 - e^{-aG} \tag{9.26}$$

So that $E[U]$ becomes

$$E[U] = \frac{aGe^{-aG}}{1 - e^{-aG}} \tag{9.27}$$

Since, in normalized time units, the busy period is always $1 + a$, its average value is simply

$$E[B] = 1 + a \qquad (9.28)$$

Now the average value of the idle period $E[I]$ is to be derived. For slotted non-persistent CSMA, an idle period always comprises an integral number of time slots $I \geq 0$. If a packet arrives during the last time slot of a busy period, then the next slot immediately begins a new busy period so that $I = 0$. When there are no arrivals during the last slot of a busy period, then the next $I - 1$ slots are vacant, until there is an arrival in the final Ith slot. The beginning of a new busy period should also be identified. To find $E[I]$ we first consider the case $I = 0$. The probability p of this occurrence is the probability of some packets arriving in the interval. The probabilities are written as

$$P(I = 0) = p = 1 - e^{-aG} \qquad (9.29)$$

The probability for $I = 1$ is the joint probability that no arrival occurs in the last slot of the busy period and that some arrival occurs in the next time slot. It is written as

$$P(I = 1) = (1 - p)p \qquad (9.30)$$

Extending to an idle period of length $I = i$, the probability, for no arrivals in i consecutive time slots followed by some arrival in the next slot, is

$$P(I = i) = (1 - p)^i p \qquad (9.31)$$

This describes a geometrically distributed random variable V with a mean value of

$$E[V] = \sum_{i=1}^{\infty} i(1 - p)^i p = \frac{1 - p}{p} \qquad (9.32)$$

The average length of the idle period is a times $E[V]$ so that from equations (9.30) and (9.32)

$$E[I] = aE[V] = \frac{ae^{-aG}}{1 - e^{-aG}} \qquad (9.33)$$

Substituting we have for the slotted non-persistent CSMA case

$$S = \frac{E[U]}{E[B] + E[I]} = \frac{aGe^{-aG}}{1 - e^{-aG} + a} \qquad (9.34)$$

9.1.2.3 1-Persistent CSMA

The steps used in 1-persistent CSMA are given below [2,7]:

Step-1: If medium is idle, transmit
Step-2: If medium is busy, continue listening until the channel is idle; then transmit immediately.

The 1-persistent protocol avoids the situations in which a station waits before transmitting even though an idle channel. In 1-persistent scheme, a packet is transmitted with unit probability if the channel is sensed idle. If the channel is busy for a station having a packet to transmit, the station waits until the channel goes idle and then transmits immediately with probability one. When two or more stations waits to transmit, a collision is guaranteed as each station transmits immediately at the end of the busy period.

The performance of the 1-persistent CSMA protocol depends on the channel delay time [7]. After a station begins transmitting, another station has a packet to transmit and listens if the channel is idle. If the packet from the first station has not yet arrived at the second station, the latter is assumed to be channel idle and transmits its packet making a collision. As the delay time becomes longer, the performance of the protocol degrades.

A simplified approach using three-state Markov chain analyzes throughput of 1-persistent CSMA [7]. This model considers an infinite number of bursty users (users that transmit bursts of information for short periods of time) generating Poisson traffic at a rate G packets per transmission time. The users are considered to be attached to a worst-case star LAN. The analysis has a sequence of subbusy periods that takes place within each busy period, as shown in Figure 9.3 corresponding to the successful and unsuccessful transmission periods given in Figure 9.1c. All stations make packets while sensing the channel busy during the jth subbusy period and sends those to the start of the $(j + 1)$subbusy period. As soon as a busy period is over, an idle period begins when there is no transmission at the start of a subbusy period.

These are the following three types of subbusy periods:

1. Idle periods represent the time interval having zero packets
2. Subbusy periods have exactly one packet
3. Subbusy periods containing more than one packet provide a collision

These periods represent states $i = 0$ and 2 for a three-state Markov chain embedded at the start of the subbusy periods. These states and the possible transitions are shown in Figure 9.4. The Markov chain is finite, aperiodic, irreducible and therefore ergodic. Some of the transition probabilities are shown in the state transition diagram [2]. The transition probability P_{01} from state 0 (the idle state) to state 1 is equal to one,

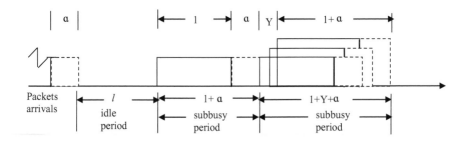

FIGURE 9.3   Channel condition in 1-persistent CSMA.

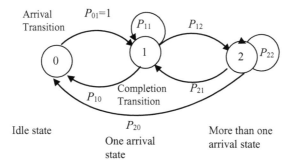

FIGURE 9.4 State transition diagram of 1-persistent CSMA (Markov-chain representation).

and for 1-persistent CSMA operation, the arrival of a packet in the idle state imme-diately begins a busy period. We also have $P_{02} = 0$. We consider the transition prob-ability P_{11}, providing the change from state 1 back to state 1, i.e., while the system is in state 1 (that is, a packet has arrived and is being transmitted), there is one arrival in the busy period. This packet in turn initiates another busy period after the first packet has been sent out.

For the analysis, $E[T_i]$ is considered to be the average time the protocol spends in state i and $\pi = \{\pi_0, \pi_1, \pi_2\}$ be the stationary probability distribution of the embedded chain. The throughput indicates the ratio of the average successful packet transmis-sion time in a cycle to the total cycle time

$$S = \frac{\pi_1 e^{-aG}}{\displaystyle\sum_{i=0}^{2} E[T_i]\pi_i} \tag{9.35}$$

Here in the numerator π_1 is the probability of the system being in state 1 during a successful transmission and, in equation (9.35), G is the probability of no arrival in the time interval a. The factor in the denominator is the state of the system during an entire cycle time.

In star topology having the 1-persistent scheduling method, all stations become ready in jth subbusy period and begins transmitting at exactly the same time in the $(j + 1)$th subbusy period. In addition, since there is an infinite population of users, the number of these transmission in the $(j + 1)$th subbusy period has no influence on the random variable Y, which indicates the last packet arriving in the vulnerable period a Therefore, the duration of subbusy periods starting with one packet (state 1) or with more than one packet (state 2) are identically distributed. So we can write

$$P_1 i = P_2 i \quad \text{for} \quad 0 \leq i \leq 2 \tag{9.36}$$

and

$$E[T_1] = E[T_2] \tag{9.37}$$

The $E[T_0]$ is just the mean idle period, which is given by $1/G$ and using the relationship $\Sigma i.\ \pi_i = 1$. The probability distribution $\{\pi_i | i \geq 0\}$ is a stationary distribution for Markov chain if

$\pi_i = \sum_{j=0}^{\infty} \pi_i P_{ij}$, and considering this probability, we can write

$$S = \frac{\pi_i e^{-aG}}{\pi_0 / G + (1 - \pi_0) E[T_1]} \tag{9.38}$$

To find π_0 and π_1, we consider a stationary distribution and we can write

$$\pi_0 = \frac{P_{10}}{1 + P_{10}} \tag{9.39}$$

$$\pi_1 = \frac{P_{10} + P_{11}}{1 + P_{10}} \tag{9.40}$$

The expression of $E[T_1]$ and its derivation are identical to $E[B]$. We thus have

$$E[T_1] = 1 + 2a - \frac{1 - e^{-aG}}{G} \tag{9.41}$$

We can find P_{10} and P_{11}. The P_{10} is the sum of the following joint probabilities

$$P_{10} = P\{\text{no arrival in time interval 1, success}\}$$

$$+ P\{\text{no arrival in time interval } 1 + Y, \text{collision}\}$$

$$= e^{-G} e^{-aG} + \int_0^a e^{-G(1+y)} G e^{-G(a-y)} \, dy \tag{9.42}$$

$$= (1 + aG) e^{-G(1+a)}$$

Similarly,

$$P_{11} = P\{\text{one arrival during interval 1, success}\}$$

$$+ P\{\text{one arrival during interval } 1 + Y, \text{collision}\}$$

$$= Ge^{-G} e^{-aG} + \int_0^a G(1+y) e^{-G(1+y)} G e^{-G(a-y)} \, dy \tag{9.43}$$

$$= \left(G^2 \frac{(1+a)^2 - 1}{2} + G \right) e^{-G(1+a)}$$

So, the throughput is written as

$$S = \frac{Ge^{-G(1+2a)}\left[1+G+aG\left(1+G+aG/2\right)\right]}{(1+a)\left(1-e^{-aG}\right)+ae^{-G(1+a)}}$$

(9.44)

Similar to the non-persistent case, a slotted version of 1-persistent CSMA is taken by using slot in the time axis and synchronizing the transmission of packets. For the slotted 1-persistent CSMA the throughput equation is written as

$$S = \frac{G\left(1+a-e^{-aG}\right)e^{-G(1+a)}}{(1+a)\left(1-e^{-aG}\right)+ae^{-G(1+a)}}$$

(9.45)

For propagation time is very small

$$\lim_{a\to 0} S = \frac{Ge^{-G}\left(1+G\right)}{G+e^{-G}}$$

(9.46)

At low values of G (very few stations sending), the throughput is low. For G becoming very large, $S \to Ge^{-G}$ which tends to zero.

9.1.2.4 *p*-Persistent CSMA

The steps used in *p*-persistent CSMA are given below:

> **Step-1:** If medium is idle, transmit with probability p
> **Step-2:** If medium is busy, continue listening until channel is idle; then transmit with probability p

To reduce the interference resulting from collisions for improvement of the throughput, the *p*-persistent CSMA scheme is used. The 1-persistent CSMA protocol applies to slotted channels. In this protocol when a station is ready to send the packets and it senses the channel to be idle, it either transmits with a probability p or the transmission is deferred by one time slot, which is typically the maximum propagation delay with a probability $q = 1-p$. If the deferred slot is idle, the station either transmits with a probability p or defers again with probability q. This process is repeated until either the packet is transmitted, or the channel becomes busy. When the channel is busy, the station acts as though there had been a collision; it waits for a random time and then starts the transmission attempt again. In case the channel was originally sensed to be busy, the station waits until the next slot and applies the above procedure.

The analysis and the resulting expression for the throughput of this protocol are rather involved. We consider this method further here as it has no advantage over non-persistent CSMA [8].

9.1.3 CSMA/CD: IEEE Standard 802.3

Considering all CSMA schemes it is evident that till there will be collision, collision cannot be avoided. To deal with these collisions, there should be CD, and once collisions are detected, steps should be taken to be retransmitted. A considerable performance improvement in the basic CSMA protocols is obtained by means of the CSMA/CD technique [9–12]. The techniques mainly use the listen-while-talk (LWT) protocol. The development of the CSMA/CD method of randomly accessing a LAN was made by a joint effort of Digital Equipment Corporation, Intel and Xerox into a detailed specification for a system called Ethernet as shown in Figure 9.5. The IEEE 802.3 CSMA/CD Standard for LANs considers the Ethernet specification.

The CSMA/CD protocols [11,12] use the concept of CSMA with the addition of the CD feature. While a CSMA/CD station senses an occurrence of collision, it immediately ceases transmitting its packet and sends a brief jamming signal to notify all stations of this collision. Collisions are detected by monitoring the analog waveform directly from the channel. When signals from two or more stations are present simultaneously, the composite waveform is distorted from that of a single station. This is noticeable in the form of larger-than-normal-voltage amplitudes on the cable. The steps are given below:

Step-1: If medium is idle, transmit immediately

Step-2: If medium is busy, continue listening until channel is idle; then transmit

Step-3: If the collision is detected, transmit a jamming signal to all the station to cease further transmission, Wait for a random amount of time and repeat steps 1 and 2

Step-4: Repeat steps 1–3 till the transmission is successful.

The basic operating characteristics/flowchart of the CSMA/CD protocol are outlined in Figure 9.5. When there is no packet waiting to be transmitted, even the CSMA/CD media-access control (MAC) sublayer [2] monitors the physical medium for traffic. When the station is ready to transmit, the behavior of the deference mechanism depends on which the protocol variation is being used. In particular, suppose the channel is idle.

The CMSA/CD is of three types – p-persistent, non-persistent and 1-persistent CSMA/CD [2]. In case of p-persistent CSMA/CD, the packet is transmitted with probability p or wait for the end-to-end propagation delay with probability $(1 - p)$. If the channel is busy, then the packet is backed off and the algorithm is repeated. In non-persistent case, the station defers transmission until the channel is sensed idle and then immediately transmits in the 1-persistent case. In a busy situation, when there is a packet ready to be sent out, the MAC sublayer postpones to the passing frame by delaying the transmission of its own waiting packet. After the last bit of the frame from the other station has passed by, the MAC sublayer remains to postpone for a certain time period called an inter-frame spacing. Transmission of any waiting packet is initiated at the end of this time. When the transmission is over the MAC sublayer monitors the carrier-sense signal. Once the postponing at a station is finished and transmission begins, collisions still occur acquisition of the network.

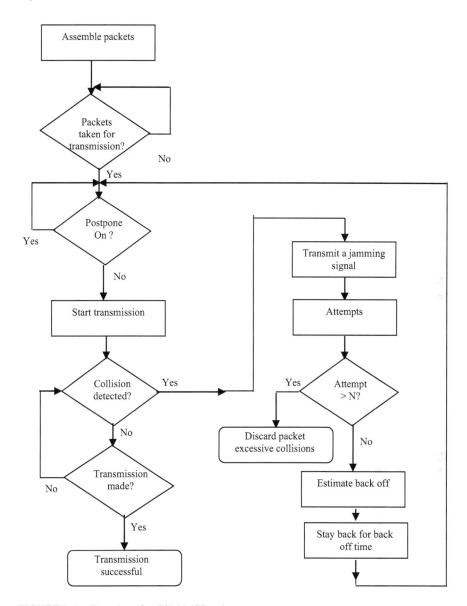

FIGURE 9.5   Flowchart for CSMA/CD scheme steps.

If a collision is detected, the station aborts the packet being transmitted and sends a jamming signal. After the jamming is transmitted, the stations involve in colliding packets and wait random amounts of time and then try to send their packets again. By using a random delay, the stations are involved in the collision that is not likely to have another collision on the next transmission attempt. However, the back-off technique maintains stability, a method known as truncated binary exponential back off used in Ethernet.

9.1.3.1 Throughput Analysis for Non-Persistent CSMA/CD

For throughput analysis of unslotted non-persistent case, Poisson arrivals are taken from an infinite population. The CD time δ is infinitesimal and negligible in comparison to the packet propagation time and treats retransmissions as independent Poisson arrivals. Figure 9.6 shows the sequence of events in a collision. Starting at time $t = 0$, station A transmits a packet. During the vulnerability time period a, station B sends its packet at time $t = y$ since it is not aware that station A is busy. At time $t = a + \delta$, station B detects the message from station A, immediately ceases its packet transmission and sends out a jamming signal of length. The transmission started by station B at $t = y$ will take time units to reach station A. At time $t = y + a + \delta$, station A detects the collision, stops sending its packet and sends out its own jamming signal of length b. The channel then becomes idle at time $t = 2a + y + b + \delta$. In the following analyses we let $\delta \rightarrow 0$.

The throughput is determined by using $E[U]$, $E[I]$ and $E[B]$, where $E[U]$ is an expected value of idle period, $E[I]$ is the average value of time during a cycle used without collisions and $E[B]$ is the expected value of the busy period. A busy period is basically either a successful transmission period or an unsuccessful contention period in which more than one station attempts to transmit. Here when any active station senses a collision, it first broadcasts a jamming tone for a time b to make aware of all other stations of the collision. In this protocol the first station transmits. The length of a successful period is $(1 + a)$, whereas that of a contention period is $(2a + b + Y_1)$. Thus, the expected duration of the busy period can be found by the expression [2], [9]

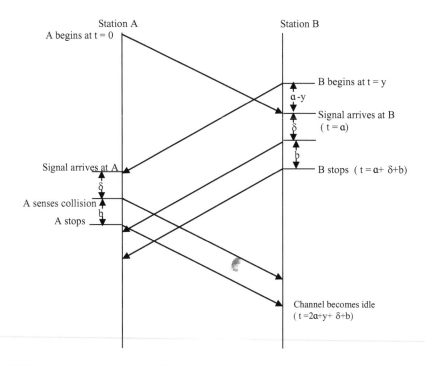

FIGURE 9.6 Time diagram of CSMA/CD.

$$E[B] = P\{\text{successful transmission}\}E[1+a]$$

$$+ P\{\text{unsuccessful transmission}\}E[2a+b+Y_1] \qquad (9.47)$$

$$= e^{-aG}(1+a) + (1-e^{-aG})(2a+b+E[Y_1])$$

To evaluate $E[Y_1]$, its probability density function $f(y)$ is represented as the probability $P\{y \geq y_1\}$, where $0 \leq y \leq a$ as follows:

$$f(y) = P\{y \geq Y_1\} = 1 - P\{y < Y_1\} \qquad (9.48)$$

where $P\{y < Y_1\}$ is the probability of no arrivals in the interval $[0, y]$, given that there is at least one arrival in $[0, a]$. Using the basic relationships of conditional probability and independent events, $P\{y < Y_1\}$ is derived as

$$P\{y < Y_1\} = \frac{P\{\text{no arrivals in}[0,y]\text{at least one arrival in}[0,a]\}}{P\{\text{at least one arrival in}[0,a]\}}$$

$$= \frac{e^{-yG}\left[1 - e^{-G(a-y)}\right]}{1 - e^{-aG}} \qquad (9.49)$$

Thus, from equation (9.48) we have

$$f[Y_1] = \frac{1 - e^{-yG}}{1 - e^{-yG}} \quad 0 \leq y \leq a \qquad (9.50)$$

From which we find

$$E[Y_1] = \int_0^a y f(y) dy = \frac{1}{G} - \frac{ae^{-aG}}{1 - e^{-aG}} \qquad (9.51)$$

The $E[B]$ is written as

$$E[B] = e^{-aG}(1+a) + (1-e^{-aG})\left(2a+b+\frac{1}{G} - \frac{ae^{-aG}}{1-e^{-aG}}\right)$$

The $E[U]$ and $E[I]$ are as same as those already derived earlier in unslotted non-persistent **CSMA** are written as

$$E[U] = e^{-aG}$$

$$E[B] = \frac{1}{G}$$

Considering $E[B]$, $E[I]$ and $E[U]$, the throughput S is derived as

$$S = \frac{Ge^{-aG}}{Ge^{-aG} + bG\left(1 - e^{-aG}\right) + 2aG\left(1 - e^{-aG}\right) + \left(2 - e^{-aG}\right)} \tag{9.52}$$

An expression can now be readily found for the slotted non-persistent CSMA/CD case, since its derivation is analogous to that of CSMA discussed earlier. The expressions for P_S, and $E[U]$ and $E[I]$ are the same as those derived for unslotted non-persistent CSMA respectively. The only difference in the analysis here is that the collision period needs to be considered. In particular, for slotted non-persistent CSMA/CD, packets are transmitted at the beginning of a slot, the contention period has a constant length $(2a + b)$, and the jamming time b is an integral number of slots. Thus, for the busy period, we have

$$E[B] = P\{\text{successful transmission}\}E\{1 + a\}$$

$$+ P\{\text{unsuccessful transmission}\}E[2a + b] \tag{9.53}$$

$$= P_S(1 + a) + (1 - P_S)(2a + b)$$

where P_S is as same that for unslotted non-persistent CSMA. Considering $E[U]$ and $E[I]$ we can write S for slotted non-persistent CSMA/CD,

$$S = \frac{aGe^{-aG}}{aGe^{-aG} + b\left(1 - e^{-aG} - aGe^{-aG}\right) + a\left(2 - e^{-aG} - aGe^{-aG}\right)} \tag{9.54}$$

The improvement in performance for CSMA/CD is achieved under high-offered traffic loads.

9.1.3.2 Throughput Analysis for 1-Persistent CSMA/CD

To analyze the throughput for 1-persistent CSMA/CD, 1-persistent CSMA is taken (already discussed earlier in this chapter). Considering Markov's chain/state transition diagram (Figure 9.4), we derive the expressions for the parameters P_{ik}, π_i and $E[T_i]$. The state of the channel showing the idle and subbusy periods is shown in Figure 9.7. As in Figure 9.4, the arrows on the timeline indicate possible times for initiating transmissions (i.e., packets arriving at these times either wait for transmission if the channel is busy or transmit immediately otherwise).

First, we consider that $E[T_1]$ is the mean idle period which is written as $1/G$. The expression for $E[T_1]$ is derived as same as that of $E[B]$.

$$E[T_1] = \left(1 - e^{-aG}\right)\left(2a + b + \frac{1}{G}\right) + e^{-aG} \tag{9.55}$$

To derive P_{10} and P_{11} we consider that, in the case of a collision, the packets generated in the time interval $a + b + Y$ in the current subbusy period will start in the next subbusy period.

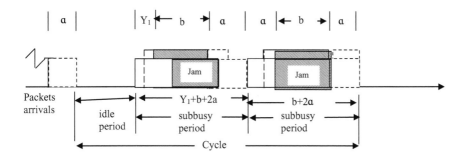

FIGURE 9.7 Channel condition in 1-persistent CSMA.

$$P_{10} = P\{\text{no arrival during interval 1, success}\}$$

$$+ P\{\text{no arrival in } a+b+y, \text{ collision}\}$$

$$= e^{-G}e^{-aG} + \int_0^a e^{-G(a+b+y)}Ge^{-Gy}\,dy$$

$$= e^{-G(1+a)} + \tfrac{1}{2}e^{-G(a+b)}\left(1 - e^{-2aG}\right) \qquad (9.56)$$

Similarly,

$$P_{11} = P\{\text{one arrival during interval 1, success}\}$$

$$+ P\{\text{one arrival in } (a+b+y), \text{ collision}\}$$

$$= Ge^{-G}e^{-aG} + \int_0^a (a+b+y)e^{-G(a+b+y)}Ge^{-Gy}\,dy \qquad (9.57)$$

$$= Ge^{-G(1+a)} + \tfrac{1}{4}e^{-G(a+b)}\left\{\left(1 - e^{-2aG}\right)\left[1 + 2G(a+b)\right] - 2aGe^{-2aG}\right\}$$

we consider that a subbusy period is generated by two or more packets (state 2). The subbusy period in right-hand side has a constant length $2a + b$, which is independent of any further colliding packets. So

$$E[T_2] = 2a + b \qquad (9.58)$$

The transition probabilities P_{2n} are then merely given by the probabilities $P_n(t)$ that exactly n packets arriving at a rate G in a time interval t. Thus, for $t = a + b$

$$P_{20} = e^{-G(a+b)} \qquad (9.59)$$

$$P_{21} = G(a+b)e^{-G(a+b)} \tag{9.60}$$

To derive π_0, π_1 and π_2 we use the following equations

$$\pi_j \sum_{\substack{i=0 \\ i \neq j}}^{\infty} P_{ji} = \sum_{\substack{i=0 \\ i \neq j}}^{\infty} \pi_i P_{ij} \tag{9.61}$$

where $P_{ij} \geq 0$ and $\displaystyle\sum_{j=0}^{\infty} P_{ij} = 1$ for $i = 0, 1, 2,\ldots$ and $P_{ij}^{n} = P\{X_{m+n} = j | X_n = i\}$ for all $m, n \geq 0$.

Using the equation (9.61), we write

$$\pi_1 = \frac{P_{20} + P_{21}}{k} \tag{9.62}$$

$$\pi_2 = \frac{1 - P_{10} + P_{11}}{k} \tag{9.63}$$

$$\pi_0 = 1 - \pi_1 - \pi_2 = \frac{(1 - P_{11})P_{20} + P_{10}P_{21}}{k} \tag{9.64}$$

$$\text{where } K = (1 - P_{10} - P_{11})(1 + P_{20}) + (1 + P_{10})(P_{20} + P_{21}) \tag{9.65}$$

Substituting all the appropriate parameters in the following expression of S, we obtain S as given below for unslotted 1-persistent CSMA/CD

$$S = \frac{\pi_1 e^{-aG}}{\displaystyle\sum_{i=0}^{2} E[T_i]\pi_i}$$

$$= \frac{(P_{20} + P_{21})e^{-aG}}{\left\{\dfrac{(1 - P_{11})P_{20} + P_{10}P_{21}}{G} + (2a + b)(1 - P_{10} - P_{11})\right.} \tag{9.66}$$

$$\left. + \left[(1 - e^{-aG})(2a + b + 1/G) + e^{-aG}\right](P_{20} + P_{21})\right\}$$

The throughput S depends on the offered load G, where the 1-persistent CSMA/CD protocol maintains throughput near capacity over a large range of loads.

9.1.4 STABILITY OF CSMA AND CSMA/CD

For the infinite population case, both CSMA and CSMA/CD channels are unstable for random retransmission delay for blocked terminals. For the finite population, CSMA channel is stable for sufficiently small retransmission probability. The stabilization is made under distributed retransmission control policies in which the retransmission probability is inversely proportional to the number of blocked stations. The Meditch and Lea (ML) model is a Markov chain model where the axis is divided into slots of length in the one-way propagation delay of the channel [12]. Here time again is normalized in terms of the packet length, so that the length of the time slot becomes a. The active stations on the network collectively make new packets in a Poisson process at a rate λ packets per slot. If a collision occurs, a blocked station reschedules its sensing of the channel in the current slot with a probability $0 < f \leq 1$. Channel stability is addressed in the ML model, considering a retransmission control policy of the form $f = \alpha/k$, where $k \geq 1$ is the number of blocked stations and $\alpha > 0$ is a constant. The transmission rate is thus given by α/a. The load G offered to the channel having the traffic from blocked and active stations is derived as

$$G = (\lambda + \alpha)/a \qquad (9.67)$$

where $0 < \lambda < 1$ is the input traffic in packets per slot and $\alpha > 0$ is the retransmission control parameter, then for channel stability,

$$\frac{\lambda}{a} < S \qquad (9.68)$$

where, as usual, S is normalized in terms of the packet transmission time t_p. From equation (9.67) channel stability is obtained in $0.05 \leq \alpha \leq 1.67$. The stable range widens if one comprises with low throughput. For reasonable values of b, however, the stable range for CSMA/CD is greater than for CSMA [12].

9.1.5   CONTROLLED-ACCESS SCHEMES

The contention-based access methods are stochastic in nature. The controlled-access technique [2] is based on deterministic/non-contention method, since a specific digital signal is used to get permission to transmit to one station at a time. The responsibility for the control signal is allotted to one station on the network (centralized control) or spread over all the nodes.

A common centralized-control method is based on polling [13]. Polling techniques determine the order in which stations can take turns to access the turns to access the network as shown in Figure 9.8, where a master station makes cycling among message source buffers. Here message arrives at N queues with queue k receiving an average rate of λk message per second considering the arrival process to be Poisson. The master station reserves each station in some prescribed order without having whether or not the station had a message to send. The polled station transmits if it had data to send, and if it does not have packets to transmit, the station sends back a special negative response. Usually the controller polls the station sequentially, but in special circumstances, important terminals may be polled several times per cycle.

Distributed-control access methods are based on token-controlled process. The techniques are used in ring or bus topologies. Tokens are special bit patterns or packets, usually several bits in length roaming from one node to the other if there is no message traffic. When a station wants to send data, it eliminates the token from the link of ring or bus and holds it. Now the station has exclusive access to the network

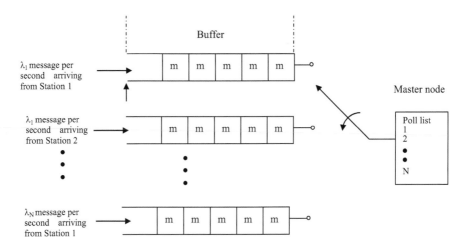

FIGURE 9.8   Diagram of polling processing.

for transmitting its message, and at the same time, other stations are continuously monitoring the messages that pass by on the network. All stations are identifying and accepting messages addressed to them. When a station ends transmitting its message, it puts the token back into circulation. This indicates to the network that the station has finished sending and gives other stations a chance to transmit. The distributed techniques are more widely used in LAN applications [2].

9.1.5.1   Token Ring: IEEE Standard 802.5

One of the control access schemes is token ring standardized as IEEE Standard 802.5 [14]. In this section, throughput is analyzed along with frame formats and the token-control access method. In a token ring the stations are connected logically in a ring as shown in Figure 9.9. Normally each station transmits the bit stream received from the previous station onto the next station. This is made after at least a one-bit delay, thereby the station is permitted to read and regenerate the incoming information before passing it. In the token ring scheme, an addressed destination

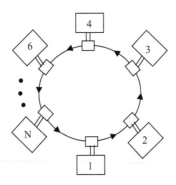

FIGURE 9.9   Token ring connected with *N* number of stations.

station copies the information as it passes, and the originating station removes it when a full loop is made around the ring. When a node wants to transmit the information, the incoming bit stream must be removed. The control is made by using an eight-bit pattern, 11111111, called a token passing from one station to another. Bit stuffing is considered to prevent this pattern from occurring in the data. Network control monitors the station in which each bit passes through the ring interface (called as a tap), which attaches the station to the ring. The operations of the tap have been already discussed in Chapter 6. Then the station wishes to transmit, it waits for the token to arrive and then changes it into another bit pattern, for example, 11111110. The station has to capture the token from the ring before transmitting immediately [2,14].

9.1.5.2   Token Bus: IEEE Standard 802.4

The token control scheme is also used in a bus topology as indicated in Figure 9.10. Here stations are with ordered identities, and they are offered service if offered one at a time in a round-robin order. The operation is thus similar to that of a ring except that whereas in a ring the token is passed implicitly for a bus an explicit token containing a specific node address is used. This results in an ordering of the station that resembles a logical ring. A token bus may be either bidirectional or unidirectional [15]. For a bidirectional bus, a logical sequential ordering of the attached stations is obtained by using an addressed idle token. When a station wishes to transmit, it waits for the free token addressed to it before accessing the channel. When the station ends transmission, an addressed idle token is sent to the next station that is in queue to access rights to the channel.

The address is selected from a predetermined list in which the stations are assigned positions in a sequence, with the fact that, in the sequence, the last station chases to the first station once a cycle is finished. Unlike the bidirectional token bus, a unidirectional token bus does not comprise an addressed token, rather it has an implicit token-passing protocol to efficiently use channel capacity. Users compete for the transmission medium according to some distributed conflict-free round-robin algorithm. The transmission medium in these unidirectional broadcast systems comprises an inbound and an outbound channel to which the stations are linked. The outbound channel exclusively transmits data, and the inbound channel reads the transmitted data. The two channels are linked so that all signals transmitted on the outbound channel are copied in the inbound channel, providing broadcast communication among the stations [15].

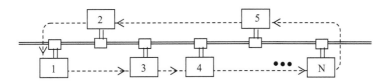

FIGURE 9.10   Token bus network having N number of stations.

9.1.5.2.1   Delay-Throughput Characteristics

This section mentions the comparison of two popular access schemes – a random-access bus (CSMA/CD) and a token ring [2] in terms of the delay-throughput characteristics of each type of LAN. Delay is measured as the average interval of time between when a packet is ready for transmission from a source station until its successful reception at the destination station. The delay time is due to the transfer time having the queuing and access delay at the sender, the transmission time of the packet and the propagation delay. The following considerations in the analyses are

- There are n stations each of which generates packets at a rate $\lambda_i, i = 1,2,...,N$
- The aggregate arrival rate of all stations is given by $\lambda = \lambda_1 + \lambda_2 + \cdots + \lambda_N$
- The packet lengths L_p are distributed in a general fashion
- A header of length L_h which contains control and addressing information is added to each packet
- The transmission rate (in bits per second) is denoted by R
- The parameter τ denotes the maximum end-to-end propagation delay for a bus or the round-trip delay for a ring

The service time of a packet is given by $T_p = \left(L_h - L_p\right)/R$

The performance model for a token ring follows a single-server queuing model, where each queue represents a station in the ring. The queues are serviced in sequence as represented by the rotating switch which corresponds to the token. The service time of a packet is represented by and the time required to pass the token from station i to station $i + 1$ is indicated by a constant switch-over delay τ_i. The analysis of Bux begins with the mean-queuing-delay modal derived by using a discrete-time approach. In this model equal arrival rates λ_i and equal switch-over delays τ_i of token are considered. Considering the case of the discrete-time interval approaching zero, the mean delay time D is written as

$$D = \frac{\rho E\left[T_p^2\right]}{2(1-\rho)E\left[T_p\right]} + E\left[T_p\right] + \frac{\tau(1-N)}{2(1-\rho)} + \frac{\tau}{2} \qquad (9.69)$$

where $\rho = \lambda$. $E[T_p]$ indicates the traffic intensity and $E[T_p]$ is the average value of T_p. Under delay model, all stations form the same amount of traffic, since no explicit analytical expression exists if the stations form unequal amounts of traffic.

The performance analysis of CSMA/CD networks is formulated by Lam, Tobagi and Hunt [4,5]. The access-control comparison of Bux considers a modified result of Lam in which, analytical analysis assumes a slotted channel with slot length 2τ. This assumption implies that even if the network utilization approaches zero, packets have to wait an average time τ before they are transmitted. Since this does not occur in a non-slotted system, Bux modified Lam's delay formula by reducing the mean delay time by the factor τ. A delay time for a CSMA/CD bus is formulated as

$$D_{\text{CSMA/CD}} = \frac{\lambda\left\{E\left[T_p^2\right] + (4e+2)\tau E\left[T_p\right] + 5\tau^2 + 4e(2e-1)\tau^2\right\}}{2\left\{1 - \lambda\left(e\left[T_p\right] + \tau + 2e\tau\right)\right\}} + E\left[T_p\right] + 2\tau e$$

$$+ \frac{\tau}{2} - \frac{\left[1 - \exp(-2\lambda\tau)\right]\left(2/\lambda + 2\tau/e - 6\tau\right)}{2\left\{F_p\exp\left[-(\lambda\tau+1)\right] - 1 + \exp(-2\lambda\tau)\right\}} \tag{9.70}$$

where F_p = Laplace transform of the probability density function of the packet service time T_p.

The maximum mean data rate is an important parameter for token rings, token buses and CSMA/CD networks [2]. The station is always ready to transmit. As soon as it transmits, it has another message waiting to be sent out. In this case it is expected that the limit on the transmission speed would be created by the station and not by the network. The total time required to transmit a message is

$$T_m = T_e + T_4 \tag{9.71}$$

where T_e and T_4 are the times needed to transmit the control and data bits, respectively.

Token Ring: The most important performance parameters for the token ring are the propagation time a and the interface time T_{iR} representing the time needed at a station to detect the token and to decide whether to pass it on or hold it. This time is generally equal to one bit transmission time [16,17]. When only one station is active on the token ring, the maximum mean data rate R_{max} is the reciprocal of the time required to transmit plus the time it takes the token to pass through other idle stations and the active station

$$R_{\text{max}} = \frac{1}{T_m + T_{\text{token},R}} \tag{9.72}$$

$$T_{\text{token},R} = a + NT_{iR} \tag{9.73}$$

In this case, all N stations are active, and R_{max} is the reciprocal of the time required to transmit plus the time needed to pass the token t to the next station. We consider that the stations are equally spaced, and the token-passing time is a/N, so that

$$R_{\text{max}} = \frac{1}{T_m + T_{iR} + a/N} \tag{9.74}$$

Unlike the token ring in which the interface delay is one bit, on a token bus, each station interface has to accept the entire message preamble before it makes a decision whether to pass the token onto the next station or to hold it for transmitting its own message. This interval is the time needed for a signal bit to travel from one end to the other, plus the time that a station needs to perform the associated interface signal processing. This bus interface delay T_{iB} is ~4 µs. When only one of the N stations on a token bus is active, the maximum mean throughput rate is the reciprocal of the time required for the station to transmit, plus the time requiring the token to pass through N idle stations and the active station

$$R_{max} = \frac{1}{T_m + T_{token,B}}$$

where

$$T_{token,B} = a + NT_{iB}$$

Here a is the propagation time from one end of the bus to the other.

Considering all N stations to be active, R_{max}, the reciprocal of the time required for one station to transmit and pass the token to the next station plus the interface delay, is given by

$$R_{max} = \frac{1}{T_m + T_{iB} + a/N} \tag{9.75}$$

CSMA/CD Bus: In a CSMA/CD bus topology [2,15], a station does not transmit senses a carrier waveform on the bus (since this indicates that another station is active). If a carrier is sensed, a station will defer transmission for a time interval called an inter frame gap T_{if}. When a transmitting station detects a collision (a carrier waveform from another station), it stops transmitting information time. The interface delay for the CSMA/CD bus is the same as for the token bus, that is T_{iB}. When only one out of N stations is active, each successful transmission needs a message transmission time T_m and an interframe gap time T_{if} for circuit and propagation transients. The maximum mean data rate is written as

$$R_{max} = \frac{1}{T_m + T_{if}} \tag{9.76}$$

Considering the rule of thumb of $2e$ collisions per successful transmission, the maximum mean data rate is written as

$$R_{max} = \frac{1}{T_m + T_{if} + (2e - 1)(T_{slot} + T_{jam})} \tag{9.77}$$

The token ring is the least sensitive to workload, whereas CSMA/CD has the shorted delay under light-load conditions. It is most sensitive to variations in load when the traffic load is heavy. CSMA/CD is more sensitive to propagation effects than the token bus, whereas the token ring is the least sensitive.

9.2 OPTICAL ACCESS NETWORK

This section describes optical access network architectures that are based on Passive Optical Network (PON) [18–22]. The PON provides high bandwidths in access networks. Here we discuss the Ethernet PON (EPON) [20,23], ATM-based PON (APON) [24], Broadband PON (BPON) and Generalize Framing Procedure (GFP) PON (GFP-PON [25]). We also mention wavelength-division multiplexing (WDM)-based PON. The access network is also represented as the "first-mile" network, connecting central offices (COs) to the subscribers.

9.2.1 ISSUES IN OPTICAL ACCESS ARCHITECTURE

The issues in optical access network are to develop high-capacity backbone networks in which backbone network operators provide the accessing of high-capacity OC-192 (10 Gbps) links (recently OC-768(40 Gbps). The recently developed access network technologies are Digital Subscriber Loop (DSL), which has the limitation to provide broadband services such as video-on-demand, interactive games and video conferencing to end users. The predominant broadband access solutions are the DSL and Community Antenna Television (cable TV)-based networks. However, both of these technologies have limitations because these were originally built for carrying voice and analog TV signals, but their carrying data are not optimal. DSL has another restriction that the distance of any DSL subscriber to a CO must be less than 18,000 ft due to having signal distortions.

Cable television networks can also support Internet services by dedicating some Radio Frequency (RF) channels in a co-axial cable for data. Cable networks are mainly built for delivering broadcast services. At high load, the network's performance is not used for satisfying end users.

The access networks bring fiber to the home with FTTx models – Fiber-to-the-Home (FTTH) [22], Fiber-to-the-Curb (FTTC) [26], Fiber-to-the-Building (FTTB) [26], etc. These models offer the access bandwidths to end users. These technologies provide fiber directly to the home or very near to the home. The FTTx models are mainly based on the PON. The major developments on PON in recent years are EPON, APON, GFP-PON and WDM-PON which are discussed in this section.

9.3 SIMPLE FIBER-OPTIC ACCESS NETWORK ARCHITECTURES

Optical fiber transmits up to 50 km or beyond in the subscriber access network. Figure 9.11 shows an FTTH access network in which a logical method accessing optical fiber in the local access network uses a point-to-point (PtP) topology. Three configurations are – dedicated fiber set from the CO to each end-user subscriber for

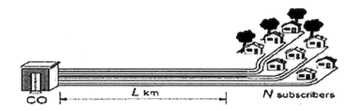

FIGURE 9.11   FTTH deployment scenarios: direct to [26].

bandwidth-intensive, integrated, voice, data and video services: (a) simple architecture having connector termination space curb switch connected to a Local Exchange (called as CO) with bidirectional fiber in which N subscribers are connected with a curb switch, (b) architecture having passive optical power splitter connected to the Local Exchange (called as CO) with bidirectional fiber in which N subscribers connect the curb switch at an average distance L kilometers from CO. A PtP structure has $2N$ transceivers and $N \times L$ total fiber length (assuming that a single fiber is used for bidirectional transmission).

Since the curb switch is expensive, the active curb-side switch can be replaced with an inexpensive passive optical split as shown in Figure 9.12b. PON problem [21] can reduce the number of optical transceivers, CO terminations and fiber deployment. A POM is a point-to-multipoint optical network with no active elements in the signal path from the source to the destination. The interior elements used in a POM are passive optical components, such as optical fiber, splices and splitters. An access network based on a single-fiber PON only requires $N + 1$ transceivers and L kilometers of fiber.

9.4 COMPONENTS OF PON TECHNOLOGIES

The key components of PON are optical splitter/couplers, burst mode switches, MAC, etc. These components have been discussed in detail in Chapters 3 and 6. In this section, we try to give an overview of these components and how these components have been related to PONs.

9.4.1 OPTICAL SPLITTERS/COUPLERS

A PON needs a passive device to divide an optical signal (power) coming from one fiber into several fibers and at the receiver to combine optical signals from multiple

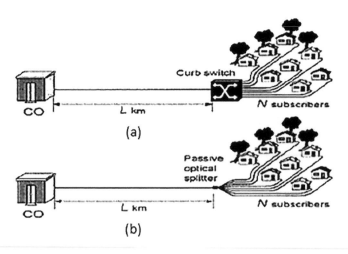

FIGURE 9.12 FTTH deployment scenarios: (a) via curb switch and (b) passive optical switch [26].

fibers into one. This device is an optical coupler/splitter. Section 3.1 already discusses the different optical couplers in detail. An optical coupler is made by using two fibers fused together in which signal can coupler from one fiber to the other with the principle of evanescent wave principle. The $N \times N$ couplers are made by staggering multiple 2×2 couplers or by using planar waveguide technology. Couplers are characterized by the following parameters:

1. **Splitting loss:** Power level at a half power 2×2 fiber coupler, and this value is 3 dB.
2. **Insertion loss:** Insertion loss arises due to imperfections of the coupler's fabrication process and coupling loss from 0.1 to 1 dB.
3. **Directivity:** Some amount of input power coming out as a leakage from one input port to another input port. The directivity ~40–50 dB.

The couplers made to have only one input to more outputs are referred to as a splitter. A coupler with more inputs and only one output is a combiner. Sometimes, 2×2 couplers are made highly asymmetric (with splitting ratios 5/95 or 10/90). This kind of coupler is used to branch off a small portion of the signal power for monitoring purposes. Such devices are called tap couplers.

9.4.2 PON TOPOLOGIES

Since a CO has multiple subscribers, multipoint topologies are suitable for the access network, including tree, tree-and-branch, ring or bus (Figure 9.13a). The use of 1:2 optical tap couplers and 1:N optical splitters, PONs are flexible in ring, bus and tree topologies. All transmissions in a PON are carried out between an Optical Line Terminal (OLT) [26] and Optical Network Units (ONUs) [26], as shown Figure 9.13.

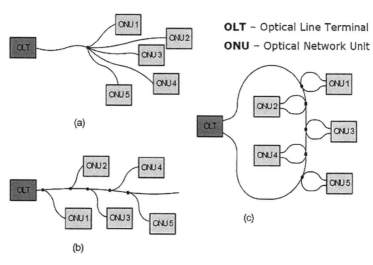

FIGURE 9.13   (A) PON topologies: (a) tree topology, (b) bus topology and (c) ring topology [26].

(Continued)

The OLT stays in the CO connecting the optical access network to the MAN or WAN [26], which is a long-haul network. The ONU is located at the end-user location (FTTH and FTTB) or at the curb requiring FTTC architecture.

The advantages of using PONs in access networks are numerous:

- PONs permit transmission for long distances (up to 20 km or beyond).
- PONs make minimization of fiber deployment.
- PONs support higher bandwidth due to the use of fiber.
- PONs permit for video broadcasting either as Internet Protocol (IP) video or video.
- PONs remove the necessity of installing active multiplexers.

9.4.3 BURST-MODE TRANSCEIVERS

Having unequal distances between CO and ONUs, optical signal attenuation in the PON is not uniform for all ONUs. The ONU's signal power is small at the OLT because of its transmission for longer distance. If the receiver is tuned to properly receive high-power signal from a close ONU, it receives a weak signal from a distant ONU. If the receiver is trained on a weak signal, it differentiates zeros or ones when receiving a strong signal.

Time slot is a period of time during which certain data are transmitted by specific regulations in the network. Time slot is not a periodic time allocation concept as in TDM here. Time slot is a "time slice" that is assigned to a node for transmission of its backlogged packet slot, i.e., it should operate in burst mode. A burst-mode receiver is required only in the OLT. The ONUs study a continuous bit stream (data or idle bits) sent by the OLT and do not need to readjust quickly.

Another approach has ONUs to adjust their transmitter powers where power levels were received by OLT from all ONUs. This method is critical particularly for transceiver design as it makes the ONU hardware more complex, requiring a special signaling protocol for feedback from the OLT to each ONU.

9.5 EPON ACCESS ARCHITECTURE

EPON transmits data traffic encapsulated in Ethernet frames (IEEE 802.3 standard using 8 bit/10 bit line coding), operating at standard Ethernet data rates. Ethernet is used due to low-cost line cards and widely used in LANs [20,23]. Since access networks are focused towards end users and LANs, high-speed Gigabit Ethernet deployment is used. The 10-Gigabit Ethernet products are easily available. Ethernet provides a much more efficient MAC protocol in comparison to ATM imposing a considerable overhead on variable-length IP packets.

9.5.1 OPERATION OF EPON

In the downstream (OLT to ONUs), Ethernet frames are transmitted by the OLT pass via a 1:N passive splitter and reach each ONU. Typical values of N are between 8 and 32. Packets are broadcast by the OLT and extracted by their destination ONU based on a Logical Link Identifier (LLID), which the ONU is assigned when it registers with the network. Figure 9.13b shows the downstream traffic in EPON [23].

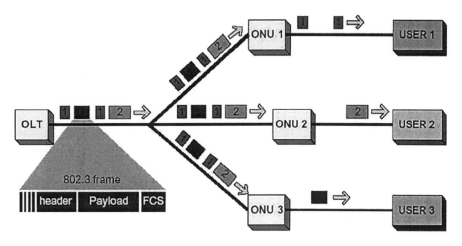

FIGURE 9.13 (CONTINUED)   (B) Downstream operation in EPON [26].

(Continued)

In the upstream, data frames originate from an ONU transmitting to the OLT and never send to any other ONU due to the directional properties of a passive optical combiner. The function of EPON is similar to that of a PtP architecture. However, in EPON, data frames from different ONUs transmitted simultaneously may make a collision. Thus, in the upstream direction a contention-based media-access mechanism is required (similar to CSMA/CD), and it is not easy to implement because ONUs are not able to detect a collision in the fiber from the combiner to the OLT due to directional properties of the combiner. An OLT can find a collision and inform ONUs by sending a jam signal; however, the relatively large propagation delay in a PON (where the typical distance from the OLT to ONU is 20 km) greatly reduces the efficiency of such a scheme. Figure 9.13c illustrates the operation of upstream, time-shared, data flow in an EPON [26].

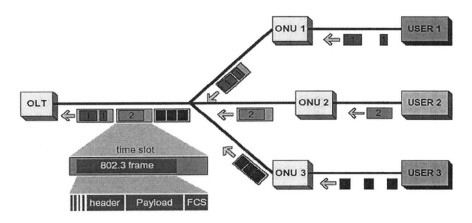

FIGURE 9.13 (CONTINUED)   (C) Upstream operation in EPON [26].

Each ONU is assigned a time slot in which all ONUs are synchronized to transmit each time slot capable of carrying several Ethernet frames. When its time slot come as per term wise, the ONU would burst all stored frames at full channel speed. If there are no frames in the buffer to fill the entire time slot, an idle pattern is transmitted. The time-slot assignment is a very crucial step. The possible time-slot allocation schemes could range from a static allocation (fixed time-division multiple access (TDMA)) to a dynamically adapting scheme based on instantaneous queue size in every ONU (statistical multiplexing scheme). In the dynamic scheme, the OLT can play the role of collecting the queue sizes from the ONUs and then issuing time slots. Although this approach leads to a significant signaling overhead between the OLT and ONUs, the centralized intelligence may lead to more efficient use of bandwidth. More advanced bandwidth-allocation schemes are also possible, including schemes utilizing notions of traffic priority [27].

9.6 MULTI-POINT CONTROL PROTOCOL (MPCP)

The multi-point control protocol (MPCP) provides a dynamic time-slot allocation scheme [26]. The MPCP is standardized in the IEEE 802.3ah. The MPCP has a signaling protocol between the OLT and ONUs but does not have any bandwidth provisioning scheme. The MPCP has three functions.

1. **Discovery processing:** An ONU is registered in the network while compensating for the round-trip time (RTT).
2. **Report handling:** ONUs generate REPORT messages through which bandwidth requirements and messages are sent to the OLT. The OLT needs to process the REPORT messages so that it can make bandwidth assignments.
3. **Gate handling:** Gate messages are takaen by the OLT to assign a time slot at which the ONU can start transmitting data. Time slots are estimated at the OLT while making bandwidth allocation.

9.6.1 DISCOVERY PROCESSING

Discovery is the technique where newly connected or offline ONUs register in the network. The steps are shown in Figure 9.14.

1. **OLT:** The OLT periodically is taken as a discovery time window during which the offline ONUs provide the opportunity to register with the OLT. A DISCOVERY-GATE message having the starting and ending time of the discovery window is transmitted to all ONUs.
2. **ONU:** Any offline ONU for registering, waits for a random amount of time within the discovery window, and then transmits a REGISTERREQ message. The REGISTERREQ message has the ONU's MAC address. The random delay is required to reduce the probability of REGISTERREQ messages transmitted by multiple ONUs from colliding.

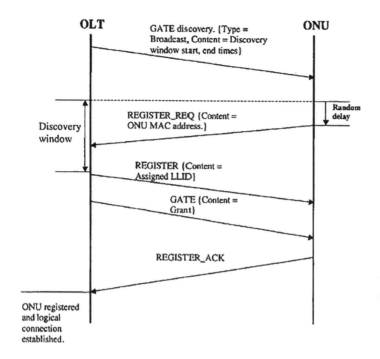

FIGURE 9.14   Timing diagram of discovery-phase message exchanges [26].

3. **OLT:** The OLT, after receiving a valid REGISTERREQ message, registers the ONU and allocates it to an LLID. The OLT now sends a REGISTER message to the newly discovered ONU that has the ONU's LLID.

4. **OLT:** The OLT now sends a standard GATE message, representing a time slot to transmit data.

5. **ONU:** Upon receiving the GATE message, the ONU sends with a REGISTERACK message in the assigned time slot. Upon receipt of the REGISTERACK, the discovery process is complete and the normal operation is started now.

At each step, timeouts are monitored at the ONU and OLT. If an expected message does not arrive before a timeout, then the OLT transmits a DEREGISTER message making the ONU register again.

9.6.2   REPORT HANDLING

REPORT messages are sent by ONUs in their assigned transmission windows along with data frames as shown in Figure 9.15. A REPORT message is made in the MAC control client layer and is time-stamped in the MAC controller, as shown in Figure 9.15.

Typically, REPORT message is of the desired size of the next time slot based on ONU's queue size. REPORT messages are made periodically, even when no request for bandwidth is being made. This opposes the OLT from deregistering the ONU. Thus, for the proper operation of this mechanism, the OLT must allow the ONU for

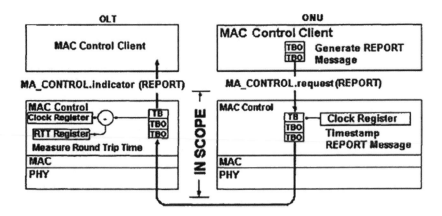

FIGURE 9.15   Multi-point control protocol REPORT operation [26].

a transmission of window periodically. At the OLT, the REPORT is processed, and the data are used for the next round of bandwidth assignments [26].

9.6.3   GATE HANDLING

The transmitting window of an ONU is expressed in the GATE message from the OLT. Upon receiving a GATE message matching the ONU's LLID, the ONU will program its local registers with the transmission start and transmission length times shown in Figure 9.16. When the time at the local clock of the ONU comes to the transmission start, the ONU begins transmitting data [26].

9.6.4   CLOCK SYNCHRONIZATION

The correct operation of MPCP is based on clock synchronization between the OLT and the ONU for compensation of the RTT that differs for each ONU to ONU as these are located at different distances from the OLT. Clock synchronization compensates

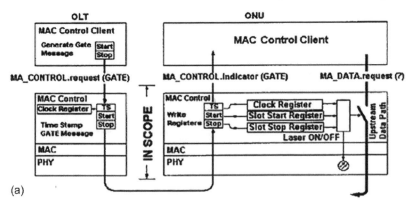

(a)

FIGURE 9.16   (a) Multi-point control protocol GATE operation [26] and (b) Message exchange process of REPORT and GATE handling [26].

(Continued)

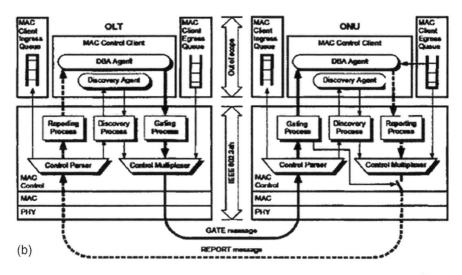

FIGURE 9.16 (CONTINUED) (a) Multi-point control protocol GATE operation [26] and (b) Message exchange process of REPORT and GATE handling [26].

for RTT is required because the OLT does not keep track of the different RTTs of different ONUs, when it transmits time slots in GATE messages. Figure 9.16 shows the message exchange process of REPORT and GATE handling.

If the differences in values go beyond a predefined threshold, the ONU **loses** its synchronization and will switch itself into offline mode. The ONU then attempts to register again using the next discovery process. The ONU also checks that the time (according to the local clock at the ONU) for arrival of the GATE message is close to the time-stamp value.

If the ONU receives an MPCP message, it lays down its local time from the time-stamp of that message, and as soon as the OLT gets an MPCP message, it determines the RTT as the difference between its local time and time-stamp of the message. Any significant change in RTT implies that the OLT and the ONU clocks are not synchronous any more, and the OLT now transmits a DEREGISTER message for the same ONU, which will then attempt to register in the network again through the discovery process.

9.7 DYNAMIC BANDWIDTH ALLOCATION (DBA) ALGORITHMS IN EPON

EPON is applied for broadcast in the downstream direction where packets are broadcast by the OLT and received by their destination ONUs via the MAC address. In the upstream direction the ONUs allocate the channel capacity and resources. The OLT assigns time slots through which the ONUs are allowed to send data. The allotment of static time slots for each ONU is obtained by using synchronous TDMA. Since the OLT has to poll the ONUs no longer and schedule the time slots, avoiding the need for REPORT messages in the MPCP protocol altogether, it becomes cost effective. But the TDMA approach lacks statistical multiplexing. Moreover, the network traffic causes the flow of some time slots even under very light load. This makes packets being delayed for several time-slot periods, while a large number of slots are unused. So there is a need for Dynamic Bandwidth Allocation (DBA) algorithms, in which the OLT schedules the time slots in which the ONUs send the time slots, is to achieve statistical multiplexing. There is Interleaved Polling with Adaptive Cycle Time (IPACT) which is used in this direction.

9.7.1   IPACT

The IPACT protocol is given below:

1. The OLT tracks the earliest scheduling time by a variable T_{schedule}. T_{schedule} altered after each allocated time slot.
2. Whenever a REPORT message having the requested time slot arrives at the OLT, the DBA agent at the OLT estimates the start time for the next transmission time slot for that ONU. To maintain high utilization in the upstream channel, the DBA agent assigns the next time slot immediately adjacent to the already allocated time slot with only a guard time interval separation. The start time is written as

$$T_{\text{start}} = T_{\text{schedule}} + T_{\text{guard}}$$

The DBA agent also ensures that the ONU has enough time to process the GATE message before the granted time slot is scheduled to start the local time, and $T_{\text{processing}}$ denotes the processing time. The start time T_{start} is then written as

$$T_{\text{start}} = T_{\text{local}} + T_{\text{processing}}$$

3. Maximum scheduling time slot: If the OLT permits to each ONU to send its entire buffer contents in one transmission, the ONUs with high data volume controls the entire bandwidth, and subsequently the average delay in the network is very large. To avoid this situation, the OLT restricts the maximum transmission size. This is described as a limited-service scheme, in which every ONU is assigned a time slot to transmit as many bytes as per its request, but no more than some upper limit, i.e., the Maximum Scheduling Time Slot. There are various schemes for specifying the limit based on a Service-Level Agreement (SLA) for each ONU, or dynamic based on network conditions.

The constant time slot is used to transmit a REPORT message be T_{REPORT}, and maxLength denotes the Maximum Scheduling Time Slot. Ignoring other overheads, such as light source on/off times and synchronization times, the length of the time slot under the limited service scheme is formulated as

$$t_{\text{length}} = \text{REPORT } t_{\text{length}} + T_{\text{REPORT}}, \text{ if } t_{\text{length}} \geq \text{maxLength}$$

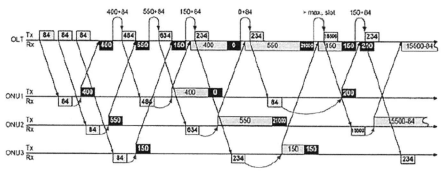

FIGURE 9.17   Timing diagram of the limited-service scheme in IPACT [26].

4. Once the above are estimated, the corresponding GATE message is transmitted by the OLT. The T_{schedule} is modified as

$$T_{\text{schdule}} = T_{\text{short}} + t_{\text{length}}$$

Figure 9.17 shows a timing diagram for the limited-service scheme in IPACT simplicity consisting of only three ONUs. Upon completion of the discovery procedure
in MPCP, the OLT sends individual GATE messages to all the ONUs. The ONUs
send their REPORT messages in their respective time slots. When the REPORT
from an ONU is reached to OLT, the OLT starts scheduling time slots. We consider
that when ONU_2 requests a time slot of 21,000 bytes, the *Maximum Scheduling Time
Slot* of 15,500 bytes is granted. If the ONU vacates its buffer completely, it shall
report 0 bytes to the OLT. Correspondingly, in the next cycle, the ONU assigns only
a small time slot to send a REPORT message (which is 84 bytes). The upstream
channel is almost 100% utilized (barring REPORT messages and guard times). Idle
ONUs are allocated very small time slots. Hence, if the system is under low load,
then ONUs will be polled frequently [26].

9.7.2   Services

There are few service disciplines for IPACT [26]. The size of the buffer (requested
time-slot size) at the ONU is known by the OLT with a REPORT message as considered in the MPCP [26]. The DBA agent at the OLT selects the size of the next time
slot allotted to the ONU based on the following service disciplines. A key metric is
the *cycle time*, which is the time duration between two successive time slots allotted
to the same ONU. The following are the six services [26]:

1. **Fixed service:** discards the requested time-slot size and always allows a
   fixed time slot, thus corresponding to synchronous TDMA. It has a constant
   cycle time.

2. **Limited service:** allows the requested time-slot size not more than the Maximum Scheduling Time Slot WMAX. It is the most conservative scheme and has the shortest cycle time among all the schemes.

3. **Gated service:** is limited to the Maximum Scheduling Time Slot which uses for limited service does.

4. **Constant-credit service:** connects constant credit to the requested time-slot size. The idea behind adding the credit is that x bytes arrived at an ONU between the time the ONU sent a REPORT and the time it received the corresponding GATE message with its time-slot assignment. If the granted window size equals the requested window (i.e., it has a credit of size x), these x bytes will not wait for the next GRANT to arrive.

5. **Linear-credit service** has a similar approach where the size of the credit is proportional to the requested window. Network traffic has a certain degree of predictability, i.e., specifically, for a long burst of data, this burst is likely to continue for some time into the future.

6. **Elastic service**: tries to avoid a fixed *Maximum Scheduling Time Slot* WMAX. The maximum window is granted in such a way that the accumulated size of last N Grants (including the one being granted) does not exceed $N \times$ WMAX bytes, where N is the number of ONUs).

Derivatives of IPACT include supporting bandwidth guarantees for high-priority traffic [28], deterministic effective bandwidth (DEB) admission control [20] and many others. A good survey of DBAs reported in the literature may be found in [23].

9.8   IP-BASED SERVICES OVER EPON

One of the critical issues in Ethernet into the subscriber access area is Ethernet's efficiency for delivering IP packets. Data and telecom network convergence direct to more and more telecommunication services migrating to variable-length-packet-based data networks.

Being designed with an IP layer, EPON effortlessly operates with IP-based traffic flows, similar to any switched Ethernet network. The speciality with the typical switched architecture is that, in an EPON, the user's throughput is slotted (gated), i.e., packets are not transmitted by an ONU at any time. This feature causes two issues unique to EPONs: (a) potential slot under-utilization problem due to variable-length packets and (b) slot scheduling to support real-time and controlled-load traffic classes.

9.8.1   SLOT-UTILIZATION PROBLEM

The slot-utilization problem is mainly associated with splitting of Ethernet frames, and variable-length packets cannot fill up a given slot completely. This problem represents in a fixed service when slots of constant size are given to an ONU regardless of its queue occupancy. Slots are not full to capacity also in the case when the OLT grants to an ONU a slot lesser than what the ONU requested based on its queue size.

Slot utilization is improved by using smarter packet scheduling (e.g., the bin-packing problem) [26]. Rather than stopping the transmission when the head-of-the-queue frame exceeds the remainder of the slot, the algorithm finds the buffer to pick a smaller packet for immediate transmission by using first-fit scheduling. However, first-fit scheduling is not a good approach. To understand the problem, it requires to see the effects of packet reordering from the perspective of Transmission Control Protocol/IP (TCP/IP) payloads carried by Ethernet frames. TCP will restore the proper sequence of packets and an excessive reordering requires the following consequences [26]:

1. According to the fast retransmission protocol, the TCP receiver sends an immediate acknowledgment (ACK) for any out-of-order packet, whereas for in-order packets, it makes a cumulative ACK (typically for every other packet).
2. Packet reordering in ONUS causes the fact that n later packets are being transmitted before an earlier packet. It generates n ACKs ($n-1$ duplicate ACKs) for the earlier packet. If n exceeds a predefined threshold, it will trigger packet retransmission and reduction of TCP's congestion window size. Currently, the threshold value in most TCP/IP protocol stacks is set to 3.

Even if special care is taken at the ONU to restrict out-of-order packets to only 1 or 2, the rest of the end-to-end path may contribute additional reordering. While true reordering typically generates <3 duplicate ACKs and is ignored by the TCP sender, together with reordering introduced by the ONU, the number of duplicate ACKs may exceed 3, thus forcing the sender to retransmit a packet. As a result, the overall throughput of the user's data may decrease.

For the solution of the above problem, it is reasonable to assume that the traffic entering the ONU is an aggregate of multiple flows. In the case of business users, it would be the aggregated flows from multiple workstations. In the case of a residential network, we may still expect multiple connections at the same time. This is because, as converged access network, PON will carry not only data but also voice-over-IP (VoIP) and video traffic. Also, home appliances are becoming as network plug-and-play devices. The conclusion is that, if we have multiple connections, we can reorder packets that belong to different connections, and never reorder them if they belong to the same connection. Connections can be distinguished by examining the source/destination address pairs and source/destination port numbers. An ONU finds layer-3 and layer-4 information in the packets. Thus, the important trade-off is made whether it considerably increases the required processing power in an ONU to improve the bandwidth utilization or not.

9.8.2 CIRCUIT EMULATION (TDM OVER IP)

The TDM circuit-switched networks switch to IP packet-switched networks rapidly. The next-generation access network is used and optimized for IP data traffic, RF set-top boxes, analog TV sets, TDM private branch exchanges (PBXs), etc. and legacy services (T_1/E_1, Integrated Services Digital Network (ISDN), Plain Old Telephone Service (POTS), etc.) are used in the near future [26].

The issue for implementing a circuit-over-packet emulation scheme is mostly associated to clock distribution. In one scheme, users provide a clock to their respective

ONUs, which in turn is delivered to the OLT. But, since the ONUs cannot send all the time, the clock information must be carried with packets. The OLT will restore the clock using this information for recovering data. The OLT should have a clock master for all downstream ONU devices. The ONU will be able to recover the clock from its receive channel, use it in its transmit channel and distribute it to all legacy devices connected to it.

9.8.3 REAL-TIME VIDEO AND VoIP

Performance of a packet-based network are analyzed by bandwidth, packet delay (latency) and delay variation (jitter), and packet-loss ratio. Quality of Service (QoS) requires the bounds on some or all of these parameters [26]. Further characterization of statistical QoS is required from guaranteed QoS. The guaranteed QoS refers to a network architecture parameters to stay within the specified bounds for the entire duration of a connection. A network is required to provide QoS (i.e., bounds on performance parameters) to obtain proper operation of real-time services – video-over-packets (digital video conference, VoD), VoIP, real-time transactions, etc. QoS for higher-layer services must be retained in all traversed network segments, including the access network portion of the end-to-end path.

The CSMA/CD MAC protocol may not give surety of QoS. All connections (traffic flows) were treated equally and were given best-effort service from the network. The first major step in allowing QoS in Ethernet was an introduction of the full duplex mode. Full duplex MAC can transmit data frames at any time, and this eliminated non-deterministic delay in accessing the medium. In a full duplex link (segment), once a packet is provided to a transmitting MAC layer, its delay, jitter and loss probability are known or predictable all the way to the receiving MAC layer. Delay and jitter may be exaggerated by head-of-line blocking when the MAC port is busy transmitting the previous frame at the time when the next one arrives.

The enabling QoS in Ethernet has two extensions – (a) 802.1~ Supplement to MAC Bridges: Traffic Class Expediting and Dynamic Multicast Filtering (later merged with 802.1D) and (b) 802.1Q Virtual Bridged LANs. 802.1Q having a frame-format extension allowing Ethernet frames to carry priority information. The standard distinguishes the following traffic classes [26]:

1. **Network control:** characterized by a must-get-there requirement to maintain and support the network infrastructure.
2. **Voice:** characterized by <10-ms delay, and hence maximum jitter (one-way transmission through the LAN infrastructure of a single campus).
3. **Video:** characterized by <100-ms delay.
4. **Controlled load:** important business applications subject to some form of admission control, be that pre-planning of the network requirement at one extreme to bandwidth reservation per flow at the time the flow started at the other.
5. **CEO's best effort (BE):** the best-effort-type services that an information services organization would deliver to its most important customers.
6. **BE:** LAN traffic.

7. **Background:** bulk transfers and other activities are permitted on the network but that should not impact the use of the network by other users and applications.

Without admission control, each priority class may intermittently degrade to best-effort performance. Here, EPON has a simple and robust method to perform admission control. MPCP relies on GATE messages sent from the OLT to ONUs to allot the transmission window. A very simple protocol modification may allow a single GATE message to obtain multiple windows, one for each priority class. The REPORT message is extended to report queue states for each priority class. The admission control has higher-layer intelligence in ONUs. When the next transmission window reaches, it will plan packets for transmission of different classes of service (CoS) accordingly.

9.8.4 Performance of CoS-Aware EPON

The priority queuing provides a delay-bound service. In simulation setup data arriving from the user has three priority classes. The queues are then serviced in order of their priority [29].

BE: class is given the lowest priority. This priority level is used for non-real-time data transfers. There is no delivery or delay guarantees in this service. The BE queue in the ONU is served only if higher-priority queues are empty.

Assured Forwarding (**AF**): class is given higher priority than the BE class. The AF queue is served before the BE queue. The AF traffic consisted of a VBR stream with an average bit rate of 16 Mbps – three simultaneous MPEG-2-coded video streams. The AF traffic is also highly bursty (LRD).

Guaranteed Forwarding (**GF**): is given the highest priority to emulate a T_1 line in the packet-based access network. The GF can dislodge the BE and AF data from their queues if there is not enough buffer space to store the GF packet. The GF queue is served before the AF and BE queues. The T_1 data arriving from the user is packetized in the ONU by placing 24 bytes of data in a packet.

We can observe that the BE traffic suffers the most when the ambient load increases. The simulation results showed that some packets were discarded when the network load exceeded 80%. The AF data also experienced an increased delay, but no packet losses were observed. The increased delay in the AF traffic can be attributed to long bursts of data. The GF data experiences a very slight increase in both average and maximum delays. This is due to the fact that the packets were generated with a constant rate, i.e., no data bursts. The maximum delay is equal to the maximum observed cycle time [26].

9.8.5 Light-Load Penalty

The default priority queuing with a simple polling mechanism in an EPON shows an improvement. As the load decreases from moderate (0.25) to very light (0.05), the average delay for the lowest priority class (P_2) increases significantly. The average packet delay at a load of 0.05 corresponding to a bit rate of 5 Mbps is 17.8 ms, and a load of more than 1200% provides higher than 1.4-ms delay at a load of 0.25 (25 Mbps). This is called as *light-load penalty,* where *at* the end of every time slot, an ONU creates a REPORT message having the number of bytes that remain in the queue. Newly arrived packets may have higher priority than some packets already stored in the queue, and are sent in the next transmission slot before the lower-priority packets. Since these new packets were not reported to the OLT before, the given slot cannot accommodate all the stored packets. This causes some lower-priority packets to be left in the queue. This situation may repeat over many cycles, causing some lower-priority packets to be delayed for multiple cycles. A lower-priority packet will finally be transmitted when more lower-priority packets (bytes) accumulate (and are reported to the OLT) behind a given packet than higher-priority packets cut in front of it. As the load increases, the queue behind a lower-priority packet grows faster and the light-load penalty decreases. At a load of 0.25, the average delay for P_2 packets is only about 4.6 cycles.

One way to eliminate the light-load penalty is to implement a two-stage queue in an ONU [29]. In a two-stage system, stage-1 consists of multiple-priority queues and stage-2 consists of one First-Come First-Serve (FCFS) queue. When a time slot arrives, data packets from stage-2 are sent to the OLT, thereby vacating the queue; simultaneously, data packets from stage-1 are advanced into vacant spaces in the stage-2 queue. At the end of the current time slot, the ONU informs to the OLT to have a corresponding slot size in the next cycle.

9.9 OTHER TYPES OF PONS

Besides EPON, there are other technologies for PON – APON, GFP-PON and WDM-PON.

9.9.1 APON

APON is based on Asynchronous Transfer Mode using the MAC layer Protocol [26]. Figure 9.18 shows downstream frames having 56 ATM cells (53 bytes each) for the basic rate of 155 Mbps, scaling up to 224 cells for 622 Mbps. It has two Physical Layer Operation, Administration, and Maintenance (PLOAM) cells, starting one at the beginning of the frame and one in the middle. The remaining 54 cells are data ATM cells. The upstream transmission in Figure 9.19 is made in the form of bursts of ATM cells, with a 3-byte physical overhead appended to each 53-byte ATM cell to permit for burst-mode receivers. Burst-mode receivers are used at the OLT to synchro-nize the different ONUs situated at different distances from the OLT; and, hence, the

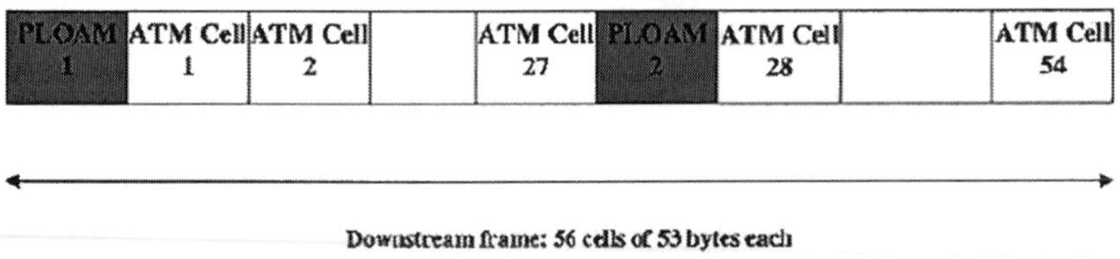

PLOAM 1	ATM Cell 1	ATM Cell 2		ATM Cell 27	PLOAM 2	ATM Cell 28		ATM Cell 54

Downstream frame: 56 cells of 53 bytes each

FIGURE 9.18   Downstream frame format.

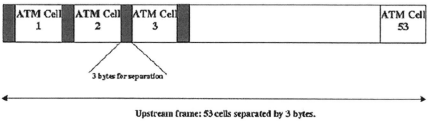

Upstream frame format.

FIGURE 9.19   Upstream frame formats of APON.

power received at the OLT from different ONUs is different. The ATM cell is either an ATM data cell or a PLOAM cell. In downstream direction, the PLOAM cells carries grants from the OLT to the ONU. Each grant becomes one-time permission for the ONU to transmit payload data in an ATM cell. The 53 grants for the 53 upstream frame cells are used for PLOAM cells. During its operation, the OLT transmits a continuous stream of grants to all the ONUs in the PON. Thus, the OLT has the portion of upstream bandwidth assigned to each ONU. In the upstream direction, the PLOAM cells are taken by the ONUs for transmission of these queue sizes to the OLT.

Initially, APON supports only ATM-based services, and it is also known as BPON. BPON is also represented by the Internal Telecommunication Union (ITU) specification G.983.1. BPON provides overlay capabilities for services such as video and Ethernet traffic [26].

9.9.2   GFP-PON

The GFP-PON is also represented by the ITU in specification G.984.x. It gives bit rates of up to 2.5 Gbps made by using hierarchical time division multiplexer STS-48. It provides higher efficiency while carrying multiple services over the PON. It reports a protocol based on GFP [30]. GFP uses a generic mechanism to adapt traffic from higher layers (Ethernet MACIIP), over a transport layer such as Synchronous Optical Network/Synchronous Digital Hierarchy (SONET/SDH). Other functionalities such as dynamic bandwidth assignment, operation and maintenance are made by using APON.

But both APON and GFP-PON have the disadvantage of a complex protocol and implementation, relative to EPON [30].

9.9.3   WDM-PON

The WDM in PON is already reported in [31,32]. In this section we discuss WDM-PON.

9.9.3.1   Need for WDM in PONs

The basic form of PON uses only a single optical channel, and the available bandwidth is restricted to a maximum bit rate of 1 Gbps at optical transceivers. The attenuation causes due to having splitting restriction and the maximum number of ONUs to 64 kbps. This restricts the network's scalability. The installation cost of laying fiber in the access network is high but for maintaining a high bandwidth it is required.

Many telecom vendors use PONs considering an FTTx model to accommodate converged IP video, voice and data services (called as "triple-play"). Although PON offers a higher bandwidth than traditional copper-based access networks, there is a need for further increasing the bandwidth of the PON by employing WDM so that multiple wavelengths are accommodated in either or both upstream and downstream directions. Such a PON is known as a WDM-PON. In WDM-PON, each ONU has its own unique dedicated wavelength. However, we can consider the mode of operation of WDM-PONs, in which multiple ONUs make sharing of a wavelength, or an ONU uses multiple wavelengths for different ONUs as destinations/sink different traffic. A WDM-PON is a PtP access network (as opposed to point-to-multipoint in PON) having a separate wavelength between the OLT and each ONU.

9.9.3.2 Arrayed Waveguide Grating (AWG)-Based WDM-PON

Wavelength routing in a WDM-PON is realized through an Arrayed Waveguide Grating (AWG). The AWG is a passive device providing a fixed routing of an optical signal from a given input port to a given output port, based on the wavelength of the signal. Signals of different wavelengths coming into an input port will be routed to a different output port. Similarly, different signals of the same wavelength are transmitted to different input ports, and also routed to different output ports, as shown in Figure 9.20. Each wavelength is routed by a passive AWG (AWG). The AWG is discussed briefly in Chapter 2. In a WDM-PON, different ONUs can be supported at different bit rates. Each ONU is operated at the full bit rate of a wavelength channel. Moreover, unlike the basic PON, the WDM-PON has no power-splitting losses. Use of individual wavelengths for each ONU also facilitates privacy and reduces security concerns which the PON has. The periodic routing pattern of an AWG provides scalable WDM-PON. WDM-PONs uses WDM for upgrading to the PON in the ITU-T G.983 [32].

In the figure there is a broad-spectrum optical source entering the input port x. For the optical signals entering port x and routed to a given output port y, the AWG

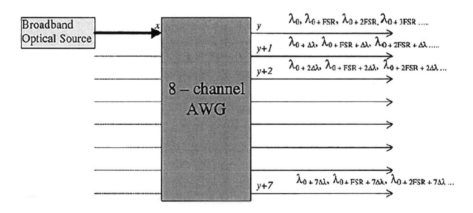

FIGURE 9.20 AWG-based WDM-PON [26].

finds the path of the wavelengths separated by a fixed wavelength interval called the free spectral range (FSR). Therefore, using a base wavelength λ_0, the output wavelengths at port y are λ_0 λ_{0+FSR}, λ_{0+2FSR} and so on. At output port, $y + 1$ number of wavelengths are $\lambda_{0+\Delta\lambda}$, $\lambda_{0+\Delta\lambda+FSR}$ and so on. This periodic routing property of the AWG helps immensely in scaling the network.

9.9.3.3 WDM-PON Architectures

Most WDM-PON architectures use a separate wavelength channel for each ONU in the downstream direction (from OLT to ONU). However, the various architectures differ in the amount of resources used in the upstream direction (from ONU to OLT). Upstream communication differs from downstream communication due to two main reasons. The ONU equipment (transmitters) must be inexpensive for the use of ONUs largely in the network. The wavelength-specific equipment is preferred at the ONU. The WDM-PON architecture uses WDM in the 1550-nm band in downstream and a single upstream wavelength in the 1300-nm band shared through TDMA. The upstream and downstream transmissions are accommodated on a single fiber through coarse WDM (CWDM). This architecture is referred as Composite PON (CPON). A single-wavelength burst-mode receiver is used at the OLT to receive the upstream signal. A burst-mode receiver is required to synchronize to the clock signals of different transmitting ONUs, which are placed at different distances from the OLT. Figure 9.21 shows the layout of a CPON [31].

CPON architecture has limitations that a single-frequency laser (DFB laser) at the ONU is expensive. The control of the wavelength changes is difficult due to temperature fluctuations at the remote (ONU) end.

The Local Access Remote Network (LARNET) architecture [33] deals with the above limitations by using a broad-spectrum source at the ONU, such as an inexpensive edge-emitting LED having slicing of spectrum by the AWG-based router in the upstream direction. When a broad-spectrum source is used into one input port of the AWG, the various constituent wavelengths are directed to different output ports. Figure 9.22 shows the architecture of LARNET [33] in which the devices are commercially available for its operation. The limitation is that spectrally slicing a

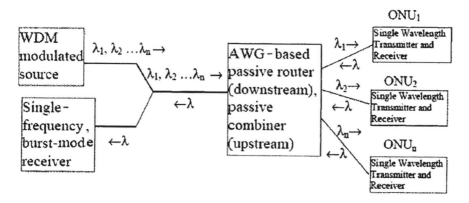

FIGURE 9.21 CPON architecture.

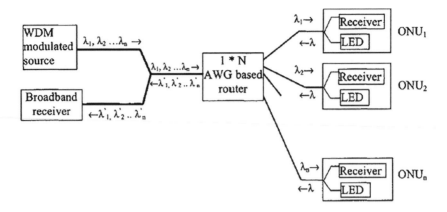

FIGURE 9.22   LARNET architecture.

broad-spectrum source by an AWG leads to a very high power loss. Therefore, the distance from the OLT to the ONU is considerably reduced in LARNET. Recently in the LARNET architecture, the upstream signal from an ONU can be looped back downstream to all other ONUs from the AWG, through appropriate wiring at the AWG [34] based on the periodic routing property of the AWG. Since the propagation delay from the ONU to the AWG is very small, a CSMA/CD MAC protocol can be used for contention resolution of upstream traffic [34].

The Remote Integrated Terminal Network (RITENET) architecture [35] avoids the transmitter at the ONU by modulating the downstream signal from the OLT and sending it back. The signal from the OLT is shared downstream and upstream through time-sharing. A frame is split into two parts – one used for downstream transmission and the other for upstream transmission. Figure 9.23 shows the architecture of RITENET which reduces the end-terminal cost at the ONU, and the distance between the OLT and the ONU is also much less, as the signal at the OLT

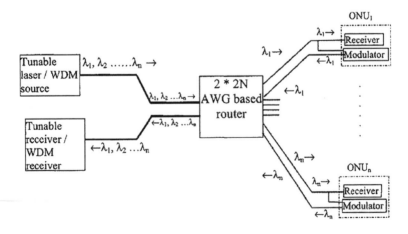

FIGURE 9.23   RITENET architecture [26].

needs to travel double the distance. Also, since the signal is now shared between the two ends, the bit rate of the PON is also doubled. Moreover the number of fibers employed is also doubled, which doubles the cost of deployment and maintenance. A WDM receiver is used at the OLT (unlike LARNET or CPON), which increases the cost of the OLT terminal equipment.

All of the above architectures use a single multi-wavelength laser source at the OLT. Commercial products generating a multi-wavelength optical spectrum is composed of many stable individual optical frequencies locked into the standard ITU grid. Figure 9.24 shows that the multi-wavelength source is modulated with independent modulators. A multi-wavelength laser source provides greater wavelength stability in the network when compared with the use of numerous DFB lasers in an array, because the single laser source can be easily and efficiently controlled for temperature variations. AWG-based routers have been used by many WDM-PON architectures. Integrated-optic technology has been used for the implementation of AWG and the number of channels that supported has also scaled very well. Outdoor temperatures may vary between −40°C and 45°C. Temperature variations may cause the pass bands at which the AWG operates to drift.

9.9.3.4 Scalability of WDM-PON

Any architecture should be scalable in order to use it. For an access network, scalability is needed to represent in terms of bandwidth and the number of end-user access points (ONUs). The scalability is obtained without much deployment overhead, and there should be reuse of AWG devices because of the high cost when scalability is desired. Since all end users can be upgraded to higher bandwidths at the same time, it should be ensured that more users can be supported while scaling the network. The combination of all the above makes scalability in WDM-PON architectures "a challenging issue". Figure 9.25 shows the architecture where additional AWGs are deployed to scale from a and wavelength, 8-ONU WDM-PON architecture of a 32-wavelength 32-ONU WDM-PON architecture. The subscript of the wavelength indicates the wavelength number while the superscript represents the source. The inheritance ONUs – ONU_I – continued to use wavelengths λ_1^1 to λ_4^1 and λ_1^2 to λ_4^2. The legacy 2×8 channel AWG and 8 new 1×4 channel AWGs are employed to scale the network. This architecture uses wavelength reuse.

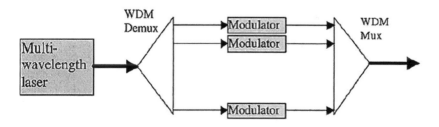

FIGURE 9.24 Modulating a multi-wavelength laser source.

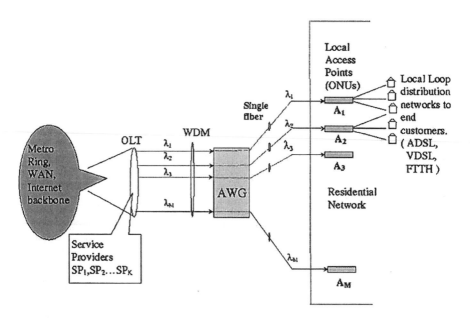

FIGURE 9.25 A WDM-PON-based FTTC network deployed as an open access network. SPs use this infrastructure to serve end users by leasing bandwidth from the access network.

9.9.4 DEPLOYMENT MODEL OF WDM-PONS

Various models have been reported for the deployment of fiber in the access network. FTTH, FTTC and FTTB [26]. In FTTC/FTTB, the ONUs are placed at the curb or in a building and serve as distribution points of bandwidth to end users. The FTTC network is a single broadband access infrastructure through which different service providers (SPs) show many services to the end user. Figure 9.26 shows a WDM-PON-based FTTC network used for an open access network. The OLT is connected to the broader Internet via a metro ring network, a wide-area LAN or a long-haul optical network. The ONUs are used as Local Access Points (LAPS) which acted **as** distribution centers for bandwidth to end users. The access network needs to be shared because it is crucial for every SP to position its own access network because of installation and operational costs. An access network is called as an *open access network* [36].

9.9.4.1 Open Access

In a broadband access network, the term "open access" means accessibility of multiple SPs to share the deployed access-network infrastructure to make services available to the end users. Multiple services are carried over a shared access channel. A PON uses a model of open access shown in Figure 9.27.

The access network has a characteristic that a single channel is shared over multiple users. An access network provides the services to the independent, non-cooperating and bandwidth-competing users to assure some minimal degree of performance for each user. A bandwidth required that the user is not affected with the

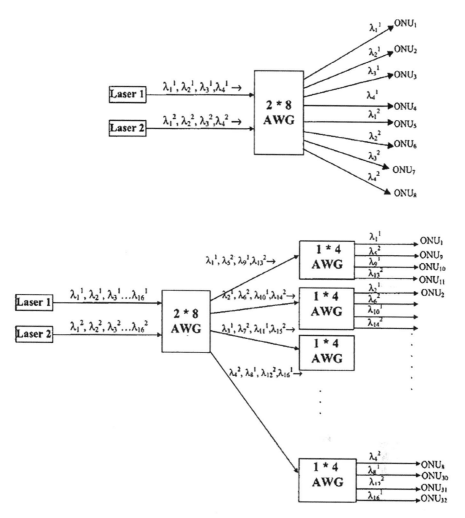

FIGURE 9.26   Scalability of WDM-PON by using AWG.

performance of other users in the network. Similarly, if two competing SPs provide the same category of services, such as IP video, the bandwidth access is required to be non-discriminatory. Most services are available on demand, and different services are obtained from different SPs at different intervals of time.

Moreover, traffic in an access network is multiplexed and several independent self-similar streams are very much burst traffic. The network has very heavy load for brief durations of time. In the best-effort traffic model, current access networks have very poor performance under heavy load.

In an MAC layer, Dual-SLA based scheduling algorithm is to use fairness in an *open access*. Dual-SLAs need two independent sets of SLAs – one for the SPs and the other for the users. The scheduling algorithm supports both these sets of SLAs simultaneously. The above scheme has the *primary* SLA that

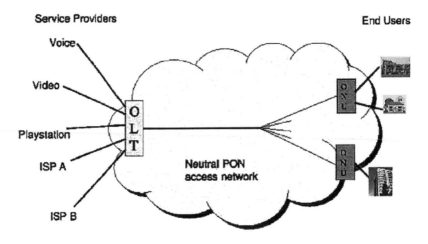

FIGURE 9.27   A PON deploying *open access.*

convenes at all instances of time and the *secondary* SLA that convenes as much as possible. The objective is to be fair with respect to the secondary SLA over a large time span.

SUMMARY

A wide variety of schemes for access to the network have been analyzed and implemented by numerous researchers. First, classic random-access scheme is ALOHA, which was used in ground-based radio networks. The broadcast concept of ALOHA is applicable to the local networking environment, and as such it is often compared to other random-access broadcast schemes used in LANs. The evolution of random-access schemes eventually led to the CSMA/CD method, which is also known as LWT. This is the basis of the popular Ethernet protocol. The controlled access to a LAN can be carried out either in a centralized or distributed fashion. The most common centralized-control method is known as polling. Distributed-control access methods are commonly referred to as token-controlled and can be employed in ring or bus topologies. Tokens are special bit patterns that circulate from node to node when there is no message traffic. When a station wants to send data, it removes the token from the line, sends its data, and then puts the token back into circulation so that the next station may gain access to the transmission medium. Two key performance considerations for a LAN are the throughput or channel utilization S and the delay characteristics.

In this chapter, the EPON is discussed. We have described the MPCP standardized for EPON. The DBA algorithms are used for EPON and the IPACT protocol. Several commercial initiatives on EPON are required as soon as standardization efforts are completed. Then, the APON (also called BPON) and the GFP-PON are also described. The EPON and WDM-PON promise for fulfilling the future end-user broadband requirements.

EXERCISES

9.1 Consider a slotted ALOHA network with 100 stations. Let $p = 0.002$ and $\alpha = 0.05$.

    a. Plot the equilibrium-throughput curve and the load line to show that the network is stable.

    b. Suppose we increase the number of stations to 200 and reduce p to 0.001. Keeping $\alpha = 0.05$, plot the load line for this case on the same graph as in part a.

    c. Show that the network in part b becomes stable if we decrease α to 0.02.

9.2 The maximum achievable throughput for an access method is called the channel capacity. This is obtained by maximizing S with respect to G. Find the channel capacity for the non-slotted–non-persistent CSMA case with $a = 0.01$.

9.3 Consider an Ethernet system that has N stations which are always ready to transmit.

    a. If each station transmits with a probability p during a contention slot, show that the probability P_A that some station acquires the channel during that slot is

$$P_A = N_p (1 - p)^{N-1}$$

    b. Show that optimum value of P_A occurs for $P = 1/N$.

    c. Find P_A for the cases $p = 0.25, 0.50,$ and 0.75 when $N = 4$.

9.4 Before a station can successfully acquire an Ethernet channel, it must wait for a mean number of slots W during a contention interval.

    a. Show that probability that the contention interval has k slots in it is

$$P_k = P_A (1 - P_A)^k$$

    b. Show that the mean number of waiting slots per contention interval is $W = (1 - P_A)/P_A$.

9.5 The difference between CSMA/CD and CSMA is that CD reduces the time that a station must continue transmitting a packet once a collision has occurred. For slotted CSMA this is a full transmission period $t_p = 1$, whereas for slotted CSMA/CD, this time is reduced to $(b + a)$.

9.6 Consider a token-passing ring network having N stations. On the same graph, make plots of the throughput S as a function of the parameter a for values of $N = 1$, 10 and 100. Let a range from 0.1 to 100.

9.7 Consider a token-passing ring network with N stations. On the same graph plot the throughput S as a function of the number of stations N for values of the parameter $a = 0.1$ and 1.0. Let N range from 1 to 25.

9.8 A token ring transmits at a rate of 5 Mbps. If the propagation speed on the medium is 2×10^{10} m/s, how many meters of cable is a 1-bit delay at a token-ring interface equivalent to?

9.9   Assume a 2-km-long token ring with 50 stations operating at a 1-Mbps. transmission rate. If the propagation delay is 5 µs/km, the header length is 24 bits, the delay per stations is 1 bit and the packet lengths are exponentially distributed with a mean length of 1000 bits, what is the mean delay time D for a traffic intensity of 0.5?

9.10 Considering the costs of deploying a PtP network, a curb-switched network and a PON are shown in Figure 9.1. Ten homes in a residential network wish to be connected to a CO via fiber, so as to receive a shared bandwidth connection of 2.5 Gbps.

   a.   Assume the following estimated costs (numbers are for example only):
      i.   Cost of fiber installation: $100/m
      ii.   Cost of a curb switch: $13,000 + $400 x number of ports
      iii. Cost of a passive optical splitter: $750 + $100 x number of ports
      iv.   Cost of a transceiver (either at CO/OLT or at ONU): $650
      v.   Distance of homes from CO: 8 km
      vi.   Possible location of curb switch/splitter: 80 m away from each home

   Compare the costs of network deployments using each of the three approaches. Please try to validate the importance of a PON.

   b.   Assume that the PON solution was chosen after analyzing the data in previous Problem 9.1. Several years after the deployment, the homes wish to upgrade their connection from a 1-Gbps shared line to a 10-Gbps shared line. We can take the following possibilities for upgrading the connection:
      i.   **Rate upgrade:** improve all the transceivers from 2.5 to 10 Gbps.
      ii.   **Spatial upgrade:** Deploy nine additional fibers, so that each ONU can now have one fiber dedicated to it, thus giving it a bandwidth of 2.5 Gbps, and making a cumulative bandwidth 10 Gbps.
      iii. **WDM upgrade:** Use WDM to support ten channels of 2.5 Gbps each on the same fiber. A passive AWG routes each channel to a different ONU.

      In addition to the costs in Problem 9.1, assume the following additional cost estimates:
      i.   Cost of AWG router: $3000 + $200 x number of ports
      ii.   Cost of a 10-Gbps transceiver: $1500
      iii. Cost of multiplexing ten channels at OLT for WDM: $900

   Compare the costs for each of the three types of upgrades. Analyze the advantages and disadvantages of each approach.

9.11 Suppose that, in an EPON, the downstream delay is equal to the upstream delay. When a REPORT message arrives from ONUi, the OLT's local time is 12. The time stamp on the REPORT message is 3. What is the current local time at ONUi? Assume negligible processing time at the OLT.

9.12 Consider IPACT with *limited service scheme.* Consider a maximum scheduling time slot W_{max} of 10,000 bytes. The number of ONUs in Find the followings:
      i.   The maximum cycle time,
      ii.   average throughput for each ONU when all ONUs are active,

    iii.   maximum throughput when only one ONU is active, while all the other ONUs are idle,

    iv.   average throughput for active ONUs when ten ONUs are active, and the remaining ten are idle.

9.13   In a PON, the downstream traffic is point-to-multipoint, whereas the upstream traffic is point to-point. Discuss the implications of such an architecture from a network safety point of view.

9.14   Once a window is granted to an ONU, there are two approaches for transmitting packets in this window. The first approach is to transmit packets in order. This means that, if the first packet in the queue is smaller than the remaining time slot, this packet will not be transmitted, which is also known as head-of-line blocking. Another approach is to use techniques such as bin-packing algorithms to ensure that as many packets may be transmitted in the allocated time slot as possible. Contrast and compare these two approaches. Write an algorithmic description for the second approach.

9.15   Draw a transition state diagram for discovery handling, GATE message handling and REPORT message handling, as defined in the MPCP protocol for an ONU.

9.16   Consider the PON network architecture in which we modify the PON architecture to build a $(1 + 1)$ protected PON, so that if any component, namely, the OLT or ONU transceivers, fiber or splitter fails, the PON network is not disrupted at all. Design a suitable architecture for such a $(1 + 1)$ protected PON.

9.17   Take the near–far problem in an EPON with four different ONUs.

       The distances of different ONUs from the OLT are: 1.5 km for ONU_1, 5 km for ONU_2, 15 km for ONUs and 20 km for ONU_4.

       Suppose each ONU's transmission power is 0.01 mW, assume that the splitter is lossless and the attenuation in the fiber is 0.15 dB/km.

9.18   Consider a PON that has 32 ONUs. Let the transmission power of the OLT transceiver be 0.02 mW, where the receiver is to detect signals of atleast 0.00015 mW for a reasonable bit-error rate. Assume that the splitter is 9% lossy. The optical fiber has an attenuation of 0.12dB/ltm. What is the maximum possible distance of an ONU from the OLT? What is the RTT corresponding to this distance

9.19   Identify whether the following statements are *true* or *false*.
    a.   The EPON is based on the CSMA/CD protocol.
    b.   Fixed time-slot assignment in a synchronous TDMA is an efficient scheme in EPON.
    c.   Dynamic slot assignment in EPON is based on a distributed approach.
    d.   The GFP-PON uses ATM cells for the MAC frames.

9.20   What is open access. State the issues in open access. How can you get solution for open access problems?

9.21   What is the *light-load penalty* in PON. Discuss the possible solutions for avoiding the light-load penalty in PON.

REFERENCES

1. G. E. Keiser, *Local Area Networks*, Tata McGraw Hill, New Delhi, 1997.
2. W. Stalling, *Data and Computer Communication*, Pearson Prentice Hall, Upper Saddle River, NJ, 1999.
3. G. Kessler and D. Train, *Metropolitan Area Networks: Concepts, Standards and Services*, McGraw Hill, New York, 1992.
4. S. S. Lam, "A carrier sense multiple access protocol for local networks," *Computer Networks*, vol. 4, pp. 21–32, 1980.
5. F. A. Tobagi and V. Hunt, "Performance analysis of carrier sense multiple access with collision detection," *Computer Networks*, vol. 4, pp. 245–259, 1980.
6. L. G. Roberts, "ALOHA packet system with and without slots and capture," *ACM SIGCOM Computer Communication Review*, vol. 5, pp. 28–42, 1975.
7. H. Takagi and L. Kleinrock, "Throughput analysis for persistent CSMA system," *IEEE Transaction on Communication*, vol. 33, pp. 627–638, 1985.
8. L. Kleinrock and H. Takagi, "Packet switching in radio channels: Part-1-carrier sense multiple access modes and their throughput delay characteristics," *IEEE Transaction on Communication*, vol. 23, pp. 891–895, 1981.
9. D. P. Heyman, "The effect of random message sizes on the performance of CSMA/CD protocols," *IEEE Transaction on Communication*, vol. 34, pp. 547–553, 1986.
10. E. J. Coyle and B. Lie, "A matrix representations of CSMA/CD networks," *IEEE Transaction on Communication*, vol. 33, pp. 53–64, 1985.
11. E. J. Coyle and B. Lie, "Finite populations CSMA/CD networks," *IEEE Transaction on Communication*, vol. 31, pp. 1247–1251, 1983.
12. J. S. Meditch and C. T. Lea, "Finite populations CSMA/CD networks," *IEEE Transaction on Communication*, vol. 31, pp. 763–774, 1983.
13. S. Tasaka, "Dyanmicbehaviour of a CSMA/CD system with a finite population of buffered users," *IEEE Transaction on Communication*, vol. 34, pp. 576–586, 1986.
14. IEEE Computer Society, "Token ring access method," ANSI/IEEE std 802.5–1985 (ISO/DP 8802/5), 1985.
15. IEEE Computer Society, "Token passing bus access method," ANSI/IEEE std 802.4–1985, 1985.
16. W. Bux, F. H. Closs, K. Kuemmerie, H. J. Keller, and H. A. Mueller, "Architecture and design of w reliable token ring network," *IEEE Journal on Selected Areas of Communication*, vol. 1, pp. 756–765, 1983.
17. H. J. Keller, H. Meyr, and H. A. Mueller, "Transmission design criteria for asynchronous token ring," *IEEE Journal on Selected Areas of Communication*, vol. 1, pp. 721–733, 1983.
18. G. Kramer, *Ethernet Passive Optical Networks*, McGraw-Hill, New York, 2005.
19. G. Kramer, B. Mukherjee, S. Dixit, Y. Ye, and R. Hirth, "Supporting differentiated classes of service in ethernet passive optical networks," *OSA Journal of Optical Networking*, vol. 1, no. 819, pp. 280–298, 2002.
20. L. Zhang, et al., "Dual DEB-GPS schedular for delay-constraint applications in ethernet passive optical networks," *IEICE Transactions in Communications*, vol. E86-B, no. 5, pp. 1575–1584, 2003.
21. G. Pesavento and M. Kelsey, "PONS for the broadband local loop," *Lightwave*, vol. 16, no. 10, pp. 68–74, 1999.
22. B. Lung, "PON architecture future proofs FTTH," *Lightwave*, vol. 16, no. 10, pp. 104–107, 1999.
23. M. McGarry, M. Meier, and M. Reisslein, "Ethernet PONS: A survey of Dynamic Bandwidth Allocation (DBA) algorithms," *IEEE Optical Communications*, vol. 42, no. 8, pp. S8–S15, 2004.

24. FSAN - Full Service Access Network, Online at http://fsan.mblast.com/default.asp.

25. E. Hernandez-Valencia, M. Scholten, and Z. Zhu, "The generic framing procedure (GFP): An overview," *IEEE Communications Magazine*, vol. 40, no. 5, pp. 63–71, 2002.

26. B. Mukherjee, *Optical WDM Networks*, Springer-Verlag, Berlin, 2006.

27. G. Kramer, B. Mukherjee, and G. Pesavento, "IPACT: A dynamic protocol for an ethernet PON," *IEEE Communications Magazine*, vol. 40, no. 2, pp. 74–80, 2002.

28. M. Ma, Y. Zhu, and T.-H. Cheng, "A bandwidth guaranteed polling MAC protocol for ethernet passive optical networks," *Proceedings, IEEE INFOCOM '03*, San Francisco, CA, pp. 22–31, 2003.

29. G. Kramer, B. Mukherjee, S. Dixit, Y. Ye, and R. Hirth, "Supporting differentiated classes of service in ethernet passive optical networks," *OSA Journal of Optical Networking*, vol. 1, no. 819, pp. 280–298, 2002.

30. E. Hernandez-Valencia, M. Scholten, and Z. Zhu, "The generic framing procedure (GFP): An overview," *IEEE Communications Magazine*, vol. 40, no. 5, pp. 63–71, 2002.

31. G. Mayer, M. Martinelli, A. Pattavina, and E. Salvadori, "Design and cost performance of the multistage WDM-PON access networks," *IEEE/OSA Journal of Lightwave Technology*, vol. 18, no. 2, pp. 121–142, 2000.

32. F. J. Effenberger, H. Ichibangase, and H. Yamashita, "Advances in broadband passive optical networking technologies," *IEEE Communications Magazine*, vol. 39, no. 12, pp. 118–122, 2001.

33. M. Zirngibl, et al., "LARNet, a local access router network," *IEEE Photonics Technology Letters*, vol. 7, no. 2, pp. 215–217, 1995.

34. B. N. Desai, "An optical implementation of a packet-based (Ethernet) MAC in a WDM passive optical network overlay," *Proceedings, OFC '01*, Anaheim, CA, p. WN5, March 2001.

35. N. J. Frigo, et al., "A wavelength-division multiplexed passive optical network with cost shared components," *IEEE Photonics Technology Letters*, vol. 6, no. 11, pp. 1365–1367, 1994.

36. I. van de Voorde, C. M. Martin, I. Vandewege, and X. Z. Oiu, "The superPON demonstrator: An exploration of possible evolution paths for optical networks," *IEEE Communications Magazine*, vol. 38, no. 2, pp. 74–82, 2000.

Index

A

Access-network operator (ANO), 51
Add-drop multiplexer (ADM), 54
    optical add-drop multiplexer (OADM), 54, 100–101
    SONET ADM (SADM), 55
    ROADM, 55
Algorithm, 205, 207, 225, 231–232, 240, 269, 271, 280–285, 289–290, 318, 327, 339, 342, 353–354
ALOHA, 213, 225, 272–275
    ALOHA/CSMA, 275
    protocol, 213
Analog to digital modulation, 41
Application layer, 2, 3–4, 215, 262
Array waveguide grating, 51, 75, 83–85, 189
ARQ, 186–189
    Go back ARQ, 188
    Select reject, 189
    stop and wait, 186
Assured forwarding (AF), 3, 4, 5
Asynchronous transfer mode
    ATM protocol, 200–201
    ATM logical connections, 201–202
    transmission ATM cells, 203
Attenuation, 21, 117–118, 199, 249, 334, 347
    low-attenuation region, 21, 33
Average hop distance, 284, 286, 289–290, 294–295

B

Bandwidth, 330, 332–336, 337–338, 339, 343, 347, 342, 351–353
    broadband, 11, 51, 206, 331, 352, 354
    narrowband, 10–11
    passband, 32, 112
    provisioning, 62, 336
Best effort (BE), 344, 353
Bi-directional, 10, 55, 58, 99, 100, 124, 128, 286, 327, 332
Bit-error rate (BER), 6, 209, 213, 247, 255
Blocking probability, 269
Broadband ISDN, 11, 206
Broadcast and select, 250, 255–256
Broader gateway protocol (BGP), 4
Bus topology
    Distributed Queue Dual Bus (DQDB), 240–242
    Expressnet, 239–241
    Fasnet, 237–239
Burstiness, 202

C

Call-blocking
    centralized control, 325
    central office (CO), 10, 331
    channel capacity, 280, 327, 339
Carrier sense multiple access (CSMA), 248–249, 272–275
    non-persistent CSMA, 313–322
    slotted Non-persistent CSMA, 322
    1-persistent CSMA, 322–324
    p-persistent CSMA, 226, 317–318
Carrier sense multiple access (CSMA)/CD, 318–325
Channel sharing, 281, 292–293
Circuit switching, 5, 10, 97
Class of service (CoS), 345
Clock synchronization, 339
Combiner, 34, 97, 256, 333, 335
Computation, 104
    on-line, 272
Connection, 4, 5, 6, 55, 57, 61, 62, 66, 69, 106, 107, 122, 201, 280, 281, 289, 343
Control information (CI), 61, 195, 216, 276
Coupled mode theory, 77
Coupler, 76–85
    directivity, 76
    insertion loss, 35, 36, 99, 111, 121, 151, 152, 164, 167, 246, 249, 333
    splitting loss, 110, 333
Cross-Connect (OXC), 55, 61–62, 68, 98–99, 251
Cross-phase modulation (XPM), 21
CSMA/ALOHA, *see* Carrier sense medium access
Cycle time, 310

D

Data link layer, 2–3
Deadlock, 261, 263, 269
Deflection routing, 279, 285
Digital subscriber loop (DSL), 331
Digital to analog modulation, 37, 40
Digital to digital modulation, 36, 37
Diode laser, 24
Directional coupler, 75, 158
Dispersion, 12, 16, 19–20
    chromatic, 19
    intermodal, 19
    material, 19
        polarization mode (PMD), 19

Distance matrix, 280
Distribution
    Boltzmann, 24
    exponential, 24
    Fermi-Dirac, 25
    Markov chain, 266
    Poisson, 303, 306, 30, 310

E

Electrooptically-tuned lasers, 27
EOMZ, 93
Error control
    Go-Back–N ARQ, 188
    Selective-Reject ARQ, 189–190
    Stop and Wait ARQ, 186

F

Fiber
    graded-index fiber, 14–16
    multimode fiber, 15
    non-zero dispersion-shifted fiber (NZDSF), 20
    single-mode fiber (SMF), 16
Fiber Bragg grating (FBG), 35
Fiber channel (FC), 238, 246
Fiber connection (FICON), 252
Fiber delay line (FDL), 67
Fibernet, 245–246
Fibernet-II, 248
Fiber-optic Cross connect (FOX), 260, 269
Fiber optic network without WDM, 237–255
    bus topology
        Fasnet, 237
        Expressnet, 239
        Distributed Queue Dual Bus (DQDB), 214
Fiber ring topology: FDDI, 242–243
File transfer protocol (FTP), 4
Fixed receiver, 258
Fixed transmitter, 293
Four-wave mixing (FWM), 22
Frame relay, 6
Free spectral range (FSR), 32–33
FTTx, 331
    FTTB, 331
    FTTC, 331
    FTTH, 332
Full-duplex mode, 344

G

GaAsInP waveguide device, 164

H

High level data link control (HDLC), 191–197

HDLC frame format, 174, 192
    examples of HDLC operations, 196
    operation of HDLC, 194–197
Hypertext transfer protocol (HTTP), 4

I

Improved random access protocols, 275
Integrated Services Digital Network (ISDN), 343
Injection-current-tuned lasers, 27
InP/GaAsInP, 159
Interleaved Polling with Adaptive Cycle Time
        (IPACT), 339
International Standard Organization (ISO), 2
Internet Protocol (IP), 3
Internet control message protocol (ICMP), 4

L

LAMBDANET, 258, 260–261
Laser, 23–27
LaserArrays, 28
LARNET, 349–350
LiNbO$_3$, 153–158
Link-capacity, 280, 283
Link control protocol
    LAPB, 197
    LAPD, 198
    LAPF, 198
    LLC/ MAC, 198
Link load, 285, 289
Local access point (LAP), 352
Local area network (LAN), 2
    ethernet, 334–336
    WDM LAN, 269
Local exchange, 332
Logical link identifier (LLID), 337–338
Logical ring, 327
LP CVD, 135

M

Mechanically-tuned lasers, 26
Media-access control (MAC), 198, 231
    round robin, 221, 238, 239, 263, 327
Mech Zehnder switch
    EOMZ based, 88, 93
    MMI coupler based MZ switch, 94
    TMI coupler based MZ switch, 94
    TOMZ switch based, 85
Micro-electro mechanical system, 99
MMI, 75, 79, 82, 94
Multi-hop WDM, 279–295
Multi-point control protocol (MPCP), 336
Multi-protocol label switching (MPLS), 62–63
Multicasting, 62

N

Narrowband ISDN, 11
Network
    multihop (*see* Multihop WDM)
    single-hop, 255–259
    throughput, 320–322
Network control, 61, 327, 344
Network element (NE), 68
Network layer, 2
Network management, 4, 173, 201
Network-to-network interface (NNI), 60
Network-to-user interface (NUI), 60
Node activity list (NAL), 276
Non reconfigurable node, 49–50
Nonlinear Refraction, 21
Numerical aperture, 14–15, 249

O

Open access, 351–352
Open shortest path first (OSPF), 4
Open System Interconnection (OSI), 2
Operation, administration, 346
Optical burst switching (OBS), 90
Optical add-drop multiplexer, 49, 54
Optical amplifier, 109–111
    amplified spontaneous emission (ASE), 110
    doped-fiber amplifiers, 115
    erbium-doped fiber amplifier, 115
    Fabry-Perot amplifier, 111
    gain equalization, 118
    Raman amplifier, 116
    semiconductor laser amplifier, 111
    unequal gain spectrum, 115
Optical delay line, 149
    wavelength conversion, 122–126, 260
Optical fiber, *see also* Fiber
    Stimulated Brillouin Scattering (SBS), 22
    Stimulated Raman scattering, 22
Optical line terminal (OLT), 333–334
Optical network unit (ONU), 334
Optical receiver, 29, 110, 248, 250
    channel equalizer, 117–118
    filter, 32–35
        acoustooptic, 34
        avalanche photodiode, 31
        electrooptic, 35
        etalon, 33
        Fabry-Perot filter, 35
        finesse, 32–33
        fixed filter, 35
        free spectral range, 32–33
        grating filter, 35
        liquid-crystal, 35
        PIN photodiode, 30
        thin-film interference filter, 36
        tunable, 36
        tuning range, 36
        tuning time, 36
    laser
        acoustooptically tuned, 27
        band gap, 24
        cavity, 24
        chirp, 25
        Distributed Bragg Reflector, 28
        Distributed Feedback (DFB), 27
        injection current tuned, 27
        laser array, 28
        multi-quantum well (MQW), 29
        on-off keying (OOK), 40
        population inversion, 111, 116
        quasi-stable, 23
Optical signal-to-noise ratio (OSNR), 101
Optical switch, 53, 58, 63, 90–91, 158, 273
Optical transmission, 12–5, 213–233
Optical transmitter, 22–28

P

Passive optical loop (POL), 270
Passive optical network (PON), 330
Passive-star coupler (PSC), 49–51, 83, 245, 256,
    269, 276, 292
    BPON, 354
    CPON, 349
    CoS EPON, 3
    Ethernet PON (EPON), 342–343
    GFP-PON, 347
    WDM-PON, 347
Point-to-multipoint (PtMP), 332
Point-to-point protocol (PPP), 8, 193
Polymeric waveguide, 164
Processing
    Ga As InP, MBE Growth System, 160
    $LiNbO_3$
        proton exchange method, 157
        thermal in Ti-diffusion method, 153
    SiO_2 GeO_2, flame hydrolysis, 147
    SiON SiO_2
        LPCVD, 137
        PECVD, 137–139
    SOI
        BESOI, 150
        SIMOX, 150
Polymeric waveguide processing, 166

R

Radio frequency (RF), 34, 137
Rainbow, 259
Random Access Protocols–I, 272
Random Access Protocols, 272

Re-amplification, re-shaping, and re-timing (3R), 110
Receiver collision, 271
Receiver collision avoidance (RCA), 275
Receiver sensitivity, 17
Regular topologies, 281
    de Bruijn Graph, 284
    GEMNET, 286, 293
    Hypercube, 286
    ShuffleNet, 281, 292
    Torus, 285
Reservation protocols, 276
Ring topology, 218
    FDDI, 242
    FDDI-I, 243
RITENET, 351
Round-trip time (RTT), 336
Routing
    adaptive, 5, 279
    OSPF, 4

S

Selection, FIFO, 123
Self-phase modulation (SPM), 21
Semiconductor optical amplifier (SOA), 110
Service provider (SP), 51, 352
Service-level agreement (SLA), 340
Single-Hop, SONATA, 260
STARNET, 269
Star topology
    Fibernet, 246
    Fibernet-II, 248
Stimulated Brillouin Scattering, 22
Stimulated Raman Scattering, 22
Session layer, 2
SiON, 139–140
SiO_2, 145–149
Silica (SiO_2)/SiO_2-GeO_2, 145–149
Simple mail transfer protocol (SMTP), 4
SONET WDM ring, 54–55
SONET, 59, 106–107
Slotted network, 66–67
Splitting ratio, 63
Subscriber access network, 331
Switch
    AWG, 256
    Clos architecture, 99
    confinement factor, 112–113
    MEMS, 122
    switching elements, 98

T

TCP, 215–216
TDM, 221, 224, 270
TDMA, 271–272, 281

Topology, 292, 294, 327, 330–331, 333, 216–222
    mesh, 221
    ring, 218, 242–243
Traffic
    arbitrary, broadcast, demands, 271
    downstream, 334
    non-uniform, 271, 279
    upstream, 350
Ti diffusion, 153
TMI, 75
TOMZ, 85–88
Transmission equipment, 62
Transport layer, 2–4, 347
Transport control protocol, 2
Tunable optical filters, 32
    acoustooptic filters, 34
    electrooptic filters, 35
    etalon, 33
    liquid-crystal, 35
Tunable receiver, 36, 258
Tunable transmitter, 258, 270–273

U

Unicast, 62
User-to-network interface (UNI), 63
User datagram protocol (UDP), 4

V

Vertical-cavity surface emitting laser (VCSEL), 28–29
Very small aperture terminals (VSAT), 6

W

Wavelength conversion, 52
    cross-gain modulation, 125
    cross-phase modulation (XPM), 21
    difference frequency generation (DFG), 124
    four-wave mixing (FWM), 21, 22, 124
Wavelength reuse, 250–251, 256
Wavelength router, 49
    non reconfigurable, 49
    reconfigurable, 53–57
Wavelength routing (WR), 53
Wavelength Conversion, 122–126
    Opto-Electronic Wavelength Conversion, 123
Wavelength Conversion Using Coherent Effects, 124
Wavelength Conversion Using Cross Modulation, 125
Wavelength-routing switch (WRS), 53, 287
WDM, demultiplexer (DEMUX), 118
Wide area network (WAN), 5, 122, 221, 275, 303
WRS, 53, 62

Printed and bound by CPI Group (UK) Ltd, Croydon, CR0 4YY

23/10/2024

01778223-0016